I0470845

FA–173/July 1997

FIRE

in the
United States

1985-1994

NINTH EDITION

This document was produced under Contract Number EMW–95–C–4717 by TriData Corporation, Arlington, Virginia. The opinions expressed in this document do not necessarily reflect the positions of the United States Fire Administration.

Federal Emergency Management Agency
United States Fire Administration
National Fire Data Center

ACKNOWLEDGEMENTS

The United States Fire Administration greatly appreciates the participation in the National Fire Incident Reporting System (NFIRS) of nearly 14,000 fire departments across the United States. The NFIRS data, on which the bulk of this report is based, are available through the work of the staffs of the various state agencies and state fire marshal's offices responsible for fire data collection and on each and every fire officer who fills out an NFIRS form. Without their painstaking efforts to collect data, this report could not exist. Although the NFIRS is wholly voluntary, the information collected on nearly one million fires each year represents the most comprehensive set of fire data and statistics in the world.

The National Fire Information Council (NFIC), a nonprofit organization of the state and metro participants in NFIRS, helps coordinate and specify requirements for NFIRS and its operation. They represent an outstanding example of local, state, and federal cooperation on a major, long-term undertaking.

The report was produced by TriData Corporation, Arlington, Virginia, for the National Fire Data Center, USFA, under Contract Number EMW−95−C−4717.

Where to get copies of this report—Copies of this report are available by writing:

U.S. Fire Administration
Federal Emergency Management Agency
Publications Center, Room N310
16825 South Seton Avenue
Emmitsburg, Maryland 21727

Documents may also be ordered on the World Wide Web: *http://www.usfa.fema.gov.*

CONTENTS

2   The National Fire Problem 29

3   Residential Properties 57

4 Non-Residential Properties 121

5 Firefighter Casualties 153

6 Special Topics 179

LIST OF FIGURES

LIST OF TABLES

EXECUTIVE SUMMARY

Fire kills thousands of Americans each year, injures hundreds of thousands, destroys billions of dollars in property, and costs tens of billions of dollars overall, but mayors and city managers, school officials, the media, and the general public still are largely unaware of the magnitude of these numbers. Their lack of awareness and failure to realize the seriousness of fire to communities and the country are factors in keeping the U.S. fire problem one of the worst in the world per capita.

PURPOSE AND SCOPE

This report is designed to arm the fire service and others with a statistical overview of the fire problem that can motivate corrective action. It can also be used to select priorities and help target fire programs, serve as a model for state or local analyses of fire data, and provide a baseline for evaluating programs.

This Ninth Edition of *Fire in the United States* covers the 10-year period from 1985 to 1994, with emphasis on 1994—the most recent year for which complete data are available at the time of printing. The primary source of data in this report is the National Fire Incident Reporting System (NFIRS), but National Fire Protection Association (NFPA) annual survey results and data from the state fire marshals are also used.

Because of the time it takes to collect data from nearly the 14,000 fire departments that participate in NFIRS, edit and obtain corrections, and analyze and display the results, the date of publication lags the date of collection.

Previous editions of *Fire in the United States* have included a state-by-state analysis and presentation of state fire statistics. This chapter has been omitted from this edition. Instead, a separate volume will be published later in 1997 that is devoted entirely (and more exhaustively than in the past) to state fire profiles.

THE NATIONAL FIRE PROBLEM

Figure 1 summarizes the national fire problem to civilians.[1] During the 10-year period 1985–1994, there were an average of 5,300 civilian fire deaths, 29,000 civilian injuries, and $9.4 billion dollar loss (adjusted to 1994 dollars) from reported fires each

---

[1] See page 25 for a discussion of how trend percentages were calculated.

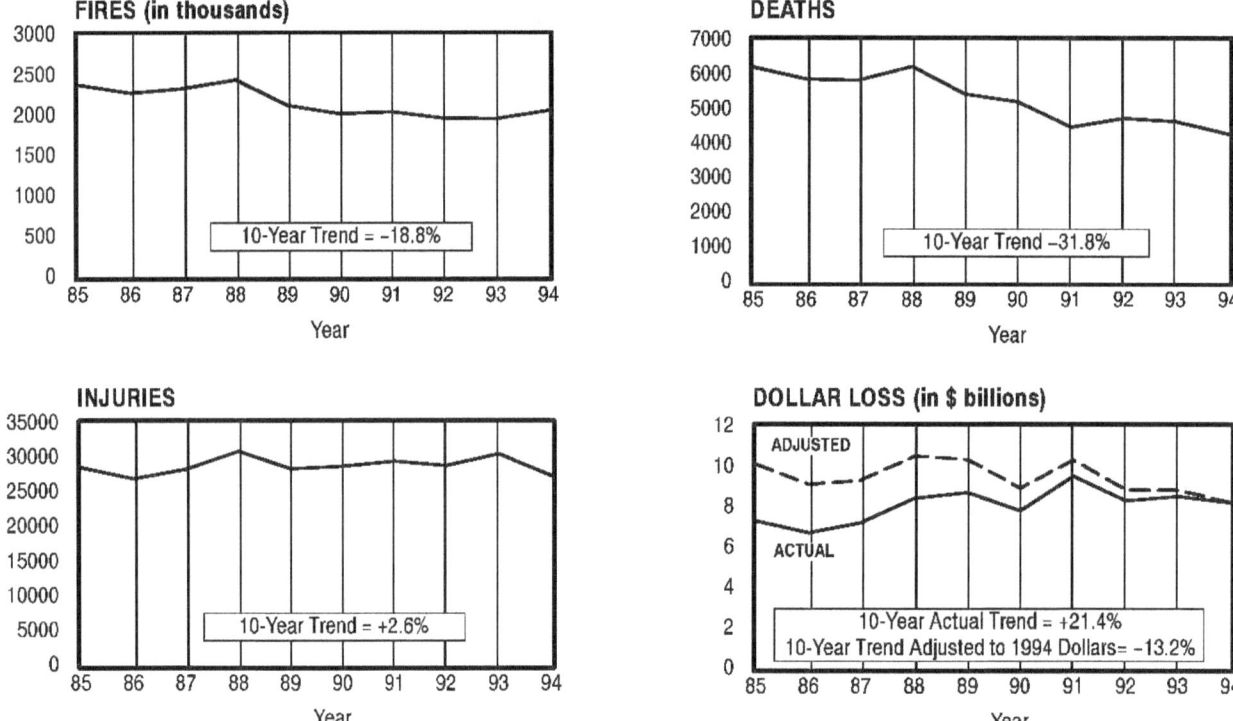

Sources: NFPA Annual Surveys and Consumer Price Index

Figure 1. Trends in Fires and Fire Losses

year. The United States had an average of 2.2 million reported fires annually during this period.

The trend over the past 10 years in the number of fires reported to the fire service has decreased 19 percent, with a noticeable drop in 1989 and continuing modest declines until 1994 when the number of fires increased to the 1991 level. Civilian deaths from fire have dropped sharply over this period (32 percent). Civilian injuries have remained steady over 10 years. The magnitude or trend of injuries from unreported fires is not known. In terms of constant (1994) dollars, losses were down a significant 13 percent over the period—much less than the rise in inflation.

On a per capita basis, the fire problem is less severe in 1994 than it was 10 years earlier, because the population increased faster than did fires and fire casualties (Figure 2). The per capita fire death trend was down 38 percent, and the per capita injury trend was down 6 percent. Although the death rate per fire in the United States has improved greatly, it remains much higher than the yearly reported fire death rates in countries such as Australia, Japan, Hong Kong, and most of the countries in Western Europe.

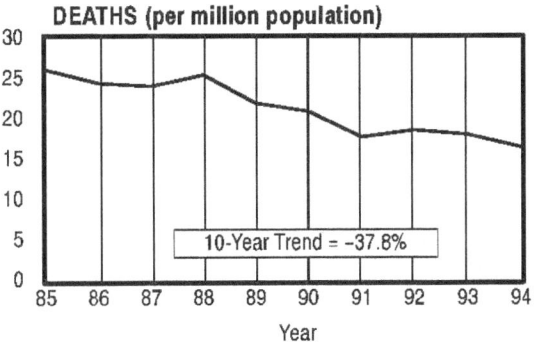

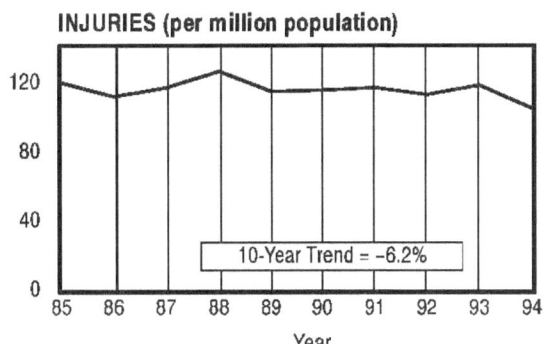

Sources: NFPA Annual Surveys, Consumer Price Index, and Bureau of the Census

Figure 2. Trends in Severity of Casualties

Regional Variations

The fire problem varies from region to region and state to state because of variations in climate, poverty, education, demographics, and other factors. The Figure 3 map shows that the fire death rate per capita is highest in the Southeast and a few isolated states. The highest death rates in 1994 were in Alaska, Mississippi, District of Columbia, and Kentucky; Alaska and Mississippi were also ranked in the top five in 1990. States with the lowest fire death rates were Hawaii, New Mexico, Utah, and New Hampshire; none of these states ranked in the lowest five in 1990.

Another important measure to examine is the absolute number of fire deaths in each state. The 11 states with the most fire deaths account by themselves for half of the national total. As expected, large-population states are at the top of this list. National totals cannot be reduced significantly unless these states reduce their fire problem.

Even though the death rate varies, the leading causes of fires (cooking, heating, and arson) and fire deaths (careless smoking, heating, and arson) are relatively similar around the nation. The rank order and magnitude of these causes vary from state to state and by whether fires, deaths, or injuries are used as the measure. Therefore, the priorities for prevention programs must be tailored to location and purpose.

Where Fire Losses Occur

The public generally does not appreciate the magnitude of the fire problem in the home nor the importance of doing its share to reduce fires in the home. Based on 1994 data, the vast majority of our civilian fire deaths (71 percent) and injuries (68 percent) continue to occur in residences, although residences have only 22 percent of the total

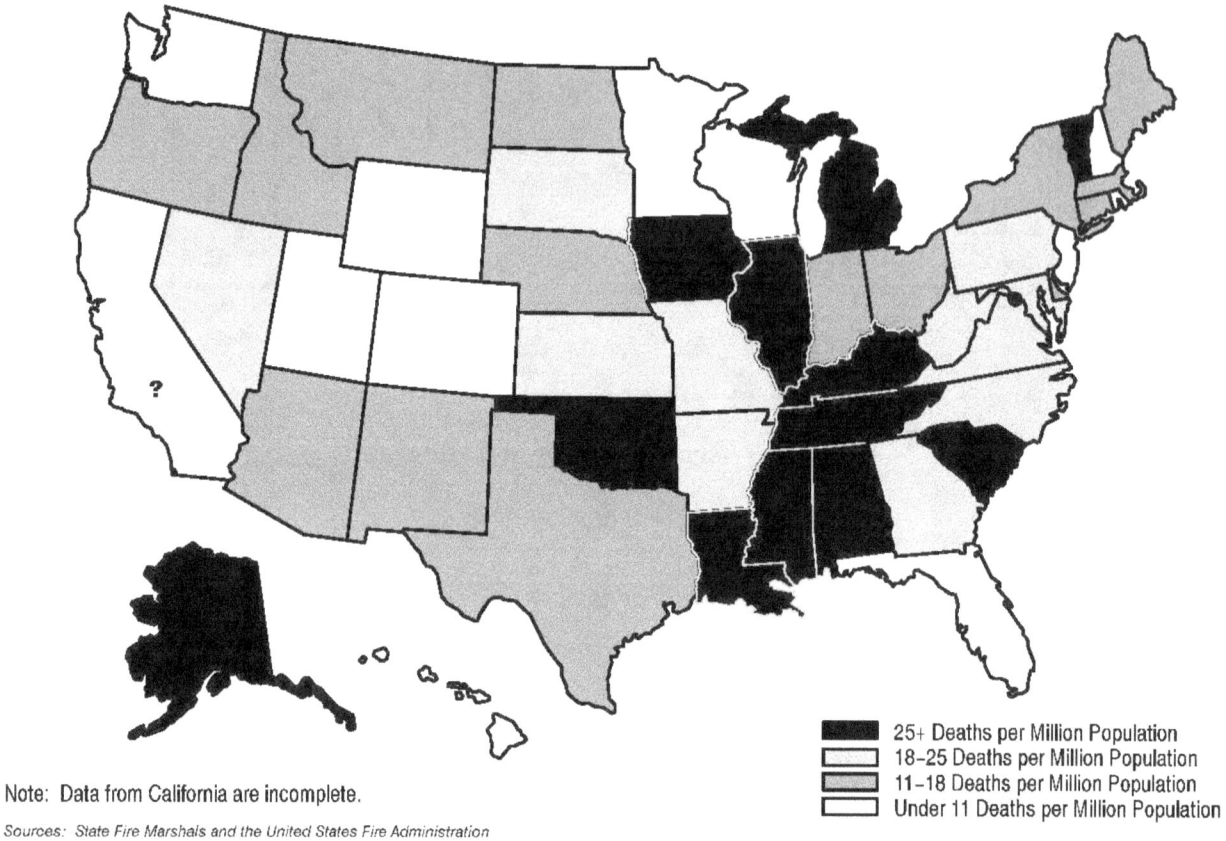

25+ Deaths per Million Population
18–25 Deaths per Million Population
11–18 Deaths per Million Population
Under 11 Deaths per Million Population

Note: Data from California are incomplete.

Sources: State Fire Marshals and the United States Fire Administration

Figure 3. Fire Death Rate by State in 1994

fires (Figure 4). More than two-thirds of injuries incurred by firefighters are in residences. And residences account for a substantial portion of the dollar loss, 44 percent. The 10-year trend mirrors the 1994 picture.

One- and two-family dwellings—where the majority of people in the United States live—dominate the fire problem. They account for 70 percent of all residential fires, 69 percent of deaths, 60 percent of injuries, and 73 percent of dollar loss. Apartments are a large category too, accounting for 32 percent of the residential injuries and 19–23 percent of the other residential fire problems. People continue to underestimate the fire problem potential in their home because large fires in hotels, high-rise office buildings, and other public buildings receive higher media attention than fires in single-family homes.

Often overlooked or ignored in the discussion of residential fires is the effect of fires in garages. In 1994, there were 17,100 fires, or 5 percent of the total in dwellings. Residential garages are coded as storage properties rather than residential properties. This method of coding garage fires has not distorted the residential fire profile in any significant way, but it does lead to understating the fire problem by 1 to 5 percent.

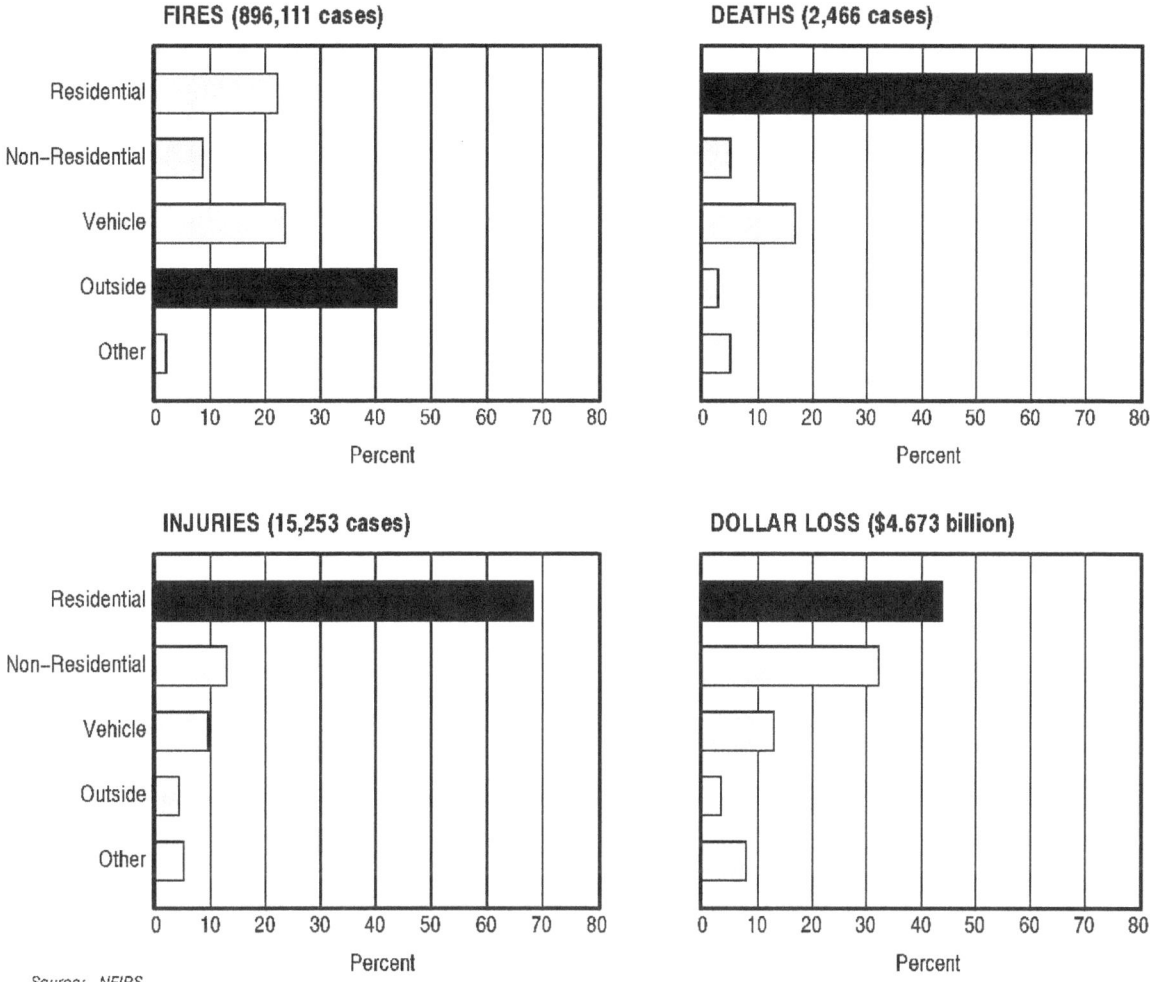

FIRES (896,111 cases)

DEATHS (2,466 cases)

INJURIES (15,253 cases)

DOLLAR LOSS ($4.673 billion)

Source: NFIRS

Figure 4. 1994 Fires and Fire Losses by General Property Type

The type of residence that is most dangerous when fire occurs is manufactured housing (mobile homes), where deaths per fire are double that of other types of residences. Hotel/motel fires account for only 1 to 2 percent of the residential fire problem in the various measures, although they tend to get a disproportionately large share of press attention when they occur.

Fires that occur outside (e.g., fields, vacant lots, wildland) are nearly double the number in any other property category, but these fires cause the fewest deaths, injuries, and dollar loss. Many of these fires to which fire departments respond are intentionally set but result in relatively little damage. Outside fires, however, are cause for concern because they may spread to structures.

One of every four calls to fire departments is to respond to a vehicle fire. The topic of vehicle fires is discussed in more detail in a separate section of this chapter.

Who Dies or Gets Injured

The elderly and the very young are the groups at highest risk. Children under 5 years of age continue to have double the national average fire death rate (Figure 5). Risk of fire death drops off sharply for children between 5 and 19, then increases slowly with age. The elderly—people over 70—have one-and-one-half to three times the national average fire death rate, depending on how old they are, with the risk increasing sharply for people over 80. However, two-thirds of the people who die in fires are neither very young nor old; the fire problem affects all age groups.

The risk of fire injury peaks at ages 20–24. Young adults have 40 percent greater risk than average. They tend to be involved in the more dangerous activities, especially involving flammable liquid and demonstrating a higher degree of bravado. People over 85 also have sharply elevated risk of fire injury.

Men are twice as likely to be killed in fires than women in 1994, a proportion that has remained relatively stable over the past 10 years. This is true for virtually every age group and has been reported every year since NFIRS started in 1975. Males also have a higher fire death rate per capita than females for all age groups except 15–19-year old females. For some age groups, the male rate is triple the female rate. Elderly men have a significantly higher fire death rate than elderly women. The male/female ratio for fire deaths is almost identical to that for fire injuries.

The reasons for the differences between the sexes in fire risk are not known for sure. Some reasons advanced are the greater likelihood of men being highly intoxicated, the more dangerous occupations of men (most industrial fire fatalities are males), the greater use of flammable liquids by men, their greater likelihood of attempting to fight fires or going back to rescue someone, or possibly that men are less safety-conscious than women.

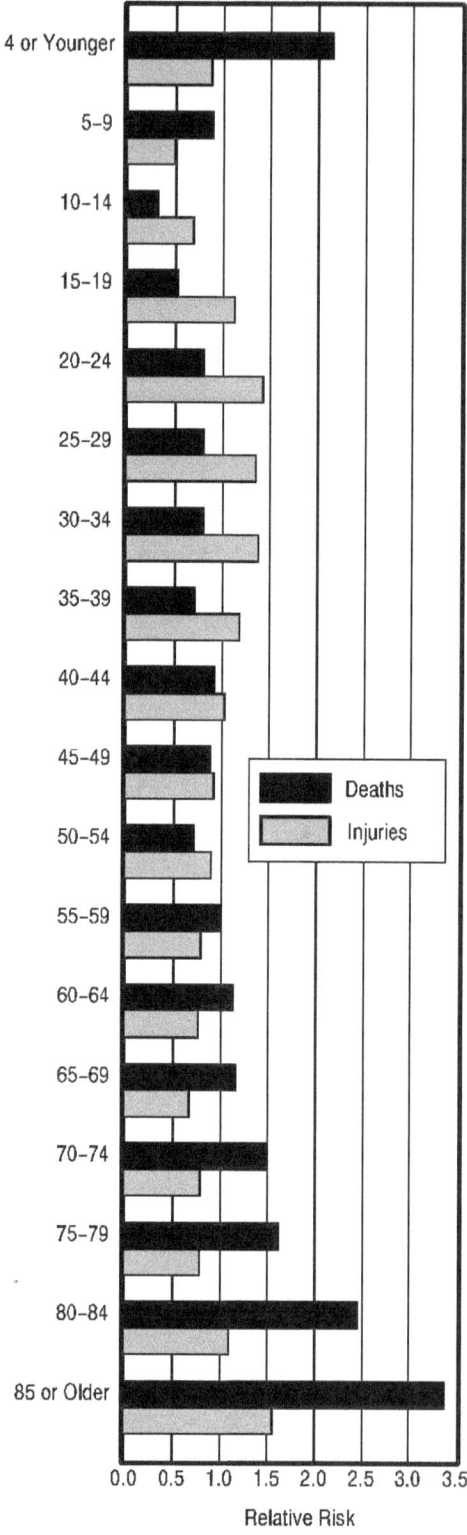

Sources: NFIRS, NFPA Annual Surveys, and Bureau of the Census

Figure 5. Relative Risk of 1994 Fire Casualties by Age

The fire problem cuts across all ethnic, economic, and regional groups. It is higher for some than for others. For example, people in rural areas and large cities have higher fire death rates than people in mid-size communities. The poor, too, suffer a disproportionate share of deaths versus the rest of the population.

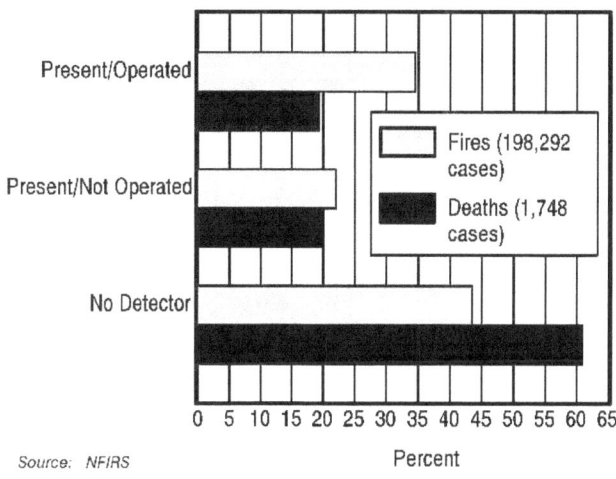

Source: NFIRS

Figure 6. Smoke Detector Performance in 1994 Residential Fires and Fire Deaths (Adjusted)

Smoke Detectors and Sprinklers

In two-thirds of household fires where the information on detectors was reported for 1994, either there was no detector or the detector did not operate (Figure 6).[2] Homes that have reported fires are less likely to have smoke detectors than the homes that do not have fires. That is, detectors are found least often in the places that need them most. Also, 22 percent (adjusted) of the installed detectors in residential fires did not operate. This does not include the situations where the fire was reported as too small to trigger the detector. Usually, detectors do not work because they have no working battery or they have been deliberately disconnected. The increasing trend of nonoperating detectors over 10 years (nearly doubling) is disturbing. A major thrust should be given to educating the populace as to the value of detectors in saving lives.

Detectors are present in a smaller proportion of houses that have fires than in apartments (57 vs. 74 percent), probably because many jurisdictions mandate the installation of detectors in multifamily dwellings. From 1990 to 1994, there was only a 1 percent increase in the presence of detectors in houses but an 8 percent increase in apartments.

Smoke detectors are much less likely to be present when there are fatalities. Detectors do indeed make a difference. Yet in 19 percent of the reported residential fire deaths in 1994, a detector did operate; in 1988, it was 9 percent. In some cases, the detector may have gone off too late to help the victim, or the victim may have been too incapacitated to react. But the percentage of deaths with detectors present, especially the upward trend, is somewhat disturbing since there is widespread belief than an operating detector will save lives. Further study is needed to show what other factors were involved in these deaths.

---

[2] The percentages discussed throughout this chapter have been adjusted to apportion the "unknowns" across the other categories.

Sprinklers are not installed in enough residences (3 percent) or apartments (6 percent) to provide meaningful insight. Sprinklers are more prevalent in hotels/motels and business, often mandated by local laws. What is not known is the portion of fires that are unreported as a result of a functioning sprinkler.

RESIDENTIAL PROPERTIES

The leading three causes of residential fires in 1994 were cooking, heating, and arson, as shown in Figure 7. This is the same ranking as reported in 1990. Cooking has been the leading cause of fires in most years, except in the 1970s when heating became the leading cause due to a surge in the use of alternative space heaters and wood heating.

Heating fires include those where the equipment involved in ignition is central heating, fireplaces, portable space heaters, fixed-room heaters, wood stoves, and water heating. The central and water heating portions of the problem have remained relatively steady over the years, while the portable space heater and wood burning stove portions of the problem, along with chimney fires, rose very sharply from the late 1970s to the early 1980s and then subsided somewhat. Although heating is the second leading cause of fires nationally, it leads cooking as the leading cause in one- and two-family dwellings both because these structures predominate in alternative heating sources and because maintenance of heating systems is handled by the homeowner rather than professionals. Heating-related fires are also the second leading cause of fire deaths and dollar loss in residences.

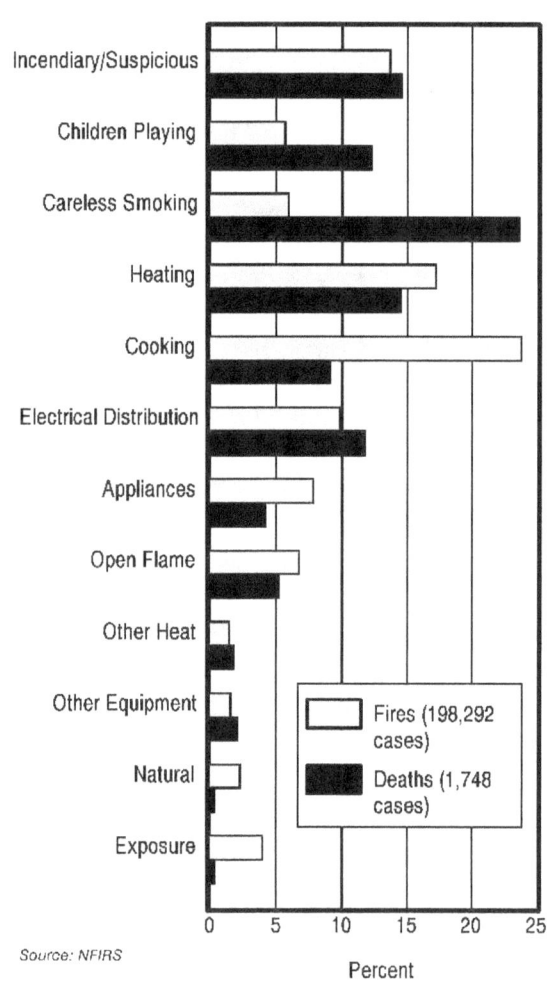

Source: NFIRS

Figure 7. Causes of 1994 Residential Fires and Fire Deaths (Adjusted)

As in all years, careless smoking continued in 1994 to be the leading cause of residential fire fatalities, followed by heating and arson. The relative magnitude of the major causes of residential fire deaths has changed somewhat over time. Deaths from careless smoking have dropped significantly (48 percent) over the 10 years. Nevertheless, one of every four fire deaths in residences is caused by careless smoking. Heating deaths peaked in 1985 but have since fallen 52 percent, moving it from second to third place. Arson has been the second leading cause of deaths since 1986. This upward trend continued until 1994, when there was a considerable decline in the number of deaths (from 570 deaths to 503). Children playing deaths dropped from 1988 to 1990 but has trended upward since then.

As in all years, cooking was the leading cause of fire injuries by a two to one margin. Most cooking fires come from unattended cooking rather than from equipment failures. Another significant cause of cooking injuries is the ignition of loose clothing such as bathrobes. Greater public awareness of these problems, coupled with information about how to quickly extinguish a cooking fire, could reduce the incidence of such injuries.

Fire deaths usually occur late at night. Nearly 50 percent of the deaths in 1994 (as in 1990) occurred in fires reported from 11:00 p.m. to 6:00 a.m. Many people are overcome in their sleep or wake up too late or too confused to escape. Fire incidence, on the other hand, peaks at 5:00 to 7:00 p.m. from the surge in cooking-related fires during the dinner period. By season, residential fires and fire deaths are most frequent during the winter when heating is a dominant cause and added to the level of year-round causes such as cooking. The residential fire rate in January is almost twice that of summer months, and the fire death rate in January is triple that of summer months.

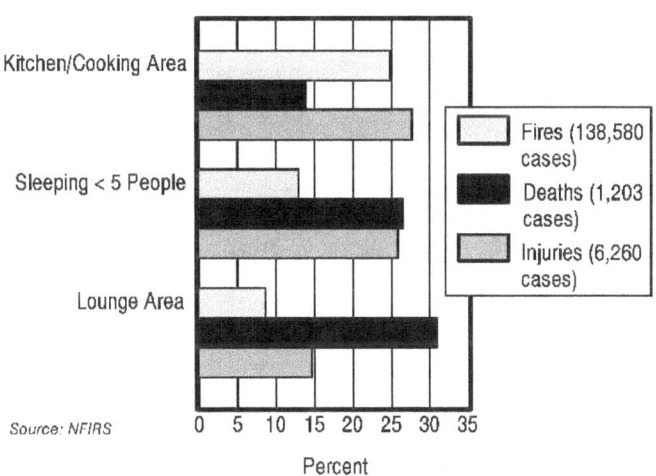

Source: NFIRS

Figure 8. Leading Rooms of Origin of 1994 One- and Two-Family Dwelling Fires and Fire Casualties (Adjusted)

The area of the home (in one- and two-family dwellings) where fires and fire injuries are most common in 1994 is the kitchen, with 25 and 28 percent, respectively (Figure 8). The second most common area for both fires and injuries is the bedroom (13 and 26 percent, respectively), where children playing, careless smoking, and intentionally set fires are the three most frequent fire causes. The third most common location for fires is the chimney (10 percent), usually because it has not been cleaned often enough or well enough (not shown because casualties from chimney fires are not large). The largest number of deaths (31 percent) occur in living rooms, family rooms, and dens; here, careless smoking is a major cause as people doze off with lighted cigarettes. This profile of fires and casualties by room is very similar to that of the preceding several years.

NON-RESIDENTIAL PROPERTIES

Non-residential properties include industrial and commercial properties, institutions, educational establishments, vacant and under construction properties, and mobile properties. Because these categories each have quite different profiles, they are examined separately.

Structures

Fire prevention efforts have focused on protecting non-residential structures and the results have been successful to a large degree. These structures accounted for 9–10 percent of all fires, 5–6 percent of fire deaths, 13–14 percent of injuries, and 32–46 percent of dollar loss. In 1994, total deaths were at a 10-year low.

Figure 9 shows the relative magnitudes of the fire problem (fires and deaths) for the non-residential property category. Half of all deaths were in the storage facility, institution, and manufacturing plant categories—the leading three property types for deaths. Fires in non-residential structures are highest from noon to 8 p.m., which correspond to the times that they are open for business and when workers may be tiring and more accident prone. There are no clear peaks for day of the week or month of year for fires in non-residential structures.

Figure 10 shows that the leading cause of non-residential fires in 1994 is arson. Arson has been the leading cause of non-residential fire deaths for 8 of the past 10 years; in 1988 and 1994, careless smoking was the leading cause of deaths. Arson is also the leading cause of fire injuries.

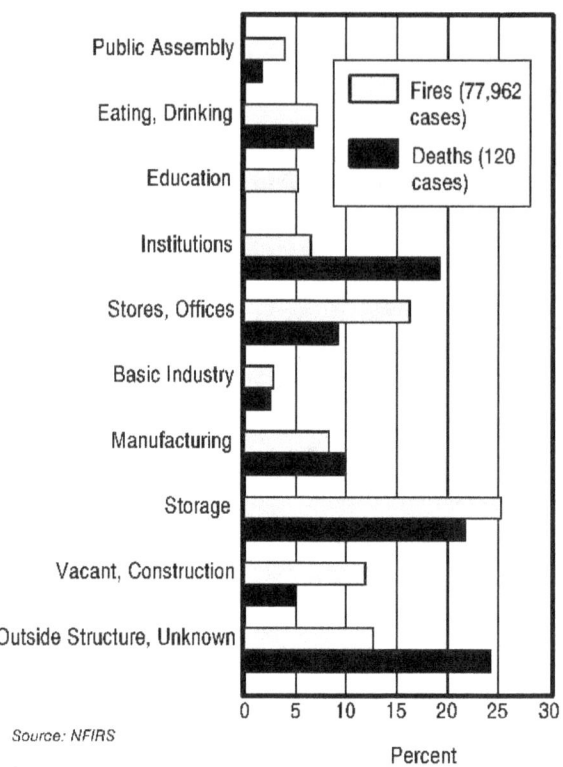

Source: NFIRS

Figure 9. 1994 Non-Residential Fires and Deaths by Property Type

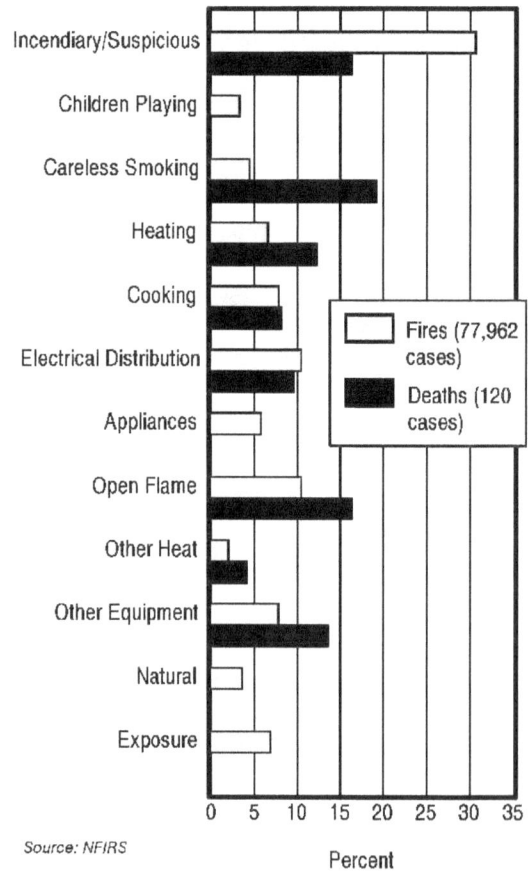

Source: NFIRS

Figure 10. Causes of 1994 Non-Residential Fires and Fire Deaths (Adjusted)

Vehicles

About one in four fires attended by the fire service involves vehicles, mainly cars and trucks. In 1994, the fire service responded to more vehicle fires than to residential fires. And this does not include the tens of thousands of fire department responses to vehicle accidents in which there was no fire.

Vehicles accounted for 17 percent of fire deaths reported to NFIRS, 10 percent of the injuries, and 13 percent of the dollar loss in 1994. The exact number of vehicle fire deaths is uncertain because of the difficulty in determining whether the impact of the accident or the subsequent fire was the cause of death in many cases. The 10-year trend in vehicle casualties has decreased steadily, perhaps due to better safety features that are being built into automobiles. However, there is no doubt that vehicles comprise a much larger segment of the fire problem than most people realize. Vehicle accidents merit more attention in fire prevention than they now receive.

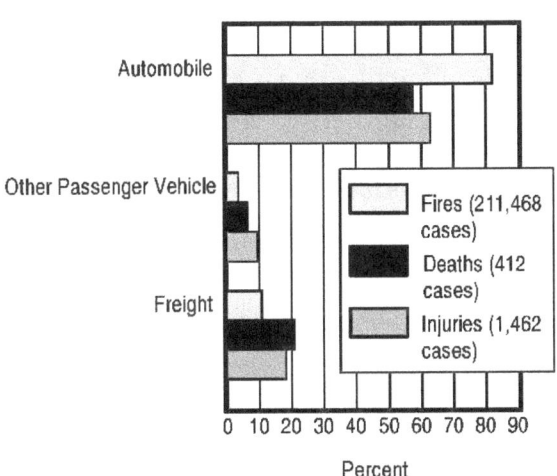

Note: Other mobile property types account for less than 4 percent of the total.

Source: NFIRS

Figure 11. 1994 Mobile Property Fires and Casualties by Major Property Type (Adjusted)

There are nearly eight times more fires associated with automobiles than with trucks (freight). However, truck fires account for a far higher proportion of deaths (one truck death for each three automobile deaths) (Figure 11).

Fires of incendiary or suspicious origin account for about one out of six mobile property fires, but may well be understated because many vehicle fires are not investigated as to cause.

Most vehicle fire deaths (63 percent in 1994) follow collisions, where preventive actions may or may not have been possible. Most of the total fires and personal injuries, however, are a result of mechanical or design problems. The overall vehicle fire loss problem is large enough to warrant adding vehicle fire prevention and possibly even accident prevention to other fire service public education programs.

11

Outside Properties

Outside fires (fires outside of structures other than vehicle) represent 44 percent of all fires in 1994 (about the same as in 1990). The 800,000 to 1 million outside fires to which fire departments respond represent a significant burden to the fire service. Although this category of property accounts for the highest number of fires, it represents the least amount of deaths (3 percent), injuries (4 percent), and dollar loss (3 percent). These numbers may not reflect the true nature of the problem because of underreporting, the difficulty in setting a price tag on outside fires, and the fact that many wildland fires are not reported to NFIRS or the NFPA annual survey.

This report examined the wildland fire problem in some detail. As urban areas continue to expand into rural and wilderness areas (called the "interface" area), the problem of fire is receiving greater attention. There has arisen a conflict between foresters, who recommend a "prescribed-burn" policy (burning within limits) in forests to remove the buildup of fuels on the forest floor, and the newly arrived populace, who are concerned about the safety of their homes and gardens. Also, as the population expands to the interface area, they may bring with them a perception that, as is the case in urban areas, local fire departments are capable of handling virtually all of the fires that occur. The reality is that the interface area presents a different set of risks than is usually faced by urban departments. The challenge is to better balance fire resources, public expectations, and prevention/mitigation efforts (e.g., better home construction and landscaping techniques) in interface areas.

FIREFIGHTER DEATHS AND INJURIES

Over the period 1985 to 1994, there has been a significant downward trend in firefighter deaths per year. A closer examination of this trend, however, shows that after a steady decline in firefighter deaths from 1988 to 1992, there has been a sharp increase in deaths—from 75 in 1992 to 104 in 1994 (Figure 12). The 1994 total includes the tragedy in which 14 firefighters were killed on Storm King Mountain in Colorado. But even if this major incident were excluded, the number of firefighter deaths increased significantly in 1994. In 1994, 78 percent of all firefighter deaths were at residential and wildland fires.

The number of firefighter injuries has changed little over the 10-year period. They averaged 100,000 per year. About 53 percent of the firefighter injuries were on the fireground. Of the firefighter injuries associated with fires, 58 percent occurred at residential fires and 25 percent at non-residential structural fires.

More firefighters get injured outside the fire building than inside. The torso, arms and hands, and legs and feet are about equally distributed as the body areas most often injured. Firefighter injuries are spread over the day, throughout the year, and across various age groups. They peak at night, in the winter, and between ages

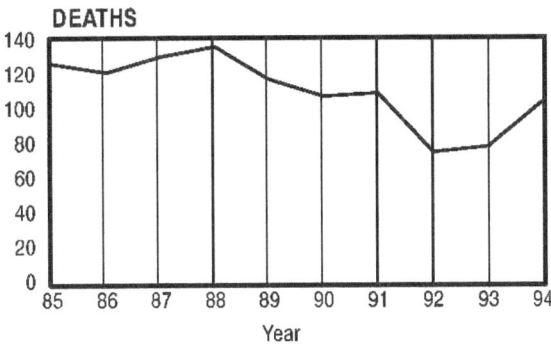
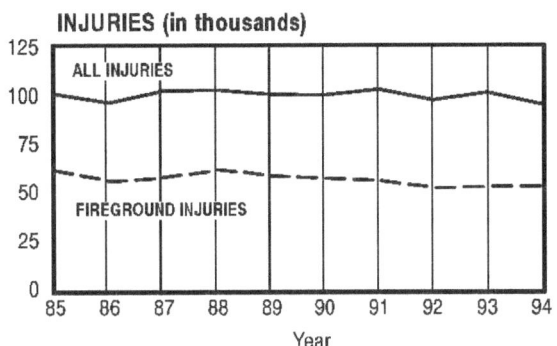

Sources: *NFPA Annual Surveys and the United States Fire Administration*

Figure 12. Trends in 1994 Firefighter Casualties

30–34. Local fire departments should consider their own profile of injuries and the reasons for any peaks in the profiles.

ARSON

Arson continues to be a pervasive problem affecting many of the major types of properties. It is the leading cause of total fires and dollar losses, primarily because it accounts for a heavy portion of loss in non-residential structures and outside properties. Annual direct cost of arson reported by the fire services is almost $3.6 billion. The true total is undoubtedly higher. Except for a spike in 1992, reported arson losses have remained somewhat steady over the 1985–1994 period. Arson is also the second leading cause of residential fire deaths and the third leading cause of fires in residences.

Arson is by far a more acute problem in urban areas than in rural locales. A 1995 study by the National Fire Protection Association shows that from 1990 to 1994, the rate for arson fires in cities of 250,000 or more was greater than twice the rate for communities of under 25,000.

In 1994, 20 percent of all arson fires occurred in structures, but these fires alone accounted for about 90 percent of all fire deaths, injuries, and dollar losses (Figure 13). But 69 percent of all arson fires are outdoor fires that cause few deaths and injuries and have relatively little impact on total dollar loss. Outdoor arson fires, however, are cause for concern because they can easily spread to structures. Although the 10-year trend for arson fires has gone down, arson is decreasing at a slower rate in residential structures than in non-residential structures. This is an important trend to watch given that more deaths and higher losses are associated with residential structures than in non-residential or commercial structures.

13

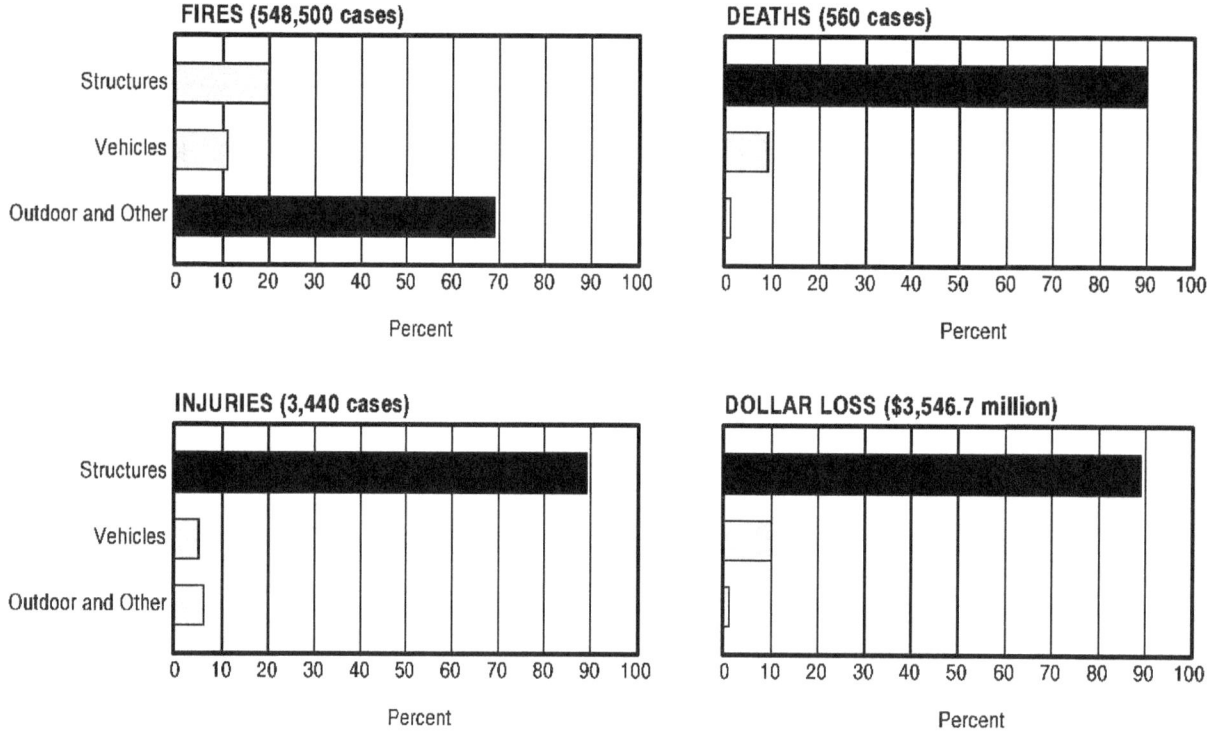

Sources: NFIRS and NFPA Annual Surveys

Figure 13. 1994 Arson Fires and Fire Losses by Major Occupancy Type

The NFPA reports that in 1994, for the first time, juvenile firesetters accounted for a majority of those who were arrested on arson charges. This growing problem is cause for serious concern and deserves more attention from policymakers.

The motives of the firesetter include vandalism, spite, revenge, intimidation, concealment of a crime, economic factors, and emotional/psychological dysfunctions. NFIRS collects data on the cause of fires, but more detailed data need to be collected on the motives of the arsonist in order to have a more complete picture of the arson problem. Furthermore, as fire investigation budgets have been reduced, many arson fires are not identified as such nor do they get fully investigated. Without thorough investigations and the willingness of local prosecutors to pursue cases against suspected arsonists, it will be difficult to make headway in the effort to reduce the incidence of arson fires in the United States.

TOTAL COST OF FIRE

The true cost of fire in the United States is much greater than just the value of property destroyed by fire—in the range of $92 to $139 billion (Table 1). The total cost includes the cost of fire services; the cost of fire protection built into buildings and equipment; the cost of fire insurance overhead; the many indirect costs, such as business interruptions, medical expenses, and temporary lodging; the value to society of the injuries and deaths caused by fire; the cost of government and private fire-related organizations; and the myriad of other related costs that add up to a very large economic impact.

Table 1. Summary of Total Cost of Fire

Cost Component	Range of Cost Estimates ($ billions)	Most Likely Estimate ($ billions)
Category A: Losses		
Residential Property	4.0	4.0
Industrial Property	4.2	4.2
Other Property	0.7	0.7
Residential Interruption	0.6–1.0	0.8
Business Interruption	6.1–8.4	8.4
Product Liability	3.5	3.5
Category B: Insurance		
Product Liability	0.1	0.1
Net Fire Insurance	5.6	5.6
Category C: Fire Service		
Paid	9.6	9.6
Volunteer Conversion	16.2–36.8	30.0
Category D: Preventative		
Built into Structures	20.7	20.7
Built into Equipment	13.5–22.5	18.0
Standards Activity	0.1–0.6	0.2
Retardants/Testing	1.9–4.0	2.5
Fire Maintenance	4.3–16.6	6.5
Disaster Recovery	0.6	0.6
Total	91.7–138.9	115.4

Source: William Meade, *A First Pass at Computing the Cost of Fire in a Modern Society,* The Herndon Group, March 1991; prepared for Center for Fire Research, National Institute of Standards and Technology, Gaithersburg, Maryland.

Because the cost of fire is perhaps 6 to 12 times higher than the direct losses alone, fire protection ranks among the larger national problems in terms of its economic impact. The full magnitude of the problem is probably underappreciated by the general public, media, and elected officials. The total cost is important to estimate when considering priorities across programs. Analysis of changes in incremental costs of the major components of the total cost of fire should be given more consideration in setting priorities than it is usually is.

U.S. FIRE VERSUS INTERNATIONAL FIRE PROBLEM

The United States historically has had one of the highest fire loss rates of the industrialized world in terms of both fire deaths and dollars loss. The fire death rates for 13 industrialized nations were compared with those in the United States. Although comparisons of total fires and total fire losses would be preferable, reliable data are not available due to diverse record keeping and fire classification practices in different countries.

Figure 14 depicts the average per capita fire death rates for 14 industrialized nations from 1979 to 1992. This figure demonstrates that the United States ranked only behind Hungary as having the highest per capita fire death rate. At a rate of 26.5 deaths per million population, the United States' fire death rate was more than five times that of Switzerland, which had the lowest rate of all the countries considered—5.2 deaths per million population.

Although the situation in the United States improved greatly between 1979 and 1992, the declining trend was international in scope. Of the countries considered, only Hungary and Denmark recorded increases in their fire death rates over that period. But the reduction of the fire death rate for the United States (46 percent, or 16.8 fire deaths per million population) was the largest absolute and relative drop of any of the countries shown.

Many people feel that there is little reason for the United States, which possesses a wealth of advanced fire suppression technologies and fire service delivery mechanisms, to lag so far behind other nations in terms of fire safety. However, most of the advanced fire technology used in the United States is installed in public places and most fire deaths occur in the home. Although statistical data are not available, the United States is widely believed to have many more residential fires on a per capita basis than the other countries studied. This higher fire rate, as well as the United States' higher fire death rates, is likely a product of several factors:

- The United States commits fewer resources, both in terms of dollars and staff time, to fire prevention activities than other industrialized countries.

- There is greater tolerance in the United States for "accidental" fires.

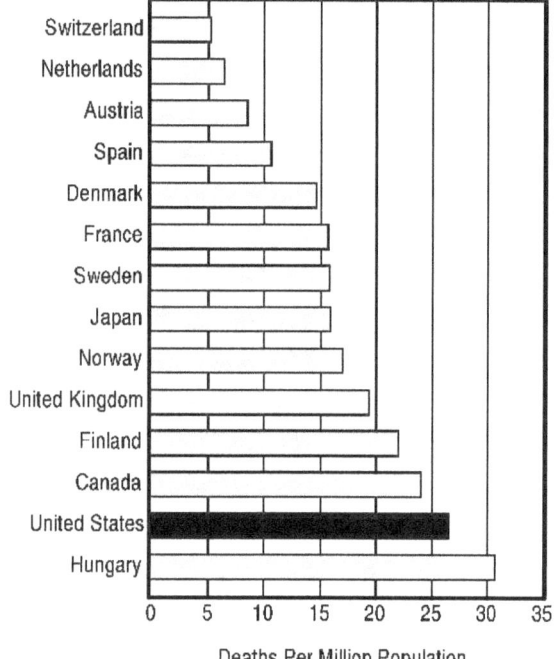

Deaths Per Million Population

Source: World Fire Statistics Centre

Figure 14. Average Fire Death Rate by Country (1979–92)

16

- Whether through ignorance or a false sense of confidence, Americans practice riskier and more careless behavior than people in other countries.

In sum, industrialized countries in Europe and Asia can provide the United States with valuable lessons on reducing the incidence of residential fires and residential fire deaths through fire prevention.

1
INTRODUCTION

The United States continues to have one of the most severe fire problems in the world relative to its population size. Most Americans are not aware of this nor of the nature of the fire problem.

This report is a statistical portrait of the fire problem in the United States over the period 1985--94. It is intended for use by a wide audience, including the fire service, the media, researchers, industry, government agencies, and interested citizens. The report focuses on the national fire problem. The magnitude and trends of the fire problem, the causes of fires, where they occur, and who gets hurt are topics that are emphasized. One specific focus is on firefighter casualties—causes, types of injuries, etc.

This document represents the sixth major edition of *Fire in the United States* published by the U.S. Fire Administration: the First Edition, published in 1978, covered 1975--76 fire data; the Second Edition, published in 1982, covered 1977--78; the Sixth Edition, published in 1987, covered 1983; the Seventh Edition, published in 1990, covered 1983--87; the Eighth Edition, published in 1991, covered 1983--90; and this Ninth Edition covers 1985--94. There were also three editions—the Third, Fourth, and Fifth—produced and used as working papers though not published.

This Ninth Edition builds on the previous edition by adding 4 years of new data. Trends and comparisons, therefore, cross a 10-year span, while profiles of causes, occurrences, age groups, sexes, and other meaningful factors are shown for 1994. Where appropriate, comparisons are made with the previous 1990 edition.

SOURCES

The report is primarily based on the National Fire Incident Reporting System (NFIRS) data, but uses other sources as well, especially the National Fire Protection Association's (NFPA's) annual survey of fire departments.

National Fire Incident Reporting System

The National Fire Incident Reporting System was started in 1975 as one of the first programs of the National Fire Prevention and Control Administration, which later became the U.S. Fire Administration (USFA). The basic concept of NFIRS has not changed since the system's inception. All states and all fire departments within them have been invited to participate on a voluntary basis. Participating fire departments collect a common core of information on fire and casualty reports using a common set of definitions. The data may be written by hand on paper forms or keyed in using computer terminals. Fire departments send these data either as a bundle of paper reports or on a computer tape to their state fire data office, which edits and collates the data. Semiannually or annually, the state's data are sent to the U.S. Fire Administration. There, the data are further validated. Data summaries and error reports may be sent back to the states to correct suspicious, incorrect, or incomplete information. Data on individual fire incidents and casualties are preserved incident by incident at local, state, and national levels.

The system has gradually grown from an initial 6 states in 1976 to 40 states and the District of Columbia in 1994. Table 2 lists the participating states over the 1985 to 1994 period. More than 25,000 fire departments provide data to the system, with more than half of these reporting to NFIRS (Table 3). Some states require their departments to participate. The future goal is voluntary participation by all states and the District of Columbia.

Corresponding to the increased participation, the number of fires, deaths, and injuries and the amount of dollar loss reported to NFIRS also grew considerably from 1975 to 1994. In 1994, over 896,000 fire incidents were collected, about 44 percent of the estimated total attended by fire departments.

There are, of course, many problems in assembling a real-world database, and NFIRS is no exception. But the enormous sample size and good efforts by the fire service allow a tremendous amount of useful information to be collected and used. Because of the enormous advances in computer technology over the past 20 years, plans for revision to NFIRS are presented in Appendix C.

Uses of NFIRS

The NFIRS data are used extensively for major fire protection decisions. At the federal level, for example, the Consumer Product Safety Commission uses the data to identify problem products and to monitor corrective actions. The Department of Transportation has used these data to identify car fire problems and has ordered recalls triggered by NFIRS data. The Department of Housing and Urban Development uses NFIRS to evaluate safety of manufactured housing (mobile homes). And of course the U.S. Fire Administration uses the data to design prevention programs, to order fire-fighter safety priorities, and for a host of other purposes. Thousands of fire departments, scores of states, and hundreds of industries have used the data. The potential for even greater use remains. One of the purposes of this report is to give some idea of the types of information available from NFIRS. The information here is highly summarized; much more detail is available.

Table 2. States Participating in NFIRS, 1985–1994

State	1985	1986	1987	1988	1989	1990	1991	1992	1993	1994
Alabama	X	X	X	X	X	X	X	X	X	X
Alaska	X	X	X	X	X	X	X	X	X	X
Arizona	X	X	X	X	X	X			X	X
Arkansas	X	X	X	X	X	X	X	X	X	X
California	X	X	X	X	X	X	X	X	X	X
Colorado	X	X		X	X	X	X	X	X	X
Connecticut	X	X	X	X	X	X	X	X	X	X
Delaware	X	X	X	X	X	X				
District of Columbia	X	X	X		X		X	X	X	X
Florida	X	X	X	X	X	X	X	X	X	X
Georgia					X	X	X	X	X	X
Hawaii	X	X	X	X	X	X	X	X		
Idaho		X	X	X	X	X	X	X	X	X
Illinois	X	X	X	X	X	X	X	X	X	X
Indiana		X	X	X	X	X	X	X	X	
Iowa	X	X	X	X	X	X	X	X	X	X
Kansas	X	X	X	X	X	X	X	X	X	X
Kentucky	X	X	X	X	X	X	X	X	X	X
Louisiana	X	X	X	X	X	X	X	X	X	X
Maine	X	X	X	X	X	X				
Maryland	X	X	X	X	X	X	X	X	X	X
Massachusetts	X	X	X	X	X	X	X	X	X	X
Michigan	X	X	X	X	X	X	X	X	X	X
Minnesota	X	X	X	X	X	X	X	X	X	X
Mississippi										
Missouri										
Montana	X	X	X	X	X	X	X	X	X	X
Nebraska	X	X	X	X	X	X	X	X	X	X
Nevada										
New Hampshire	X	X	X	X	X	X	X	X	X	X
New Jersey	X	X	X	X	X	X	X	X	X	X
New Mexico										X
New York	X	X	X	X	X	X	X	X	X	X
North Carolina	X									
North Dakota										
Ohio	X	X	X	X	X	X	X	X	X	X
Oklahoma							X	X	X	X
Oregon	X	X	X	X	X	X	X	X	X	X
Pennsylvania										
Rhode Island	X	X	X	X	X	X	X	X	X	X
South Carolina	X	X	X	X	X	X	X	X	X	X
South Dakota	X	X	X	X	X	X	X	X	X	X
Tennessee	X	X	X	X	X	X	X	X	X	X
Texas	X	X	X	X	X	X	X	X	X	X
Utah	X	X	X	X	X	X	X	X	X	X
Vermont	X	X	X	X	X	X	X	X	X	X
Virginia	X	X	X	X	X	X	X	X	X	X
Washington	X	X	X	X	X		X	X	X	X
West Virginia	X	X	X	X	X	X	X	X	X	X
Wisconsin	X	X	X	X	X	X	X		X	X
Wyoming		X	X	X	X	X	X	X	X	X
Total	40	42	41	41	43	41	41	40	41	41

Table 3. Fire Departments Reporting to NFIRS—1994

Participating State	No. of Participating Fire Departments	No. of Fire Departments in State	Fire Departments Reporting (percent)
Alabama	1	1,072	0.1
Alaska	78	253	31
Arizona	5	258	2
Arkansas	355	824	43
California	373	1,153	32
Colorado	25	400	6
Connecticut	215	274	78
District of Columbia	1	1	100
Florida	330	674	49
Georgia	151	718	21
Idaho	157	214	73
Illinois	884	1,330	66
Iowa	539	869	62
Kansas	557	674	83
Kentucky	474	794	60
Louisiana	376	700	54
Maryland	355	370	96
Massachusetts	326	364	90
Michigan	925	1,030	90
Minnesota	650	821	79
Montana	208	551	38
Nebraska	299	483	62
New Hampshire	85	253	34
New Jersey	385	788	49
New Mexico	1	359	0.3
New York	1,647	1,809	91
Ohio	906	1,300	70
Oklahoma	504	857	59
Oregon	329	325	101
Rhode Island	45	81	56
South Carolina	157	655	24
South Dakota	221	343	64
Tennessee	189	655	29
Texas	526	2,000	26
Utah	121	211	57
Vermont	113	252	45
Virginia	438	702	62
Washington	53	655	8
West Virginia	428	442	97
Wisconsin	222	901	25
Wyoming	109	252	43
Total	13,763	26,667	52

NFPA and Other Data Sources

In addition to NFIRS, this report makes use of the summary numbers for fires, deaths, injuries, and dollar loss from the National Fire Protection Association's (NFPA) annual survey of fire departments and NFPA *Fire Command* articles on firefighter casualties. It also uses data obtained from state fire marshals, the National Center for Health Statistics, the Bureau of the Census, and the Consumer Product Safety Commission. The U.S. Fire Administration gratefully acknowledges the use of their information. Sources are cited for each graph and table in the report.[1]

METHODOLOGY

An attempt was made to keep the data presentation and analysis as straightforward as possible. It was also the desire of the USFA to make the report accessible to the largest group of users, and therefore an attempt was made to avoid any unnecessarily complex methodology.

National Estimates

Most numbers in this report are national estimates or percentages, not raw totals from NFIRS. The reader does not have to scale the data.

Many of the estimates are derived by computing a percentage from NFIRS and multiplying it by the total number of fires, deaths, injuries, or dollar loss from the NFPA annual survey. For example, the national estimate for the number of residential cooking fires was computed by taking the percentage of NFIRS residential fires (with known causes) that were attributed to cooking and multiplying it by the estimated total number of residential fires from the NFPA survey.

Ideally, one would like to have all of the data come from one self-consistent data source. But because the "residential population protected" was not reported to NFIRS by many fire departments and the reliability of that data element is suspect in many other cases, especially where a county is served by several fire departments which each report their population protected independently, this data element was not used. Instead, the extrapolation of the NFIRS sample to national estimates was made using the NFPA survey for the gross totals of fires, deaths, injuries, and dollar loss.

One problem with this approach is that the proportions of residential, non-residential, mobile property, and outside fires and fire deaths differ between the large NFIRS sample and the NFPA survey sample. For example, in NFIRS, residential fire injuries in 1994 comprise 68 percent of the total, versus 73 percent in the NFPA survey. Mobile property (mostly motor vehicles) deaths in 1994 are 17 percent in NFIRS and 15 percent in NFPA. To be consistent with approaches being used by the Consumer Products Safety Commission and NFPA, we have used the NFPA estimates of fires, deaths, injuries, and dollar loss for residential, non-residential, mobile, and outside properties as a

---

[1] The term *fire losses* in captions refers to deaths, injuries, and dollar loss; the term *fire casualties* refers to deaths and injuries.

starting point. The details of the national fire problem below this level are based on proportions from NFIRS. One will not get the same numbers starting from the NFIRS proportions of residential, non-residential, etc., as from the NFPA proportions. This inconsistency will remain until all estimates can be derived from NFIRS alone.

In the future, the national estimates will be derivable solely from NFIRS if a statistically sufficient number of fire departments participating in NFIRS provide reasonably accurate estimates of their population protected.

Unknowns

On a fraction of the incident reports or casualty reports sent to NFIRS, the desired information for many data items is either left blank or reported as "unknown." The total number of blank or "unknown" entries is often larger than some of the important subcategories. For example, 44 percent of the fires in 1994 do not have sufficient data reported to NFIRS to determine cause. The lack of data, especially for fatal fires, masks the true picture of the fire problem. Many prevention and public education programs use the NFIRS data to target at-risk groups or to address critical problems, fire officials use the data in decisionmaking that affects the allocation of firefighting resources, and consumer groups and litigators use the data to assess product fire incidence. When the unknowns are large, the credibility of the data suffers. Fire departments need to be more aware of the effect of incomplete reporting.

Adjusted Percentages

In making national estimates, the unknowns should not be ignored. The approach taken in this report is to provide not only the "raw" percentages of each category, but also the "adjusted" percentages computed using only those incidents for which the cause was provided. This in effect distributes the fires for which the cause is unknown in the same proportion as the fires for which the cause is known, which may or may not be approximately right. That is the best we can do without additional knowledge of the nature of the unknowns.

To illustrate: Heating was reported as the fire cause for 9.8 percent of residential fire fatalities; another 32.6 percent of residential fatalities had cause unknown; thus, the percent of fatalities that had their cause reported was 100 - - 32.6 = 67.4 percent. With the unknown causes proportioned like the known causes, the adjusted percent of residential fire fatalities caused by heating can then be computed as 9.8 ÷ 67.4 = 14.5 percent.

Representativeness of the Sample

The percentage of fire departments participating in NFIRS varies state to state, and some states are not participating at all. To the best that USFA can determine, the distribution of participants is at least reasonably representative of the entire nation, even though the sample is not random. The sample is so large—over 40 percent of all fires—and so well distributed geographically and by size of community that there is no known major bias that will affect the results. Most of the NFIRS data exhibit stability from one year to another, without radical changes, as will be observed from the 10-year trend lines presented throughout this report. Also, results based on the full data set are generally similar to those based on part of the data, another indication of data reliability. Although the individual incident reports could and should be filled out more completely and more accurately than they are today—as can be said about most real-world data collections as large as NFIRS—the "big picture" is a reasonably accurate description of the fire problem in the United States. It is the best one we have ever had.

Trend Data

A frequently asked question is how much a particular aspect of the fire problem has changed over time. The usual response is in terms of a percentage change from one year to another. As we are dealing with real-world data that fluctuate from year to year, a percent change from one specific year to another can be misleading. This is especially true when the beginning and ending data points are extremes—either high or low. For example, in Figure 15, "Trends in Fires and Fire Losses," the percent change from 1985 of 28,425 injuries to 1994 of 27,250 injuries would be a decrease of 4 percent. Yet, if we were to choose 1986 as the beginning data point (26,826 injuries), this change would be show a 1.6 percent increase. As we are interested in *trends* in the U.S. fire problem, this edition of *Fire in the United States* reports the overall change in a data series as a trend. We have computed the best-fit trend line (which accounts for the fluctuation in the year-to-year data) and have presented the change over time based on this trend line. In this example, the overall 10-year trend is an increase in injuries of 2.6 percent—not the 4 percent decrease calculated from only beginning and ending years.

Cause Categories

The causes of fires are often a complex chain of events. To make it easier to grasp the "big picture," 13 major categories of fire causes such as heating, cooking, and children playing are used by the U.S. Fire Administration here and in many other reports. The alternative is to present scores of detailed cause categories or scenarios, each of which would have a relatively small percentage of fires. For example, "heating" includes subcategories such as misuse of portable space heaters, wood stove chimney fires, and fires involving gas central heating systems. Experience has shown that the larger categories are useful for an initial presentation of the fire problem. It then can be followed by more detailed analysis, as needed.

The cause categories used in this project are listed in the same order on each graph to make comparisons easier from one to another. The order here also is the same as used in previous *Fire in the United States* reports. The particular order chosen was a combination of the order used in the cause sorting hierarchy and a desire to put the more important causes in the top half of the charts.

A problem to keep in mind when considering the rank order of causes in this report is that suffi-cient data to categorize the cause were not reported to NFIRS for 40 percent of the fatal fires in the database. The rank order of causes might be different than shown here if the cause profile for the fires whose causes were not reported to NFIRS were substantially different from the profile for the fires whose causes was reported. However, there is no information to indicate that there is a major difference between the knowns and the unknowns, and so our present best estimate of fire causes is based on the distribution of the fires with known causes.

Fires are assigned to one of the 13 general cause groupings using a hierarchy of definitions, approximately as shown in Table 4.[2] A fire is included in the highest category into which it fits on the list. If it does not fit the top category, then the second one is considered, and if not that one, the third, and so on. For example, a fire caused by an arsonist using a match to ignite a fuse is included in the "incendiary or suspicious" category and not in the "open flame" category. If the arsonist used a cigarette to ignite the fuse, the fire still is grouped with incendiary and suspicious fires and not with "careless smoking" fires.

The NFIRS fire data can be analyzed in many ways such as by the form of the heat of ignition, the material ignited, the ignition factor, or many other groupings. The hierarchy used in this report has proved useful in understanding the fire problem and targeting prevention, but other approaches are certainly useful too. Because the NFIRS database stores records fire by fire and not just in sum-mary statistics, a very wide variety of analyses are possible.

Ratio of NFIRS to NFPA Data

There is an inconsistency between the NFIRS sample and the NFPA annual survey data: In every year, the deaths reported to NFIRS are a larger fraction of the NFPA estimate of deaths than the NFIRS fires are of the NFPA estimate of fires. NFIRS injuries and dollar loss are even larger fractions of the NFPA totals than are deaths or fires. This issue is discussed further in Appendix A.

Unreported Fires

NFIRS only includes fires to which the fire service was called. In some states, fires attended by state fire agencies (such as forestry) are included; in other states, they are not.

NFIRS does not include fires from ten states and many fire departments within participating states. However, if the fires from the reporting departments are reasonably representative, this omission does not cause a problem in making accurate national estimates for any but the smallest subcategories of data.

---

[2] The exact hierarchy and specific definition in terms of the NFIRS code may be found on pages 2--201 to 2--203 of the 1990 NFIRS System Documentation Manual, Version 4.1. The actual hierarchy involves a large number of sub-categories that are later grouped into the 13 major categories.

Table 4. Hierarchy of Cause Groupings Used in This Report

Cause Category*	Definition
Exposure	Caused by heat spreading from another hostile fire
Incendiary/Suspicious	Fire deliberately set or suspicious circumstances
Children Playing	Includes all fires caused by children playing with any materials contained in the categories below
Natural	Caused by Sun's heat, spontaneous ignition, chemicals, lightning, static discharge
Smoking	Cigarettes, cigars, pipes as accidental heat of ignition
Heating	Includes central heating, fixed and portable local heating units, fireplaces and chimneys, water heaters as source of heat
Cooking	Includes stoves, ovens, fixed and portable warming units, deep fat fryers, open grills as source of heat
Electrical Distribution	Includes wiring, transformers, meter boxes, power switching gear, outlets, cords, plugs, lighting fixtures as source of heat
Appliances (including air conditioning/refrigeration)	Includes televisions, radios, phonographs, dryers, washing machines, vacuum cleaners, hand tools, electric blankets, irons, electric razors, can openers, dehumidifiers, water cooling devices, air conditioners, refrigeration equipment as source of heat
Other Equipment	Includes special equipment (radar, x-ray, computer, telephone, transmitters, vending machine, office machine, pumps, printing press), processing equipment (furnace, kiln, other industrial machines), service, maintenance equipment (incinerator, elevator), separate motor or generator, vehicle in a structure, unspecified equipment
Open Flame, Spark (heat from)	Includes torches, candles, matches, lighters, open fire, ember, ash, rekindled fire, backfire from internal combustion engine as source of heat
Other Heat	Includes fireworks, explosives, heat or spark from friction, molten material, hot material, all other fires caused by heat from fuel-powered objects, heat from electrical equipment arcing or overloading, heat from hot objects not covered by above groups
Unknown	Cause of fire undetermined or not reported

* Fires are assigned to a cause category in the hierarchical order shown. For example, if the fire is judged incendiary and a match was used to ignite it, it is classified as incendiary and not open flame, because incendiary is higher on the list. One minor deviation: if the fire involves air conditioning or refrigeration, it is included in appliances and not electrical distribution.

An enormous number of fires are not reported to the fire service at all. Most are small fires in the home or in industry which go out by themselves or are extinguished by the occupant. These unreported fires collectively cause a great deal of property loss and a large number of injuries requiring medical attention, based on a study done in the early 1970s. CPSC commissioned a study in 1984 on unreported fires. We do not have a current study that can be used to estimate the magnitude of the problem.

Perhaps the most disturbing type of unreported fires are those not submitted by fire departments that are participating in NFIRS. Some departments submit information on most but not all of their fires. Sometimes the confusion is systematic, as when no-loss cooking fires or chimney fires are not reported. Sometimes it is inadvertent, such as when incident reports are lost or accidentally not all submitted. The information that is received is assumed to be the total for the department and is extrapolated as if it were. There is no measure of the extent of this problem at present. It does not seem to be large enough to distort the major aspects of the picture, but it needs to be monitored in the future.

ORGANIZATION OF THIS REPORT

This report is organized similarly to the Eighth Edition of *Fire in the United States*. Chapter 2 presents an overview of the national fire problem in terms of the total number of fires, deaths, injuries, and dollar loss—the four principal measures used to describe the fire problem.

Chapters 3 through 5 address residences, non-residences, and firefighter casualties, respectively. Sections within each chapter focus on subsets of the subject (e.g., within residential: one- and two-family dwellings versus apartments, vehicles, motels, etc.). Chapter 6 focuses on special topics bearing on the fire problem, including a comparison of U.S. fire statistics with 13 industrial nations, a close look at the magnitude of the arson and wildland fire problems, and identification of the cost of U.S. fires. At the end of each chapter, a section describes available resources that provide in-depth information on specific topics.

Previous editions of *Fire in the United States* have presented a state-by-state look at residential fires and deaths for 1 year and for a 3-year period. This analysis has been omitted from this edition, but will be published as a separate report in 1997. However, a quick-look, 50-state comparison chart of state fire death rates is available in Chapter 2.

Appendix A discusses the differences between NFPA and NFIRS data.

Most of the data are presented graphically for ease of comprehension. The specific data associated with the graphs are usually provided directly with the graphic. Where the data are too numerous to include in the graphic, data tables are presented in Appendix B.

The National Fire Incident Reporting System, originally designed in 1975, has become outdated and cumbersome by today's standards. Therefore, NFIRS is being redesigned. A summary of the planned changes is presented in Appendix C.

This edition of *Fire in the United States* concludes with an a comprehensive index to the topics of this report.

2

THE NATIONAL FIRE PROBLEM

OVERVIEW

The United States has a severe fire problem, more so than is generally perceived. Nationally, there are millions of fires, thousands of deaths, tens of thousands of injuries, and billions of dollar loss—which make the U.S. fire problem one of great national importance.

Although we have made much progress in the last decade, the United States continues to have one of the highest per capita fire death rates in the world. The United States has an average of 5,277 fire deaths a year from 1985 to 1994 (Figure 15). The number of deaths has been steadily trending down—32 percent over the past 10 years. In 1994, the number of deaths was 4,275.

We are less certain of the injury statistics in Figure 15 because of ambiguity about the completeness of defining and reporting minor injuries and the fact that many injured people go directly to a medical care facility themselves without going through a fire department screening. There was an average of nearly 28,700 reported civilian injuries per year from reported fires over the past 10 years and an average of 56,620 injuries to firefighters from those fires, as shown in Chapter 5, Figure 112. The actual totals for reported fires may be even higher. Furthermore, past studies suggest that the number of civilian injuries associated with fires that are not reported to the fire service might be double or more that of the number from reported fires, as discussed in Chapter 1. Fire-caused injuries to civilians trended up by 2.6 percent over the 10-year period. Injuries in 1994 were at their lowest level since 1986.

In terms of dollar loss, the estimated direct value of property destroyed in fires was $8.2 billion for 1994. The total cost of fire (direct losses, the cost of fire departments, built-in fire protection in new buildings, insurance overhead, and other annual fire protection expenditures) is much higher. The direct dollar loss increased 21 percent from 1985 to 1994, with the increase due to inflation. Using constant 1994 dollars, the loss was down by 13 percent over this period. Still, the direct dollar loss was enormously high at an average of $9.4 billion a year in adjusted 1994 dollars.

These casualties and losses come from an average of nearly 2.2 million fires a year. The trend in fire incidents has declined 19 percent since 1985, with the sharpest decline in 1989 but remaining steady since then.

On a per capita basis, the fire problem appears less severe today than 10 years ago, partially because the population has been increasing faster than have fires and fire casualties and partially

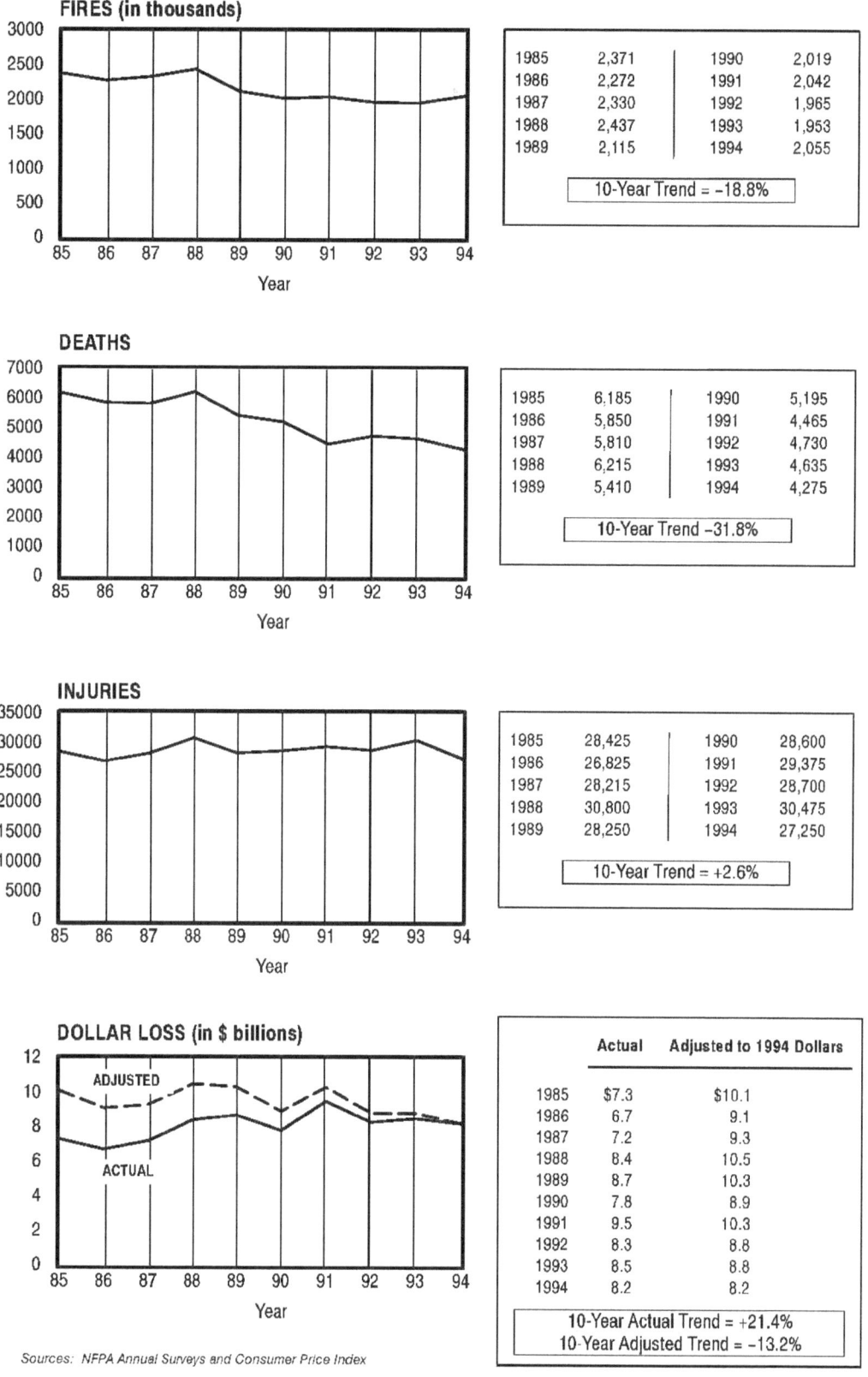

FIRES (in thousands)

1985	2,371	1990	2,019
1986	2,272	1991	2,042
1987	2,330	1992	1,965
1988	2,437	1993	1,953
1989	2,115	1994	2,055

10-Year Trend = –18.8%

DEATHS

1985	6,185	1990	5,195
1986	5,850	1991	4,465
1987	5,810	1992	4,730
1988	6,215	1993	4,635
1989	5,410	1994	4,275

10-Year Trend –31.8%

INJURIES

1985	28,425	1990	28,600
1986	26,825	1991	29,375
1987	28,215	1992	28,700
1988	30,800	1993	30,475
1989	28,250	1994	27,250

10-Year Trend = +2.6%

DOLLAR LOSS (in $ billions)

	Actual	Adjusted to 1994 Dollars
1985	$7.3	$10.1
1986	6.7	9.1
1987	7.2	9.3
1988	8.4	10.5
1989	8.7	10.3
1990	7.8	8.9
1991	9.5	10.3
1992	8.3	8.8
1993	8.5	8.8
1994	8.2	8.2

10-Year Actual Trend = +21.4%
10-Year Adjusted Trend = –13.2%

Sources: NFPA Annual Surveys and Consumer Price Index

Figure 15. Trends in Fires and Fire Losses

because of the overall decline in numbers of reported fires. Over this 10-year period, reported fires averaged 8.7 per thousand population (Figure 16).

The trend in fire death rate per million population has declined a significant 38 percent. In terms of injuries, the per capita rate was down 6 percent over 10 years. Although dollar loss per capita was $31, up 11 percent unadjusted, it trended down 26 percent over the 10 years when adjusted for inflation.

THE BROADER CONTEXT

Fires constitute a much larger problem than is generally known. Losses from all natural disasters combined—floods, hurricanes, tornadoes, earthquakes, etc.—average a fraction of the annual direct dollar losses from fire. Deaths from disasters have tended to be vastly fewer than from fires—on the order of 200 per year for disasters versus more than 4,000 for fires.

Most fires are relatively small, and their cumulative impact is not easily recognized. There are only a few fires that have the huge dollar losses that are associated with hurricanes or floods. The southern California wildland fires in the fall of 1993 resulted in over $800 million in losses. The Oakland East Bay Hills fire of October 1991 was estimated to have caused over $1 billion in losses. The Phillips petrochemical plant fire in the Houston ship channel in October 1989 caused several hundred million dollars in losses. But because most of the losses from fire are spread over the more than 2 million fires that are reported each year, the total loss is far more than the impression many people have of it from the anecdotal reporting of local fires in the media.

Fires also are an important cause of accidental deaths. The National Safety Council ranks fires as the fifth leading cause of accidental deaths, behind only vehicle accidents, falls, poisonings, and drownings

Fire-related injuries to civilians and firefighters are reported with too much uncertainty to properly rank them with confidence, but it is clear that they number over 100,000 and possibly two or three times that many when injuries from unreported fires and unreported injuries from reported fires are taken into account. Burn injuries are particularly tragic because of the tremendous pain and suffering they cause. Serious burns tend to cause psychological damage as well as physical damage, and they may well involve not only the victims but also their family, friends, and fellow workers.

U.S. Fire Deaths Versus Other Nations

The United States has one of the most severe fire problems in the industrialized nations. Although our per capita death rate is nearly half what it was in the late 1970s, and down 38 percent since 1985, current international data (1992) suggest that the United States has a fire death rate two to three times that of several European nations and at least 20 percent higher than most. In 1992,

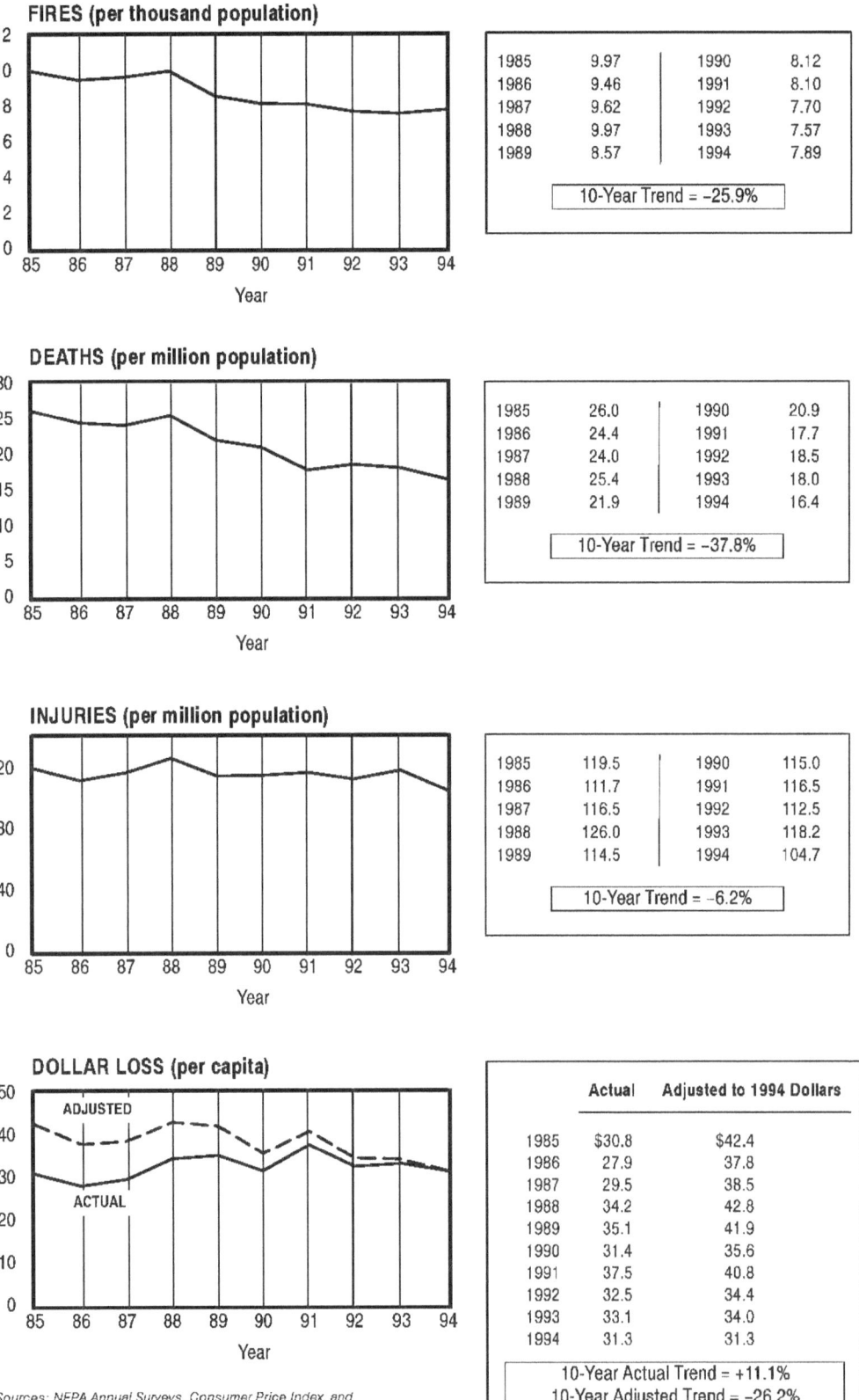

FIRES (per thousand population)

1985	9.97	1990	8.12
1986	9.46	1991	8.10
1987	9.62	1992	7.70
1988	9.97	1993	7.57
1989	8.57	1994	7.89

10-Year Trend = –25.9%

DEATHS (per million population)

1985	26.0	1990	20.9
1986	24.4	1991	17.7
1987	24.0	1992	18.5
1988	25.4	1993	18.0
1989	21.9	1994	16.4

10-Year Trend = –37.8%

INJURIES (per million population)

1985	119.5	1990	115.0
1986	111.7	1991	116.5
1987	116.5	1992	112.5
1988	126.0	1993	118.2
1989	114.5	1994	104.7

10-Year Trend = –6.2%

DOLLAR LOSS (per capita)

	Actual	Adjusted to 1994 Dollars
1985	$30.8	$42.4
1986	27.9	37.8
1987	29.5	38.5
1988	34.2	42.8
1989	35.1	41.9
1990	31.4	35.6
1991	37.5	40.8
1992	32.5	34.4
1993	33.1	34.0
1994	31.3	31.3

10-Year Actual Trend = +11.1%
10-Year Adjusted Trend = –26.2%

Sources: NFPA Annual Surveys, Consumer Price Index, and
Bureau of the Census

Figure 16. Trends in Severity of Fires and Fire Losses

our fire death rate was reported at 19.5 deaths per million population.[1] Switzerland's rate was 5.3 per million population; Canada's was 15.8. In fact, of the 14 industrial nations that are examined in detail in Chapter 6, the U.S. rate was higher than all but one—Hungary.

The declining U.S. trend in fire death rate over the past 10 years was not a singular event; all countries except Hungry and Denmark also trended downward. Furthermore, although statistical data are not available, the United States is widely believed to have many more residential fires on a per capita basis that any of the 13 countries studied.

The United States has placed greater emphasis on improving the technology in fire suppression and fire service delivery mechanisms than other nations, but these nations tend to surpass the U.S. in practicing fire prevention. The United States would be well served by studying and implementing international fire prevention programs that have proved effective in reducing the number of fires and deaths.

Total Cost of Fire

The total cost of fire to society is staggering—over $100 billion per year.[2] This includes the cost of adding fire protection to buildings, the cost of paid fire departments, the equivalent cost of volunteer fire departments ($20 billion annually), the cost of insurance overhead, the direct cost of fire-related losses, the medical cost of fire injuries, and other direct and indirect costs. Even if these numbers are high by as much as 100 percent, the total costs of fire would range from $50 to $100 billion, still enormous, and on the order of 1 to 2 percent of the gross domestic product, which was $6.74 trillion in 1994.[3] Thus from a monetary viewpoint, fire ranks among the significant national problems. A detailed examination of the total cost of fire to U.S. society is presented as a special topic in Chapter 6.

FIRE CASUALTIES BY POPULATION GROUP

The fire problem is more severe for certain groups than others. People in the Southeast, males, the old, and the very young all are at much higher risk from fires than the rest of the population.

Regional Differences

The Southeast of the United States continues to have the highest fire death rate in the nation and one of the highest in the world. Figure 17 shows the states with the highest fire death rates for

---

[1] World Fire Statistics Centre. Using NFPA estimates and Bureau of the Census data, however, the 1992 U.S. fire death rate is computed at 18.5 per million population.

[2] Meade, William P., *A First Pass at Computing the Cost of Fire in a Modern Society*, The Herndon Group, Inc., February 1991.

[3] U.S. Department of Commerce's Bureau of Economic Analysis.

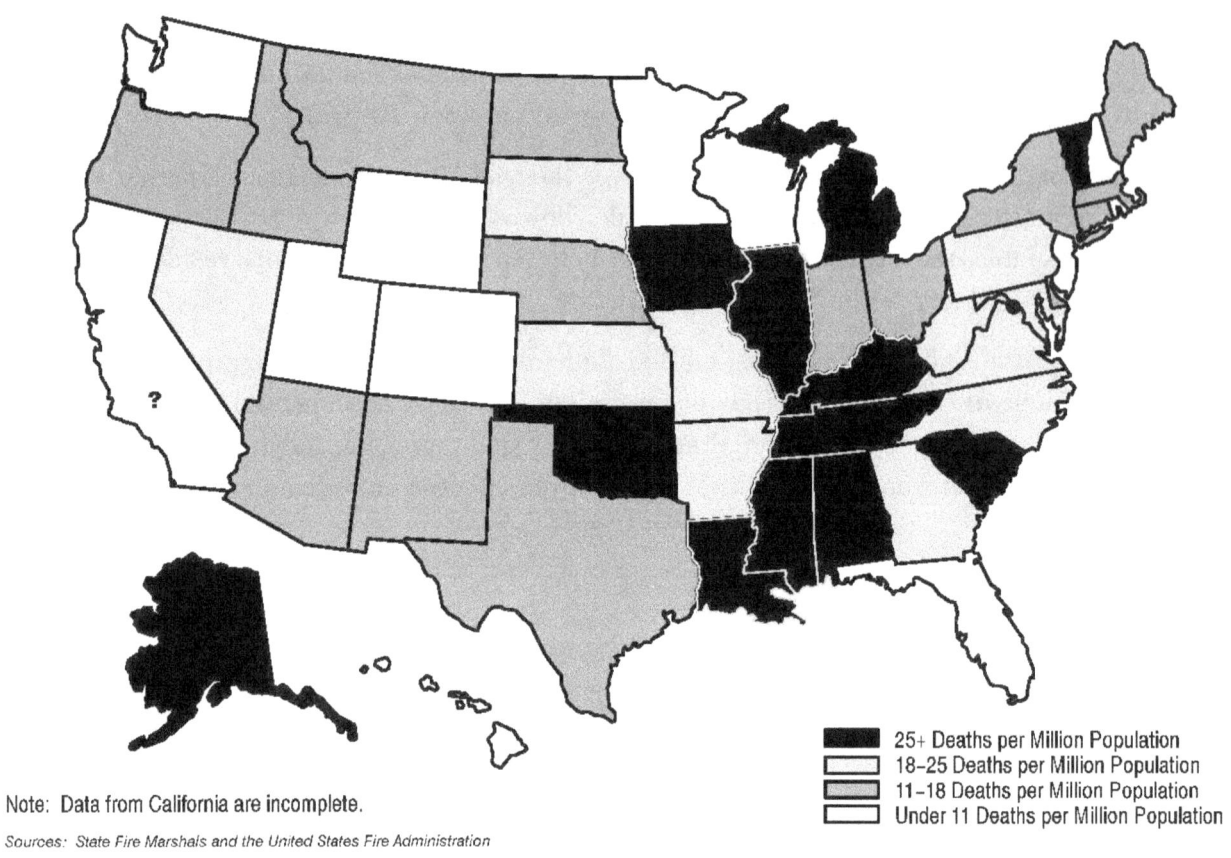

Note: Data from California are incomplete.

Sources: State Fire Marshals and the United States Fire Administration

■ 25+ Deaths per Million Population
□ 18-25 Deaths per Million Population
▨ 11–18 Deaths per Million Population
□ Under 11 Deaths per Million Population

Figure 17. Fire Death Rate by State in 1994

1994. Although considerable improvements have been seen in the death rates of the southeastern states, nearly half of these still have death rates in the highest category, here at 25 or more deaths per million population. In addition to the Southeast, the states of Alaska, Illinois, Iowa, Michigan, Vermont, and the District of Columbia were in the highest fire death rate category in 1994. The Southeast and Alaska have been consistently among the highest fire death rate areas for many years; however, their rates have been dropping along with those of the whole nation.

The next two categories of states in Figure 17, the shaded and the striped, still have fire death rates higher than many of the developed nations in Europe and the Far East. At the other extreme are the states with no shading. These "best" states are in the general range of the nations of Europe and the Far East. They tend to be the Southwest and West states, but there are some noteworthy others: Florida, Minnesota, New Hampshire, New Jersey, Rhode Island, and Wisconsin all had a low year in 1994. Florida has the lowest death rate among the high population states.[4]

Fire death rates for each state and the District of Columbia for the past 9 years are shown in Figure 18 (data for 1985 were unavailable). An overlay on each state chart represents the national

---

[4] California has had the lowest death rate. For 1994, however, the California fire death statistics are incomplete.

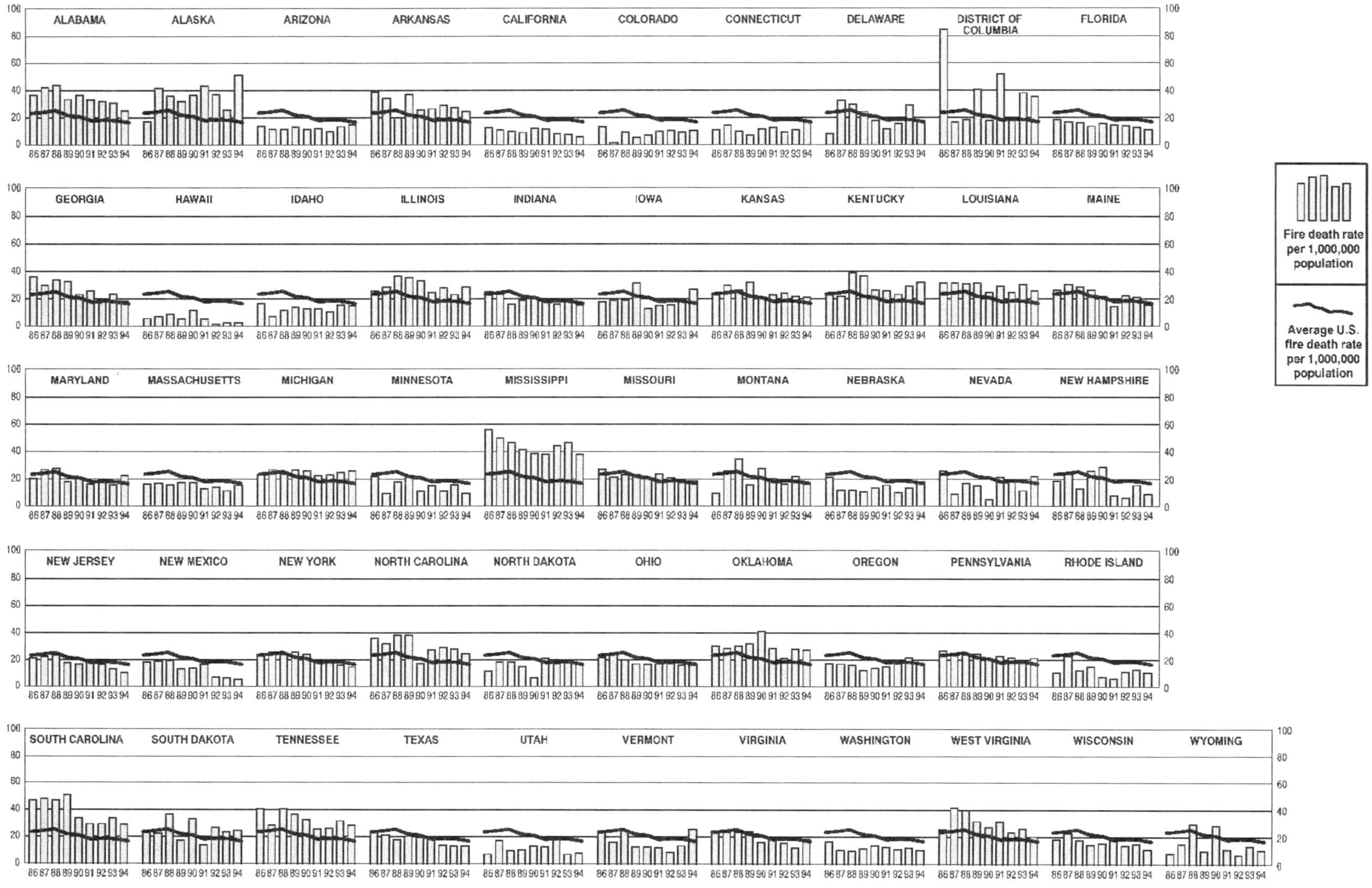

Figure 18.   9-Year Fire Death Rate by State Compared to National Average

Fire death rate per 1,000,000 population

Average U.S. fire death rate per 1,000,000 population

Source: State Fire Marshals

fire death rates. Eleven states are consistently above the national average and 18 states are consistently below it. The fire profile by state will be published separately in 1997.

The rank order of state fire death rates per million population is shown in Figure 19. States with relatively small populations may move up and down on the list from year to year as a result of only a few deaths; their death rate should be considered averaged over time. For example, both New Hampshire and Wyoming changed from one of the highest death rates from fires in 1990 to among the lowest in 1994; Iowa and Vermont were the only states to go from best to worse. The highest states were Alaska, Mississippi, District of Columbia, and Kentucky. The lowest were Hawaii, New Mexico, Utah, New Hampshire, and Minnesota.

Figure 20 shows the rank order of states in terms of the absolute number of fire deaths. Not surprisingly, large population states are at the top of the list. The 10 states with the most fire deaths account for nearly half of the national total. Unless their fire problems are significantly reduced, the national total will be difficult to lower.

The sum of the state death estimates in Figure 20 is nearly 150 deaths below the estimate of 4,275 from the NFPA survey for 1994. This difference may be due to some states underreporting their fire deaths, such as California, or an overestimate from the extrapolation of the NFPA sample of fire departments, or a combination of both. Nevertheless, the correspondence between the two sources should be considered quite good.

Gender

Men continue to have almost twice as many fire deaths as women. Figure 21 shows that the high proportion of male fire deaths has been remarkably steady over the past 10 years. The slight trend toward narrowing the gap between male and female fire deaths appears to have stopped in 1990. Males also have a higher fire death rate per capita than females for essentially all age groups. From the age of 20 on, males have twice the fire death rate as women (Figure 22).

Figure 21 also shows that the male/female ratio for fire injuries is similar to that for fire deaths except that the gender gap is narrowing. Injuries per capita for males are one and one-half to two times the female rate until age 70 (Figure 22), which can be expected because of the longer lifespan of women.

The reasons for the disparity of fire injuries between men and women are not known for certain. Suppositions include the greater likelihood of men being intoxicated, the more dangerous occupations of men (most industrial fire fatalities are males), and the greater use of gasoline and other flammable liquids by men. We do know that men have more injuries trying to react to the fire than do women.

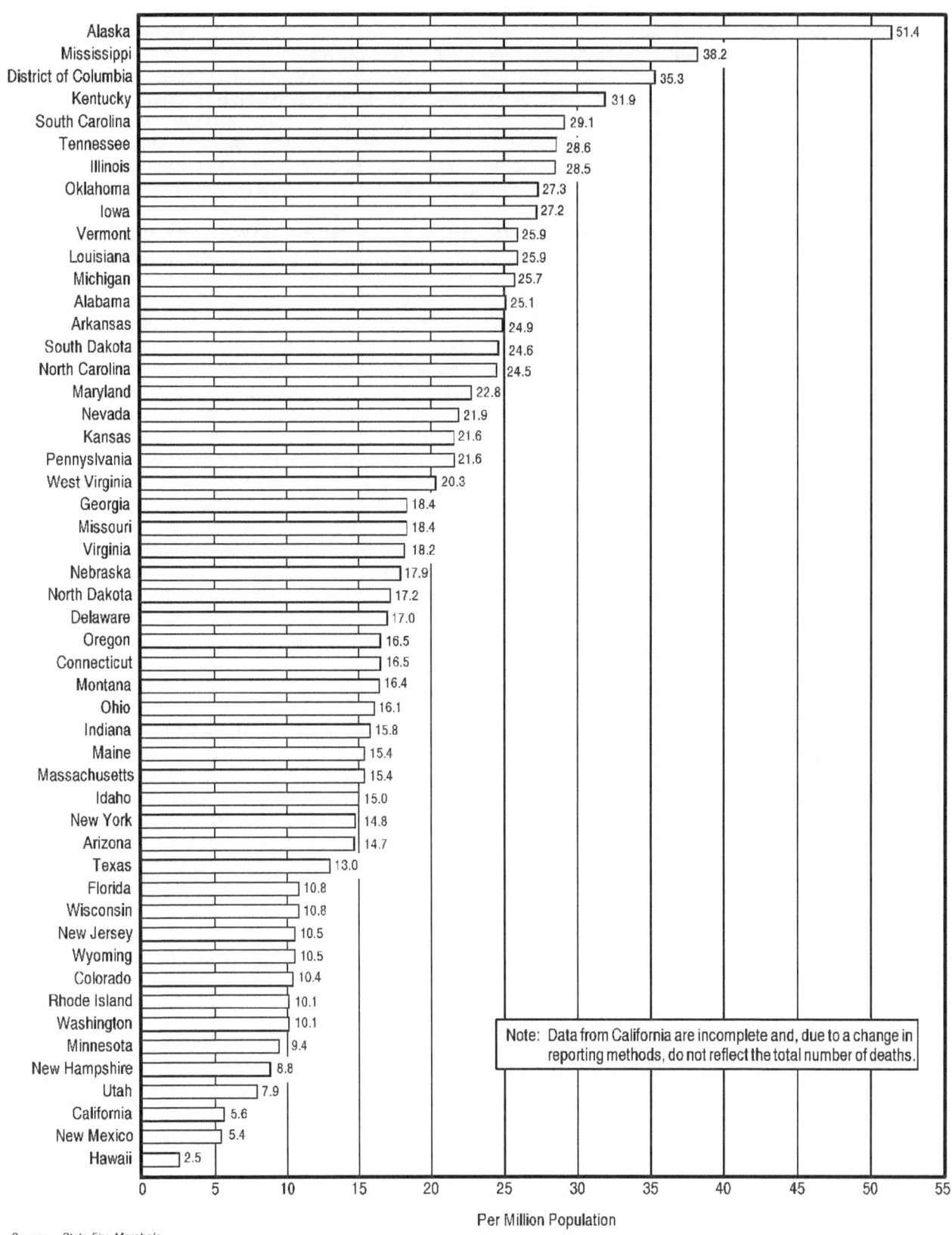

Figure 19. Rank Order of States by Severity of 1994 Civilian Deaths

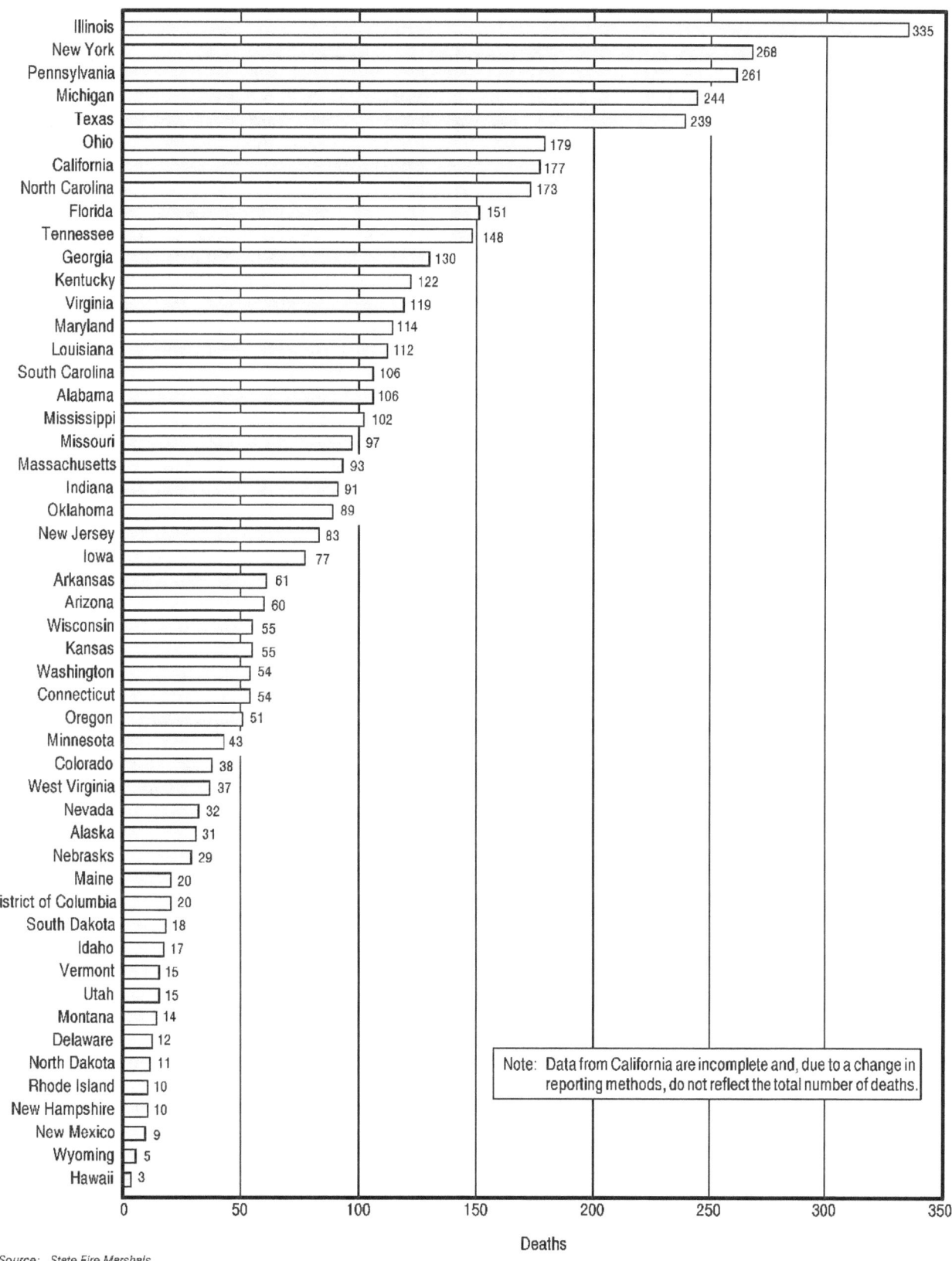

Figure 20. Rank Order of States by 1994 Civilian Deaths

Source: State Fire Marshals

39

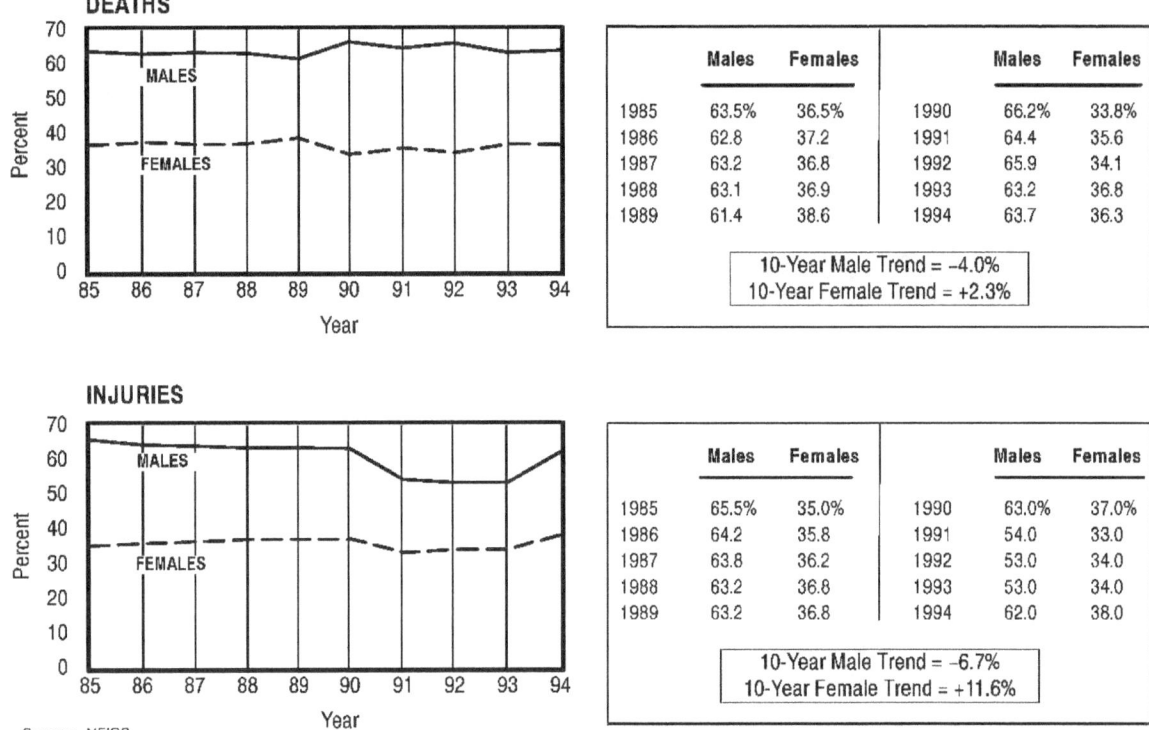

Figure 21. Trends in Male vs. Female Casualties

DEATHS

	Males	Females		Males	Females
1985	63.5%	36.5%	1990	66.2%	33.8%
1986	62.8	37.2	1991	64.4	35.6
1987	63.2	36.8	1992	65.9	34.1
1988	63.1	36.9	1993	63.2	36.8
1989	61.4	38.6	1994	63.7	36.3

10-Year Male Trend = –4.0%
10-Year Female Trend = +2.3%

INJURIES

	Males	Females		Males	Females
1985	65.5%	35.0%	1990	63.0%	37.0%
1986	64.2	35.8	1991	54.0	33.0
1987	63.8	36.2	1992	53.0	34.0
1988	63.2	36.8	1993	53.0	34.0
1989	63.2	36.8	1994	62.0	38.0

10-Year Male Trend = –6.7%
10-Year Female Trend = +11.6%

Source: NFIRS

Age

People over 70 have a much higher fire death rate than the average population, as shown in Figure 23. At the other end of the age spectrum, the very young (under 5) have a much higher than average problem too. The relative risk of dying and being injured in a fire for various age groups is shown in Figure 24. (The population as a whole has a relative risk of 1.) Children under 5 have nearly double the risk of death, children over 5 have less than average risk.[5] Risk of fire death drops off sharply between 5 and 19, then experiences minimal change until age 55. At age 55, the risk begins to increase. By age 70, there is a jump in risk; and above age 80, the risk is even higher than for the very young.

Contrary to what might be expected, the age profile of risk from injuries is very different from that for deaths. The risk of injury in a fire is highest for those aged 20–34. The risk of injury is below average for infants and children and for elderly 65–79. Over 80, the risk is above average.

Figure 25 shows the percent of 1994 fire deaths and injuries falling into each age group. (This is not the same as risk.) Those under age 5 account for 16 percent of the deaths with age reported—by far the highest proportion for any age group. Those 70 and above comprise 20 percent of the fire

---

[5] For those interested in data reliability issues, there is some concern over the coding of the ages of infants less than 1 year old. Some code them as 1, some as 0, and some to the nearest integer of 0 or 1. Also, some fire departments or states fill in blank fields with zeros. Thus, the number of casualties with age 0 has been suspect. By dropping age profiles with 0's, the difference was small; the category 0–4 still had a relative risk of nearly 2.

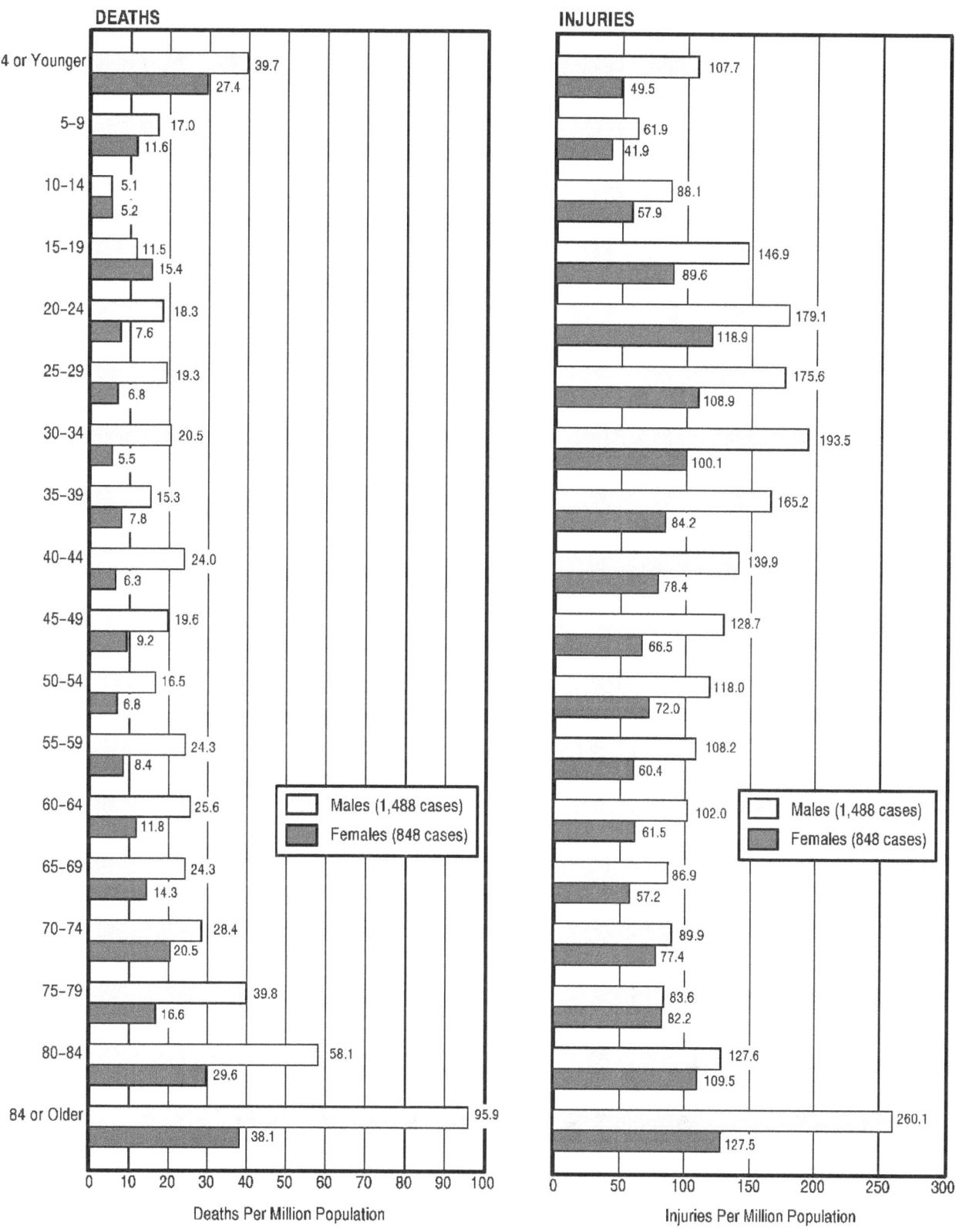

DEATHS

INJURIES

Males (1,488 cases)
Females (848 cases)

Deaths Per Million Population

Injuries Per Million Population

Sources: NFIRS, NFPA Annual Surveys, and Bureau of the Census

Figure 22. Severity of 1994 Fire Casualties by Age and Gender

41

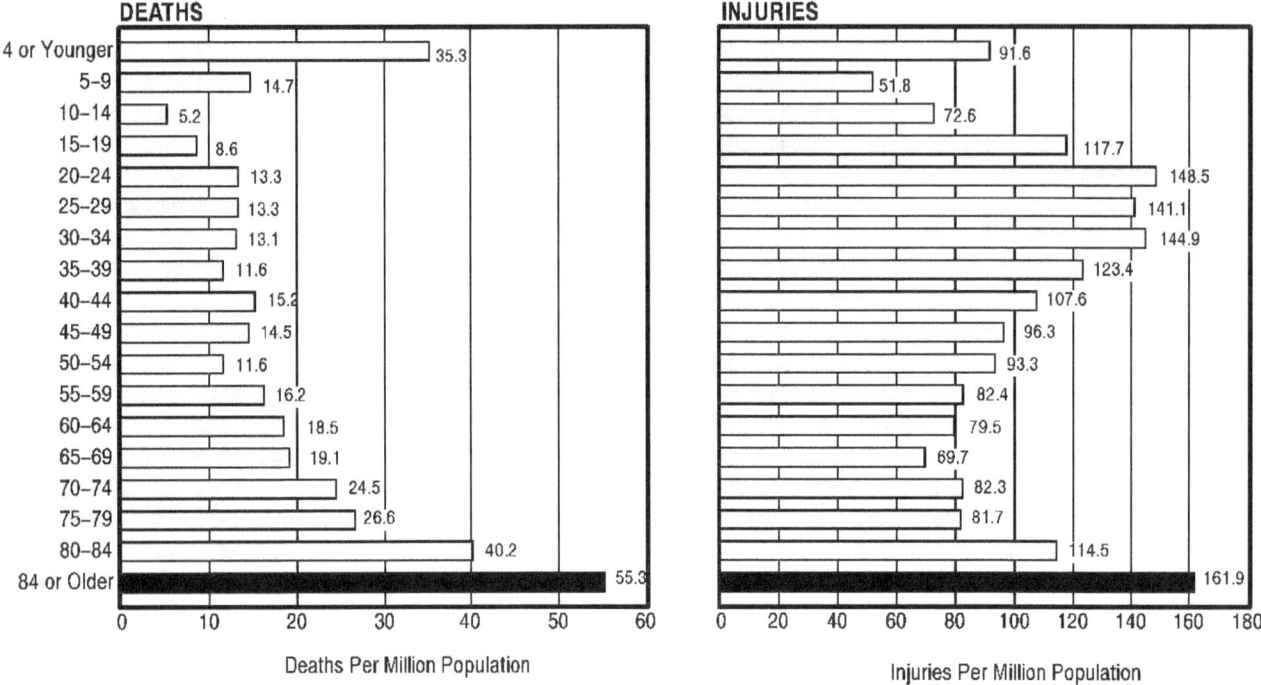

Sources: NFIRS, NFPA Annual Surveys, and Bureau of the Census

Figure 23. Severity of 1994 Fire Casualties by Age

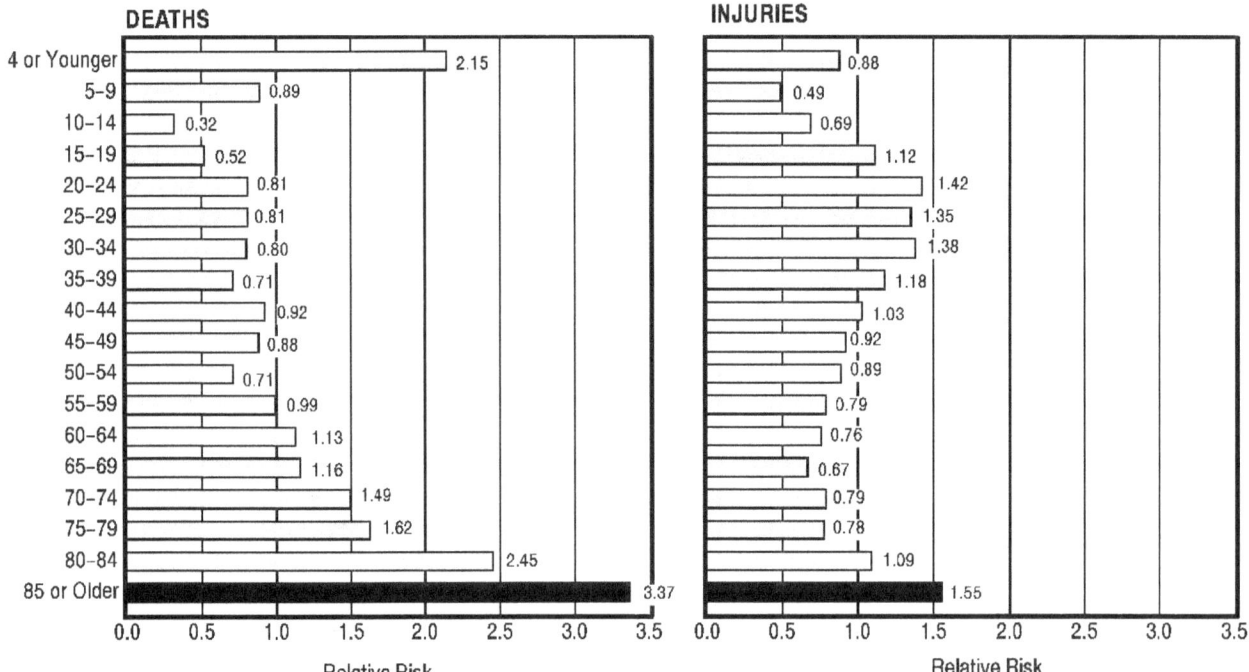

Sources: NFIRS, NFPA Annual Surveys, and Bureau of the Census

Figure 24. Relative Risk of 1994 Fire Casualties by Age

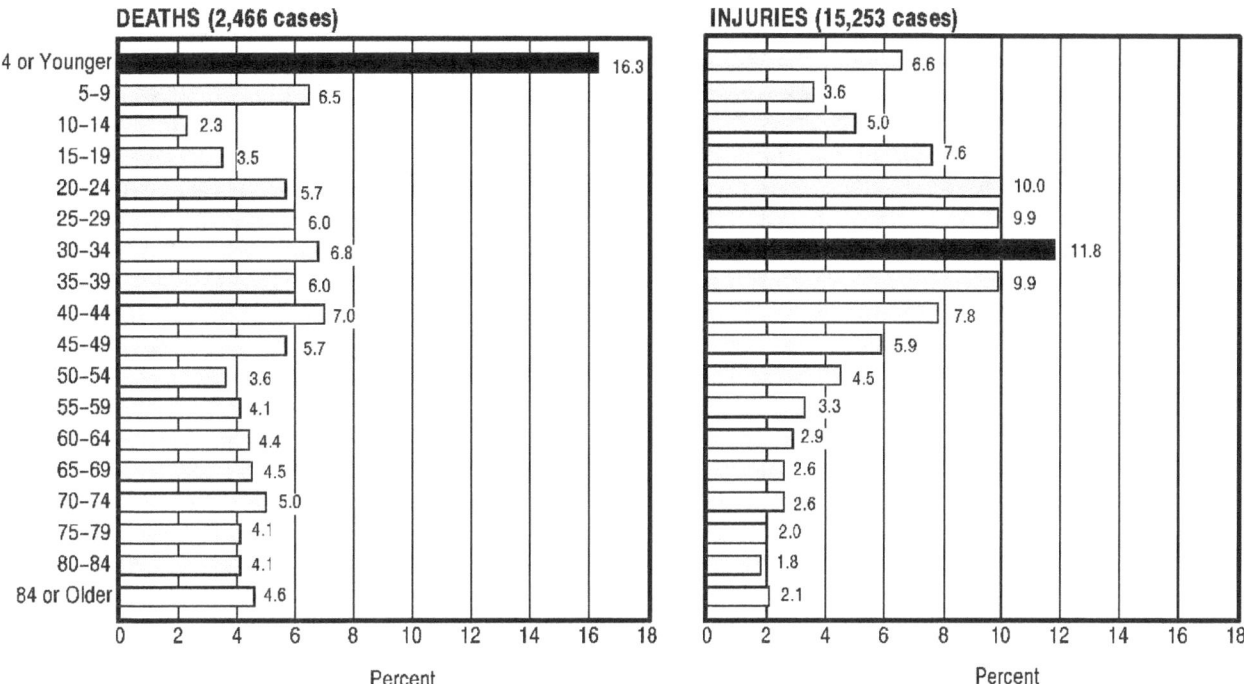

DEATHS (2,466 cases)

Age	Percent
4 or Younger	16.3
5–9	6.5
10–14	2.3
15–19	3.5
20–24	5.7
25–29	6.0
30–34	6.8
35–39	6.0
40–44	7.0
45–49	5.7
50–54	3.6
55–59	4.1
60–64	4.4
65–69	4.5
70–74	5.0
75–79	4.1
80–84	4.1
84 or Older	4.6

Percent

INJURIES (15,253 cases)

Percent
6.6
3.6
5.0
7.6
10.0
9.9
11.8
9.9
7.8
5.9
4.5
3.3
2.9
2.6
2.6
2.0
1.8
2.1

Percent

Sources: NFIRS, NFPA Annual Surveys, and Bureau of the Census

Figure 25. 1994 Fire Casualties by Age

deaths. These two peak risk groups comprise more than one-third of fire deaths and represent about equal numbers of fatalities. On the other hand, two-thirds of fire deaths fall in age groups that are not at high risk. The bulk of fire deaths occur to the not so young and not so old. Programs aimed only at the highest risk groups will not reach the majority of victims.

The injury distribution tracks closely the relative risk profile by age, except for the elderly (Figure 25). Ages 20–39 account for 32 percent of fire injuries in 1994. The very young account for 7 percent; the elderly over age 70 account for 8 percent. Although the elderly are at high risk, there are fewer of them in the total population. If their risk continues to be the same, we could expect more and more elderly fire injuries and deaths as the elderly proportion of the population increases. In the meantime, the focus for injury prevention should be on young adults 20–39. It is believed that males in this age group are greater risk takers during fires, resulting in a higher proportion of injuries.

The distribution of fire deaths by age is somewhat different for males versus females. A slightly larger proportion of female deaths in 1994 occurred in the young (through age 19) and again in the elderly (Figure 26). Male fire deaths, by contrast, are higher in the mid-life years, ages 20 to 55. Elderly females have a significantly larger proportion of injuries than males.

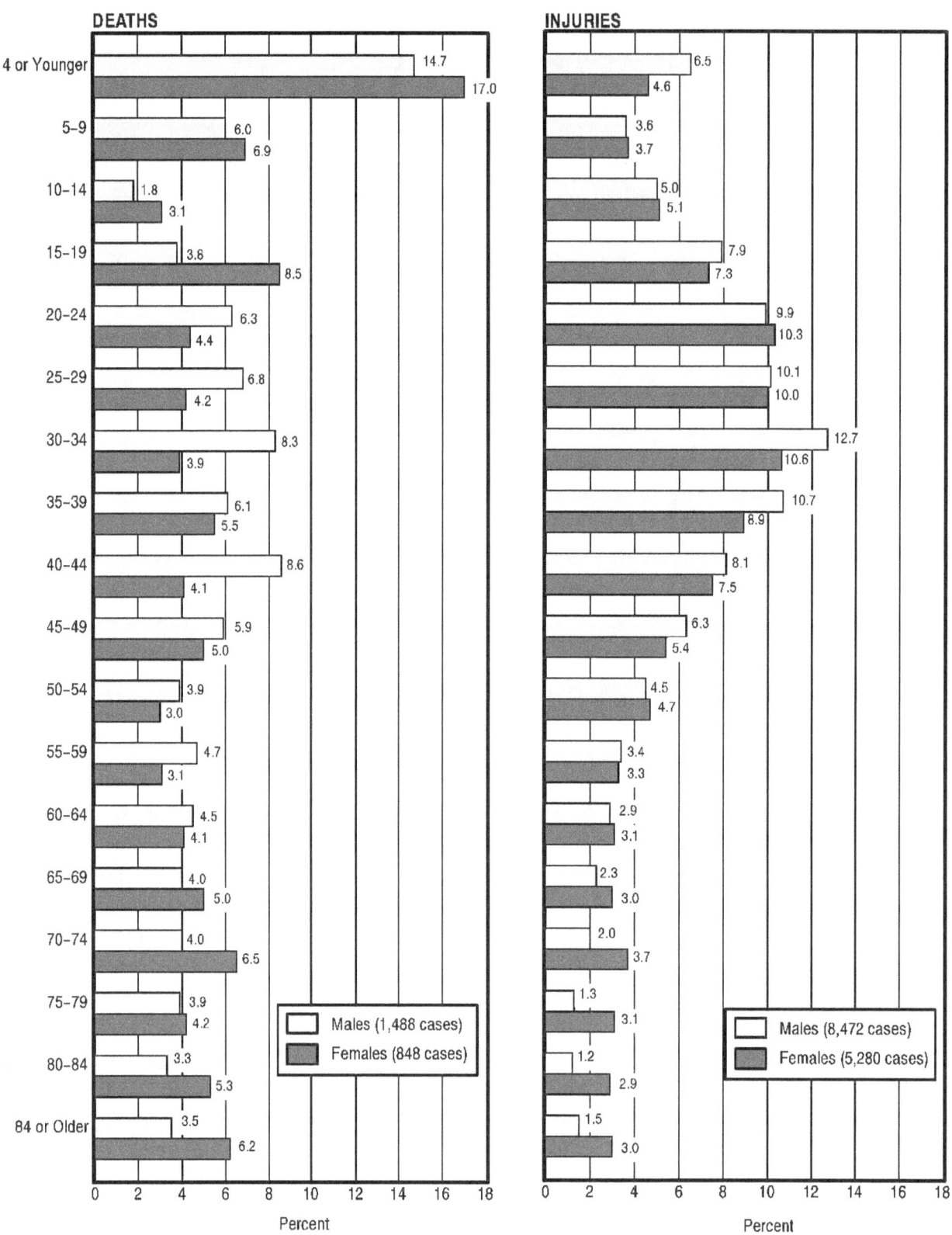

Figure 26. 1994 Fire Casualties by Gender and Age

Source: NFIRS

44

Ethnic Groups

The fire problem cuts across all groups and races, rich and poor, North and South, urban and rural. But it is higher for some groups than for others.

Data on "race" or ethnic group of victims are somewhat ambiguous in a society where many people are of mixed heritages. And many citizens, including firefighters, find it distasteful to report on race. On the other hand, there does seem to be a higher fire problem for some groups, and it can be helpful to identify their problems for use within their own communities.

KINDS OF PROPERTIES WHERE FIRES OCCUR

This section describes the proportions of the fire problem across major types of properties: residential structures, non-residential structures, vehicles, outside properties, and other or unknown properties.[6]

Property Categories

In terms of numbers of fires, the largest category in 1994 (as in all years) is outside fires (44 percent)—in fields, vacant lots, trash, etc. (Figure 27). Many of these fires are intentionally set but do not cause much damage. Residential and non-residential structure fires together comprise only about one-third of all fires. Residential fires outnumber non-residential structure fires by two and a half to one. What may surprise some is the large number of vehicle fires. In fact, one out of every four fires that fire departments attend involves vehicles. Most vehicle fires are associated with accidents. The number of vehicle fires does not include the many vehicle accidents that the fire department responds to but where there is no fire.

By far the largest percentage of deaths, 71 percent, occurs in residences, with the majority of these in one- and two-family dwellings. It may surprise some that such a large share of our fire deaths result from fires that occur in houses and apartments. Great attention is given to large, multiple death fires in public places such as hotels, nightclubs, and office buildings. But in fact, the major attention-getting fires that kill 10 or more people are few in number and constitute only a small portion of fire deaths. Firefighters generally are doing a good job in protecting public properties in this country. The area with the largest problem is where it is least suspected—in people's homes. Fire prevention efforts should be increasingly focused on this part of the overall fire problem.

Vehicles accounted for the second largest percentage of fire deaths, 17 percent. As most vehicle fires are the result of collisions, there are virtually no fire prevention programs designed to address this problem other than as part of vehicle design and as a byproduct of accident prevention in general.

---

[6] The percentage of fire deaths in the major property types differs somewhat between NFIRS and the NFPA survey. These differences are discussed in Appendix A.

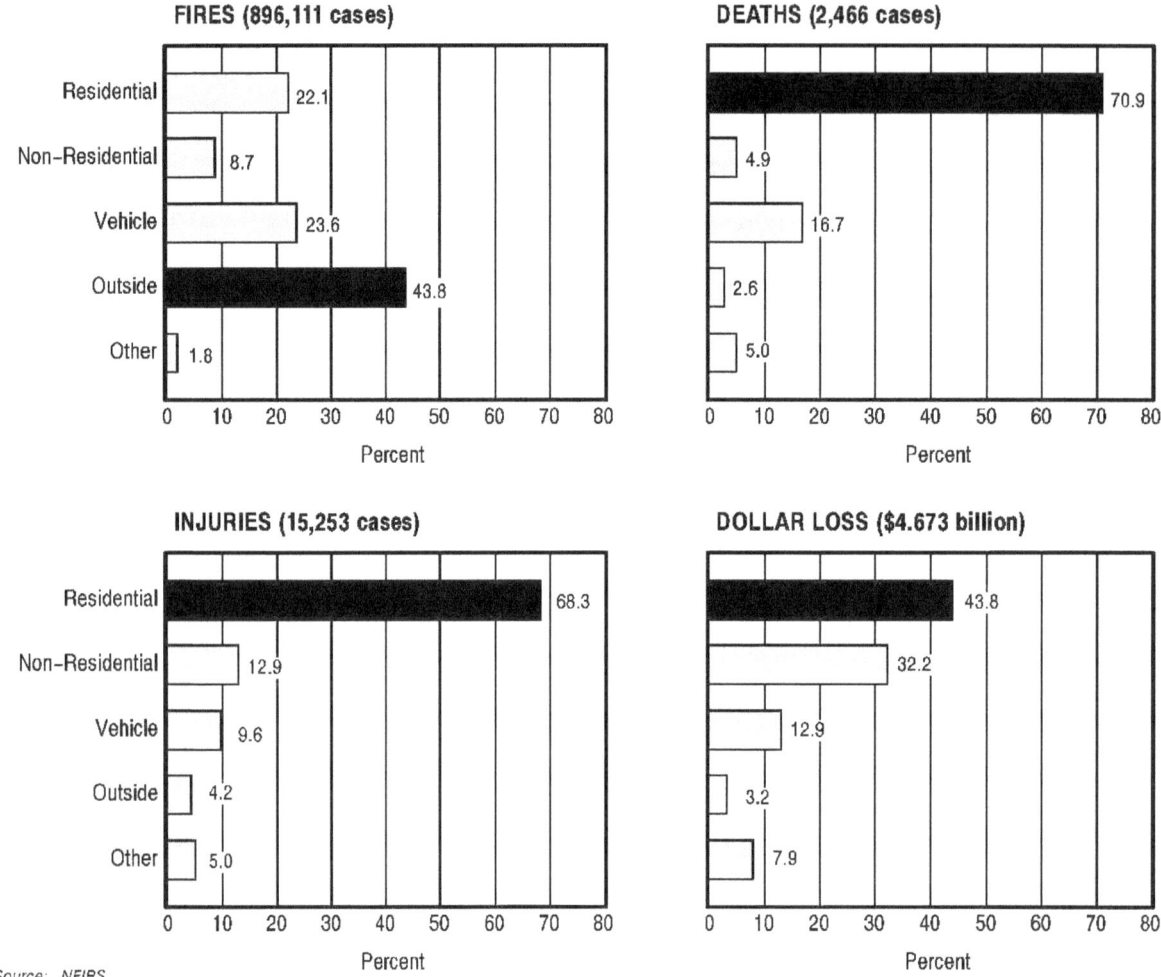

Figure 27. 1994 Fires and Fire Losses by General Property Type

Only 5 percent of the 1994 fire deaths occurred in commercial and public properties. Outside and other (unknown) fires, including wildfires, were a very small factor in fire deaths (8 percent).

As Figure 27 shows, the picture is somewhat similar for fire injuries, with more than two-thirds of all injuries occurring in residences. Non-residential structures are the location of 13 percent of all fire injuries. Vehicles account for another 10 percent. Outside and other fires account for just 9 percent of the fire injuries.

The picture changes sharply for dollar loss and fire incidence. Three-quarters of all dollar loss is from residential and non-residential structures. The proportion of dollar loss from outside fires may be understated because the destruction of trees, grass, etc., is given zero value in fire reports if it is not commercial cropland or timber.

All relative percentages for property type fires were the same in 1994 as they were in 1990.

46

Trends

The proportions of the fire problem by property type have remained quite steady over time. This is another reasonableness check for NFIRS. In terms of numbers of fires, the proportion of the problem due to outside property has increased slightly from 41 percent to 44 percent over the 10-year period, the proportion in vehicles and outside has remained constant, and the proportion in residential and non-residential structures has declined slightly (Figure 28). It has been suggested that the outside property increase might be due to an increasing number of rural departments reporting to NFIRS.

Over the 10-year period, residential property fires have ranged from 69 to 75 percent of total fire deaths, with an overall downward trend. Non-residential structures and outside fires have remained steady and represent a small proportion of deaths. The trends in vehicle fires, however, has been steadily increasing. In other words, vehicles are becoming a more important factor in fire deaths.

The trends in property types for injuries have been stable over the 10-year period, perhaps because of the much larger sample of injuries than deaths.

Dollar loss has greater trend fluctuations because this measure is highly sensitive to a few very large fires and whether they are included or omitted in the sample of fires on which estimates are based. The classic example is the 1986 pineapple fire in Hawaii ("outside" property type), which destroyed an enormous pineapple crop and caused a peak in outside fire losses that forced the other percentages downward. Similarly, in 1994 there was a huge increase in the "other" fire category.

SEVERITY OF FIRES

Figure 29 shows the severity of fires in 1994 as measured by deaths and injuries per thousand fires and by dollar loss per fire. These indicators can increase if there are more casualties or more damage per fire (the numerator) or if fewer minor fires are reported (the denominator).

As shown, residential fires have the highest death and injury rates—another important reason for prevention programs to focus on home fire safety. Usually, non-residential structure fires have the highest dollar loss per fire. In 1994, however, there was an unusually large loss in the "other" category.

The trends in severity over the 10-year period are shown in Figure 30. Residential fire severity increased slightly over the 10-year period in terms of deaths per fire and increased by 10 percent in terms of injuries per fire.

Non-residential severity remained relatively constant in both deaths and injuries for fire. Other fire (including unspecified property types) has relatively high severities but represents only small numbers of fires and fire deaths; it is a miscellaneous category.

FIRES

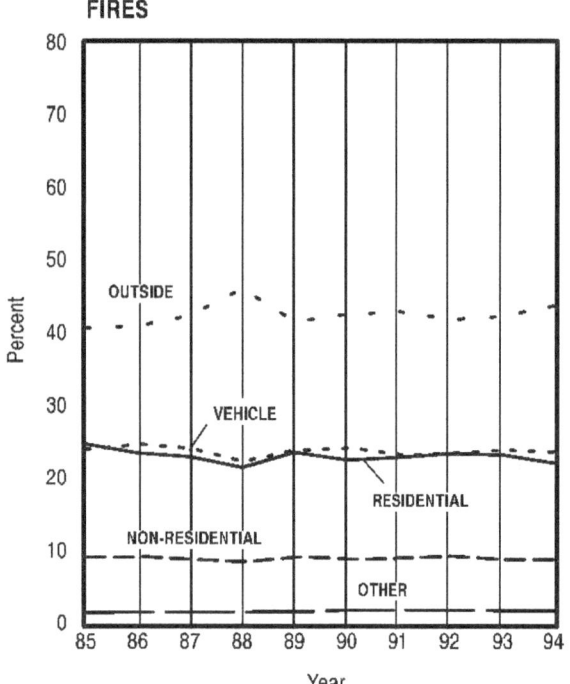

	Residential	Non-Residential	Vehicle	Outside	Other
1985	24.7%	9.1%	24.0%	40.7%	1.5%
1986	23.5	9.2	24.7	41.0	1.6
1987	23.0	8.8	24.1	42.4	1.6
1988	21.5	8.5	22.4	46.1	1.6
1989	23.6	9.1	23.9	41.6	1.7
1990	22.5	8.9	24.2	42.6	1.9
1991	22.9	9.0	23.3	42.9	1.9
1992	23.4	9.3	23.5	41.9	1.9
1993	23.2	8.8	23.9	42.3	1.8
1994	22.1	8.7	23.6	43.8	1.8

DEATHS

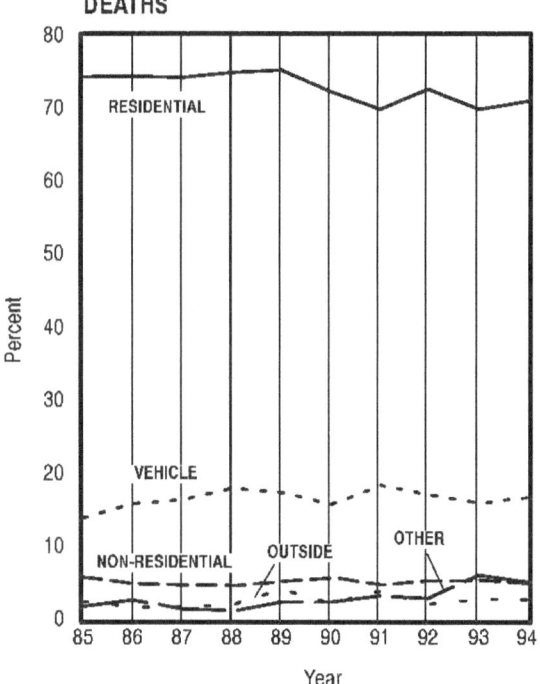

	Residential	Non-Residential	Vehicle	Outside	Other
1985	74.3%	5.7%	13.9%	2.4%	1.8%
1986	74.4	4.9	15.8	1.7	2.6
1987	74.2	4.7	16.4	1.8	1.4
1988	74.9	4.6	17.9	2.0	1.2
1989	75.2	5.2	17.3	4.1	2.4
1990	72.2	5.7	15.7	2.1	2.4
1991	69.8	4.8	18.4	3.9	3.2
1992	72.6	5.3	17.1	2.1	2.9
1993	69.8	5.4	15.9	2.8	6.1
1994	70.9	4.9	16.7	2.6	5.0

Continued on next page

Figure 28. Trends in Fires and Fire Losses by General Property Type

INJURIES

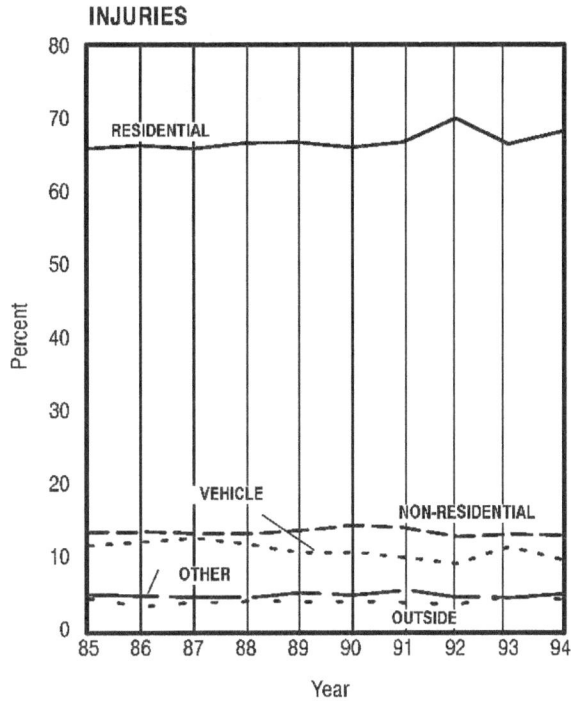

	Residential	Non-Residential	Vehicle	Outside	Other
1985	65.9%	13.4%	11.6%	4.3%	4.9%
1986	66.4	13.5	12.1	3.3	4.7
1987	65.9	13.2	12.6	3.8	4.5
1988	66.7	13.2	11.7	4.0	4.4
1989	66.8	13.6	10.6	3.9	5.1
1990	66.1	14.3	10.7	4.0	4.8
1991	66.8	14.0	9.9	3.7	5.5
1992	70.1	12.8	9.1	3.6	4.5
1993	66.5	13.1	11.4	4.6	4.4
1994	68.3	12.9	9.6	4.2	5.0

DOLLAR LOSS

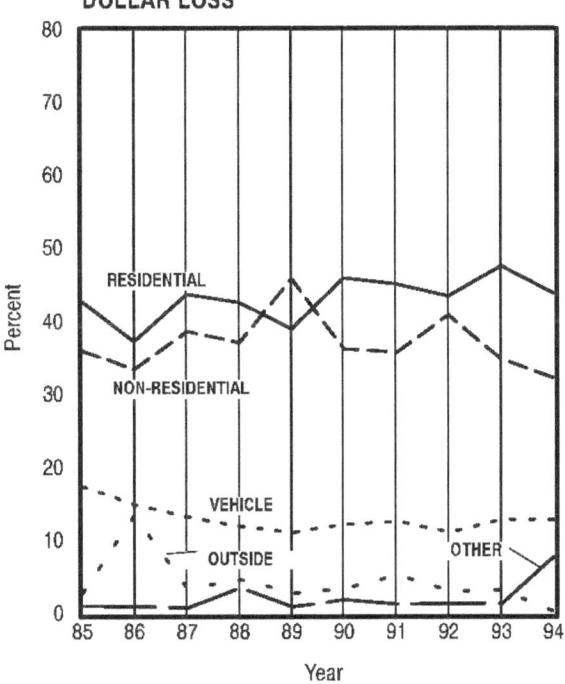

	Residential	Non-Residential	Vehicle	Outside	Other
1985	42.7%	36.0%	17.4%	2.8%	1.1%
1986	37.3	33.4	14.9	13.5	1.0
1987	43.7	38.7	13.2	3.6	0.8
1988	42.6	37.1	11.9	4.7	3.6
1989	39.0	45.9	11.1	2.9	1.0
1990	46.0	36.3	12.3	3.5	1.9
1991	45.1	35.7	12.6	5.3	1.3
1992	43.5	40.9	11.1	3.1	1.4
1993	47.6	34.8	12.9	3.3	1.3
1994	43.8	32.2	12.9	3.2	7.9

Source: NFIRS

Figure 28. Trends in Fires and Fire Losses by General Property Type (cont'd)

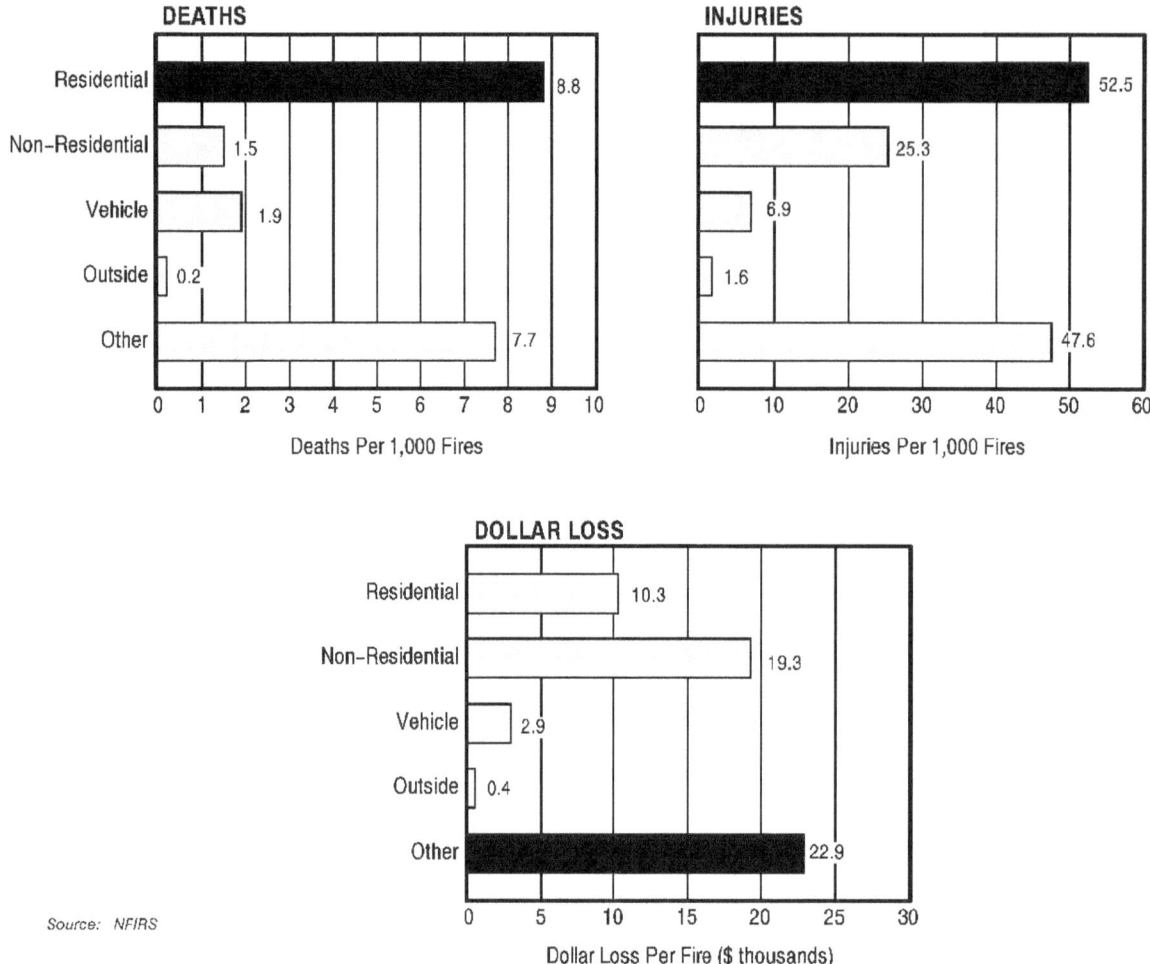

Figure 29. Severity of Fire Losses in 1994 by General Property Type

Adjusted dollar loss per fire changed significantly. Non-residential fires averaged $23,200 over 10 years per fire with wide fluctuations: from a low of $19,300 per fire in 1994 to a high of $30,300 per fire in 1990. Fire loss decreased for most categories. Residential losses increased a significant 12 percent, and from 1993 to 1994 there was an uncharacteristic increase (sixfold) in "other" fires due to one large ($300 million) explosion.

There are many reasons for increases in loss per fire in residential occupancies. It could reflect a more affluent society in part, but affluence has not increased as sharply as the losses per fire adjusted for inflation. More damage per fire also may be due to faster spreading fires. It could also be from the use of smoke detectors, where small fires are put out and the fire department is never called. One clue as to the underlying cause for the increase is that the number and percent of residential fires that spread to the whole structure (that is, were not confined to the floor of origin) increased sharply from 1985 to 1994. This is an area needing further study.

DEATHS (per 1,000 fires)

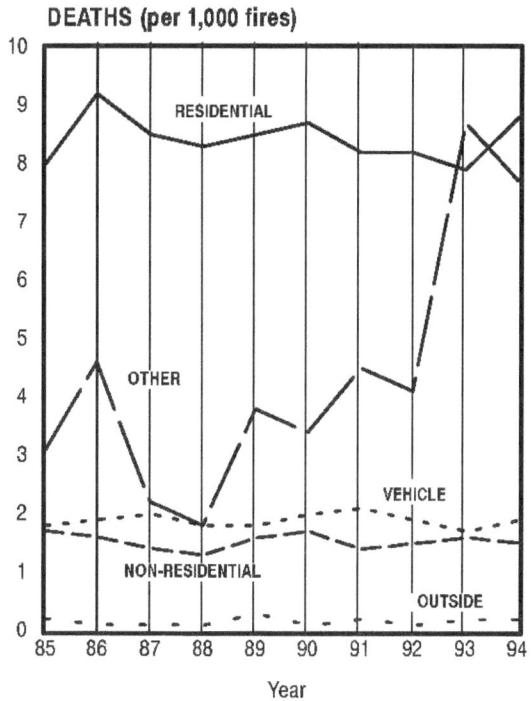

	Residential	Non-Residential	Vehicle	Outside	Other
1985	8.0	1.7	1.8	0.2	3.1
1986	9.2	1.6	1.9	0.1	4.6
1987	8.5	1.4	2.0	0.1	2.2
1988	8.3	1.3	1.8	0.1	1.8
1989	8.5	1.6	1.8	0.3	3.8
1990	8.7	1.7	2.0	0.1	3.4
1991	8.2	1.4	2.1	0.2	4.5
1992	8.2	1.5	1.9	0.1	4.1
1993	7.8	1.6	1.7	0.2	8.7
1994	8.8	1.5	1.9	0.2	7.7

INJURIES (per 1,000 fires)

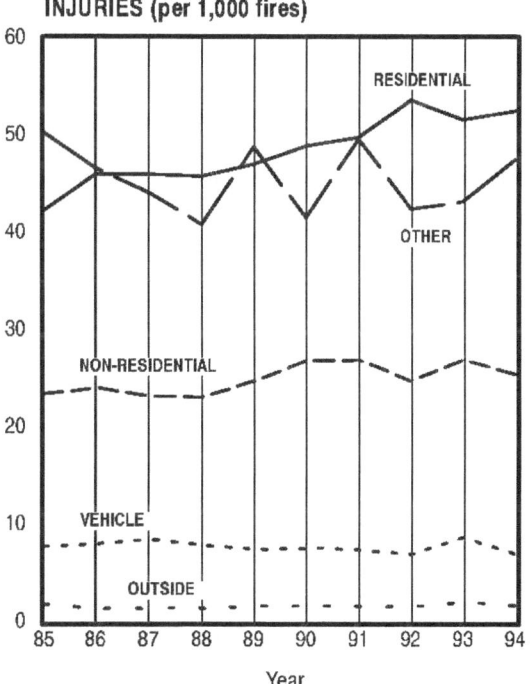

	Residential	Non-Residential	Vehicle	Outside	Other
1985	42.3	23.4	7.7	1.7	50.2
1986	46.0	24.0	8.0	1.3	46.6
1987	46.0	23.1	8.4	1.4	44.1
1988	45.8	23.0	7.8	1.3	40.8
1989	47.0	24.7	7.4	1.6	48.8
1990	48.9	26.8	7.5	1.6	41.5
1991	49.8	26.9	7.3	1.5	49.6
1992	53.6	24.7	6.9	1.5	42.4
1993	51.6	26.9	8.6	2.0	43.2
1994	52.5	25.3	6.9	1.6	47.6

Continued on next page

Figure 30. Trends in Severity of Fire Losses by General Property Type

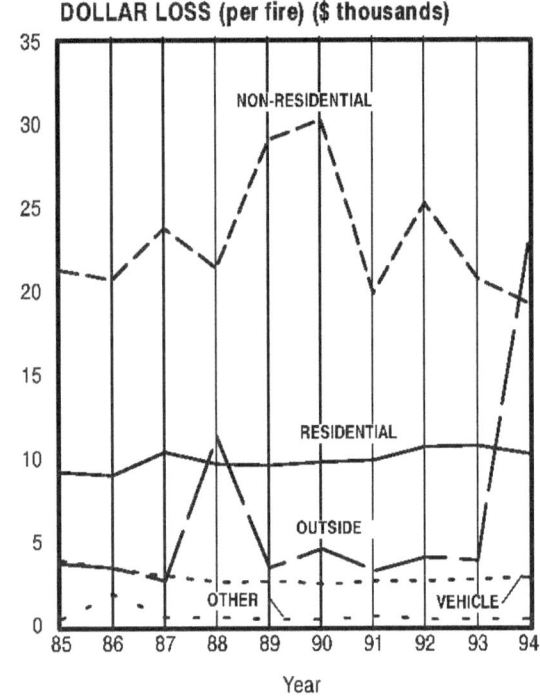

DOLLAR LOSS (per fire) ($ thousands)

	Residential	Non-Residential	Vehicle	Outside	Other
1985	$9.2	$21.3	$3.9	$0.4	$3.7
1986	9.0	20.7	3.4	1.9	3.5
1987	10.4	23.8	3.0	0.5	2.7
1988	9.7	21.4	2.6	0.5	11.3
1989	9.6	29.1	2.7	0.4	3.5
1990	9.8	30.3	2.5	0.4	4.6
1991	9.9	19.9	2.7	0.6	3.3
1992	10.7	25.3	2.7	0.4	4.1
1993	10.8	20.8	2.8	0.4	3.9
1994	10.3	19.3	2.9	0.4	22.9

Source: NFIRS

Figure 30. Trends in Severity of Fire Losses by General Property Type (cont'd)

CAUSES OF FIRES AND FIRE LOSSES

Figure 31 shows the profile of the major causes of fires, fire deaths and injuries, and direct dollar loss in 1994. Here, fire deaths occurring in all the different occupancies are grouped together. The top three causes are careless smoking (21 percent), incendiary and suspicious (or arson) (16 percent), and heating (13 percent). These percentages are adjusted, which proportionally spreads the unknowns over the other 12 causes. The leading cause of injuries is cooking (21 percent), followed by arson (14 percent) and children playing (11 percent). Careless smoking dropped from third in 1990 to fourth in 1994.

The three leading causes of fire deaths are the same for both sexes (Figure 32). The proportions of each of the remaining causes are surprisingly similar too. Males had slightly more fire deaths from open flame and other heat and females slightly more from cooking and electrical fires.

Unlike fire deaths, there is a sharp difference between the sexes in their injury cause profiles. While the leading cause, cooking, is the same, the relative role that the remaining causes play in fire injuries between men and women differs greatly in 1994. Moreover, the leading cause for both sexes is cooking, but nearly twice as many women are injured in cooking fires as men (28 vs. 17 percent). For women, children playing and arson are the second and third leading causes, although combined they account for fewer injuries than cooking. For men, arson is the second leading cause of fire injuries, followed by careless smoking. Arson is by far the leading cause of fires and direct dollar loss.

Causes of fire casualties are discussed in more detail by occupancy type in Chapters 3 and 4.

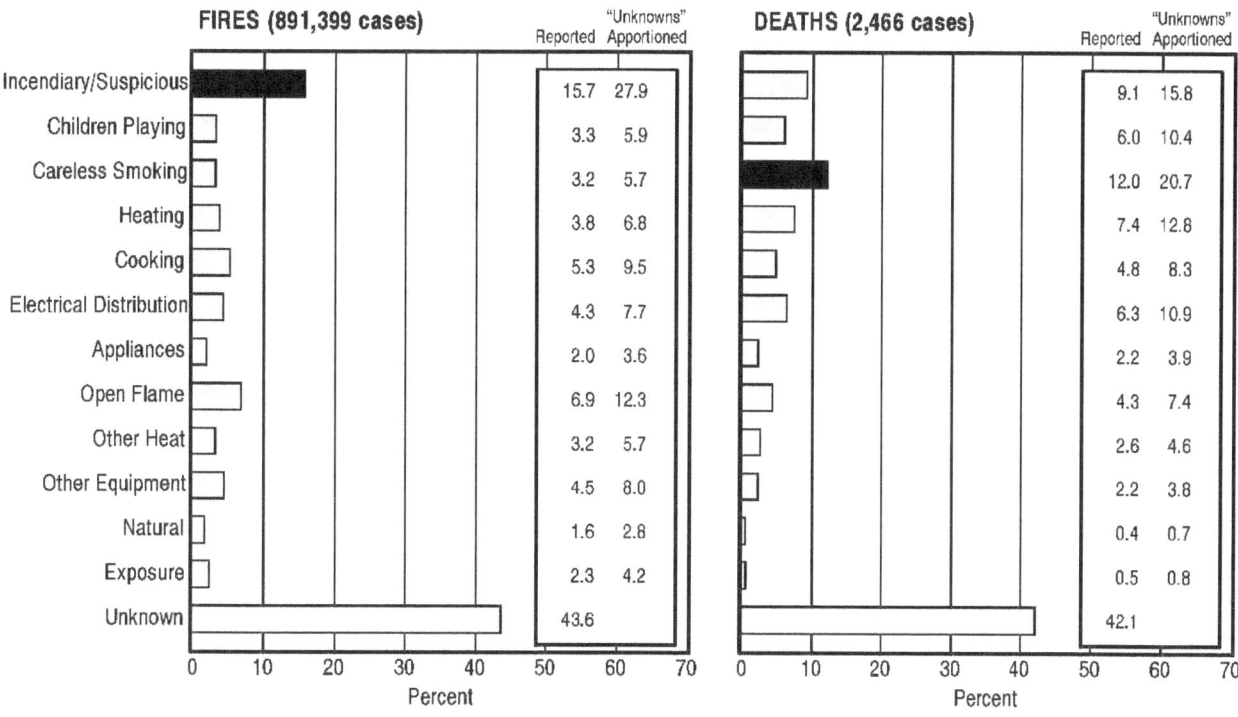

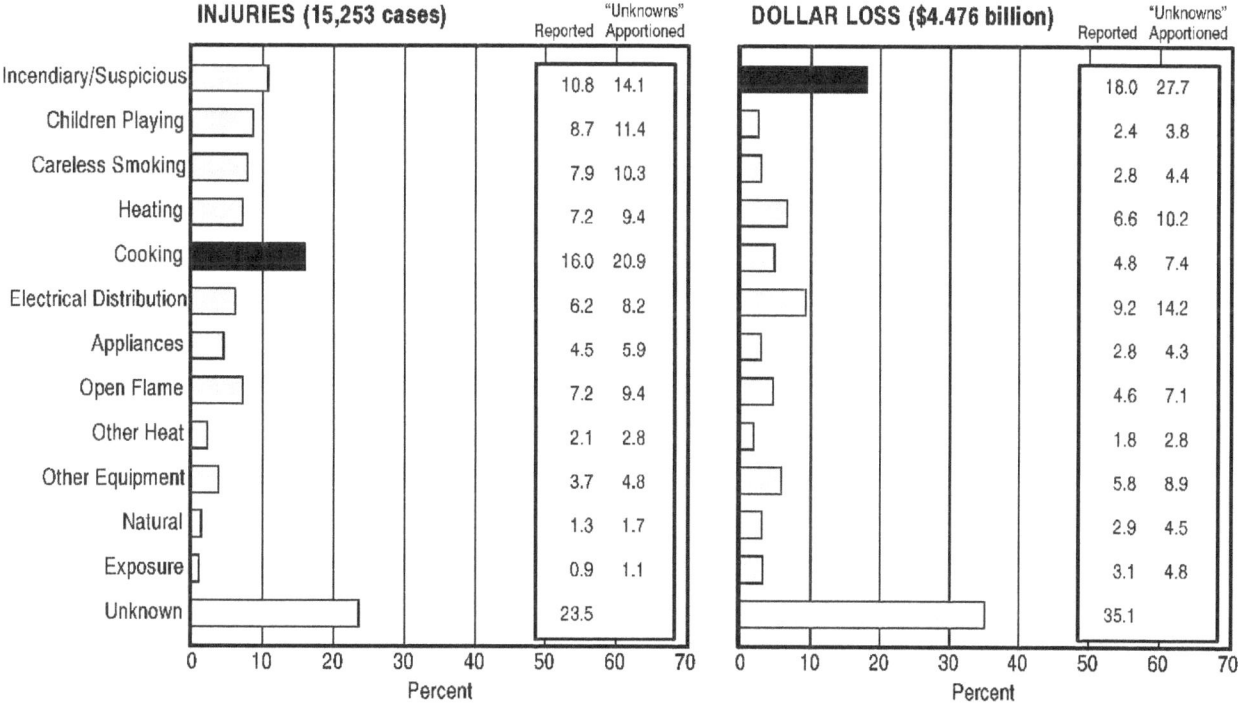

Source: NFIRS

Figure 31. Causes of 1994 Fires and Fire Losses

Source: NFIRS

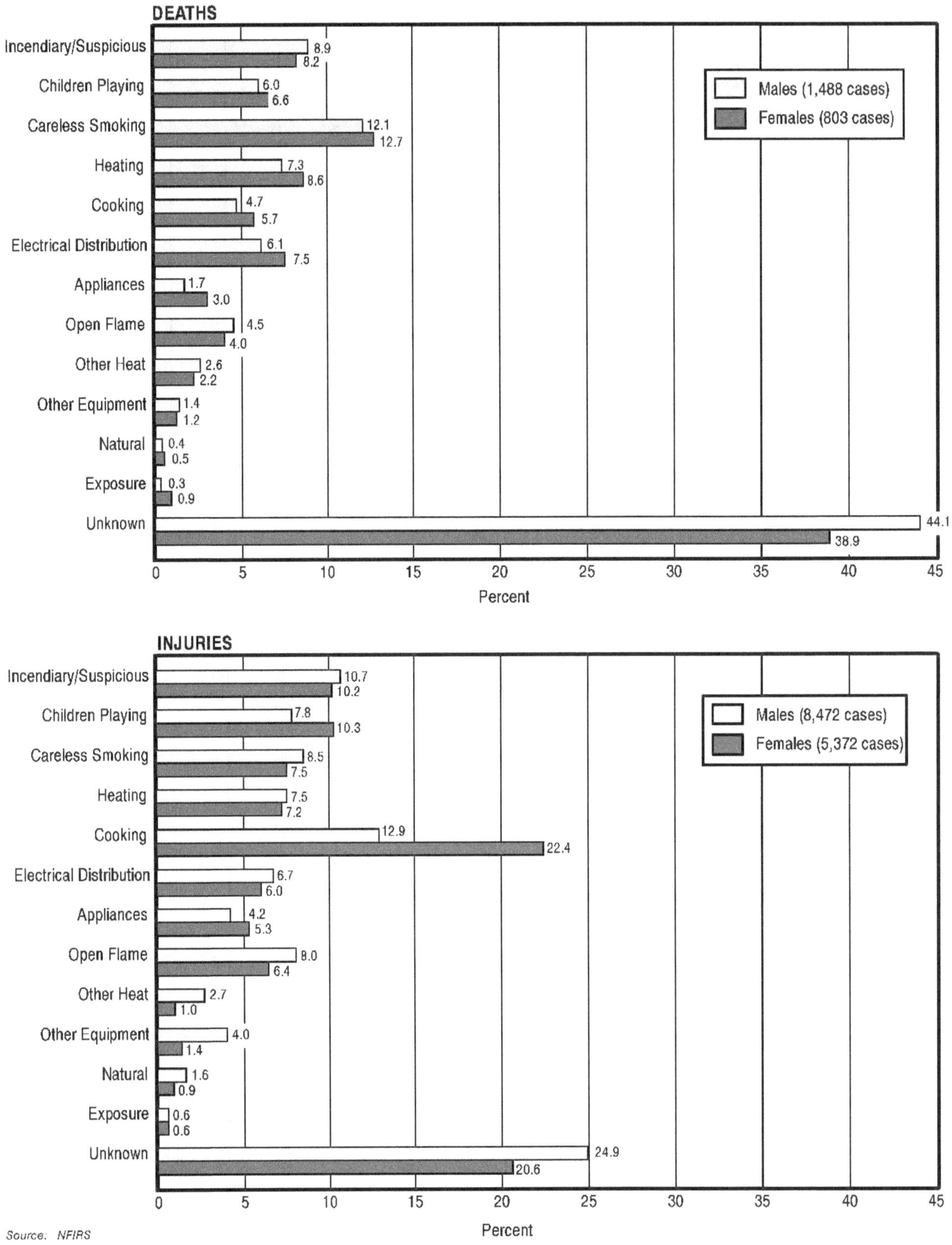

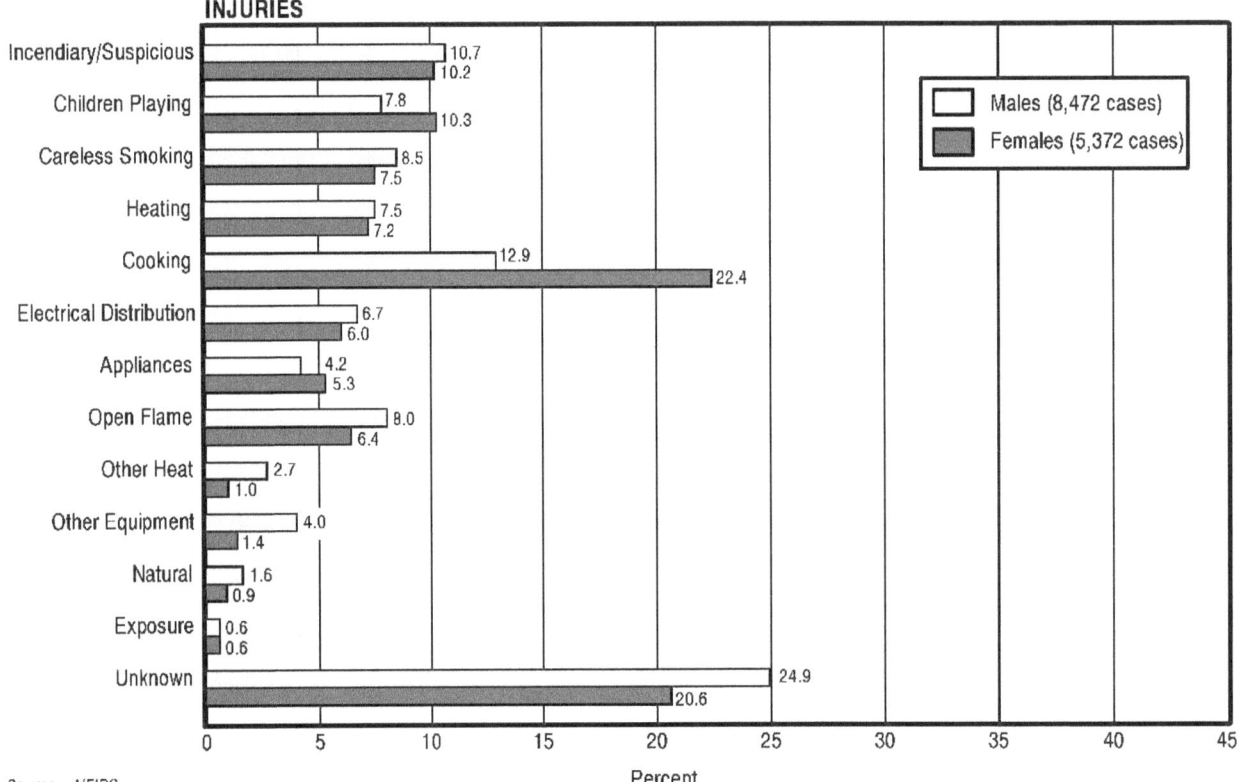

Figure 32. Causes of 1994 Fire Casualties by Gender (Unadjusted)

USFA RESOURCES ON THE NATIONAL FIRE PROBLEM

The U.S. Fire Administration has issued two reports that have attracted nationwide attention. *American Burning* is probably the most widely quoted fire protection publication. This report set the stage for national consciousness-raising about the need for as much focus on fire prevention as on fire suppression. *Fire Death Rate Trends: An International Perspective* explores the magnitude and the nature of the fire death problem in the United States. It provides a statistical portrait of fire death rates for 14 industrialized nations, and presents observations about key institutional and attitudinal differences between the U.S. and industrialized countries with significantly lower fire death rates. Chapter 6 in this document provides data extracted from this report.

These publications are available by writing to:

U.S. Fire Administration
Federal Emergency Management Agency
Publications Center, Room N310
16825 S. Seton Avenue
Emmitsburg, MD 21727

Documents may also be ordered via the World Wide Web: *http://www.usfa.fema.gov*. USFA publications are free.

3

RESIDENTIAL PROPERTIES

OVERVIEW

The residential portion of the fire problem accounts for 71 percent of fire deaths and 68 percent of the injuries to civilians. It also accounts for more firefighter injuries than any other occupancy category. This section reviews the residential problem overall, and subsequent sections present details of the fire problem for major subcategories of residential properties (one- and two-family dwellings, apartments, hotels and motels, and other types.)

The term *residential* as used in NFIRS includes what is commonly referred to as homes, whether they are one- or two-family dwellings or multifamily apartment buildings. It also includes manufactured housing, hotels and motels, residential hotels, dormitories, and much of what might be considered "halfway houses" for the care of people with problems but able to operate in the community. The term does not include "institutions" such as prisons, homes for the elderly, juvenile care facilities, or hospitals, though many people may reside in them for short or long periods of time.

Figure 33 shows the 10-year trend in residential fires, deaths, injuries, and dollar loss. Both numbers of fires and deaths have trended downward dramatically over the past 10 years (28 percent and 32 percent, respectively), although there has been relatively little change over the past 4 years. Injuries trended up to 6 percent, and dollars loss trended down 8 percent when adjusted for inflation. These results are based on the NFPA annual surveys of fire departments.[1]

Types of Residences

Figure 34 shows the relative proportions of fires, fire deaths, injuries, and dollar loss among the residential categories in 1994.

One- and two-family dwellings, where the majority of the U.S. population lives, dominate the residential statistics: 70 percent of residential fires, 69 percent of residential deaths, 60 percent of residential injuries, and 73 percent of residential dollar loss.

---

[1] A second approach to making these estimates is to use the percentage of fires that are residential from NFIRS (shown in Figure 27, Chapter 2), scaled up (multiplied by) the NFPA estimate of total fires. The results are somewhat different from those using the NFPA subtotals. We have used the NFPA residential totals for scaling residential fires because they are consistent with the total number of fires from NFPA. Better estimates from NFIRS will not be available until more of the participating NFIRS departments provide accurate "population protected" data.

57

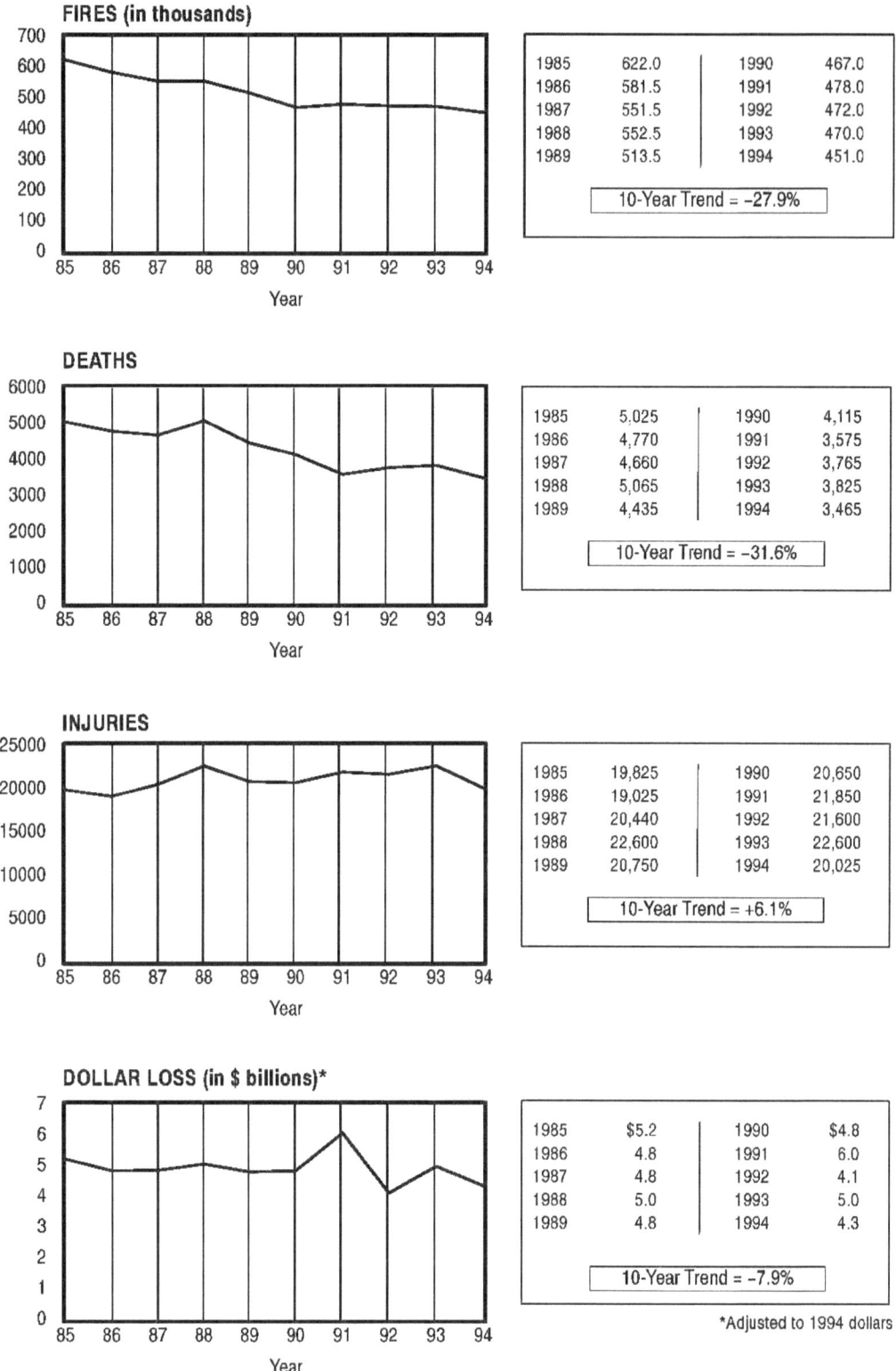

1985	622.0	1990	467.0
1986	581.5	1991	478.0
1987	551.5	1992	472.0
1988	552.5	1993	470.0
1989	513.5	1994	451.0

10-Year Trend = -27.9%

1985	5,025	1990	4,115
1986	4,770	1991	3,575
1987	4,660	1992	3,765
1988	5,065	1993	3,825
1989	4,435	1994	3,465

10-Year Trend = -31.6%

1985	19,825	1990	20,650
1986	19,025	1991	21,850
1987	20,440	1992	21,600
1988	22,600	1993	22,600
1989	20,750	1994	20,025

10-Year Trend = +6.1%

1985	$5.2	1990	$4.8
1986	4.8	1991	6.0
1987	4.8	1992	4.1
1988	5.0	1993	5.0
1989	4.8	1994	4.3

10-Year Trend = -7.9%

*Adjusted to 1994 dollars

Sources: NFPA Annual Surveys and Consumer Price Index

Figure 33. Trends in Residential Fires and Fire Losses

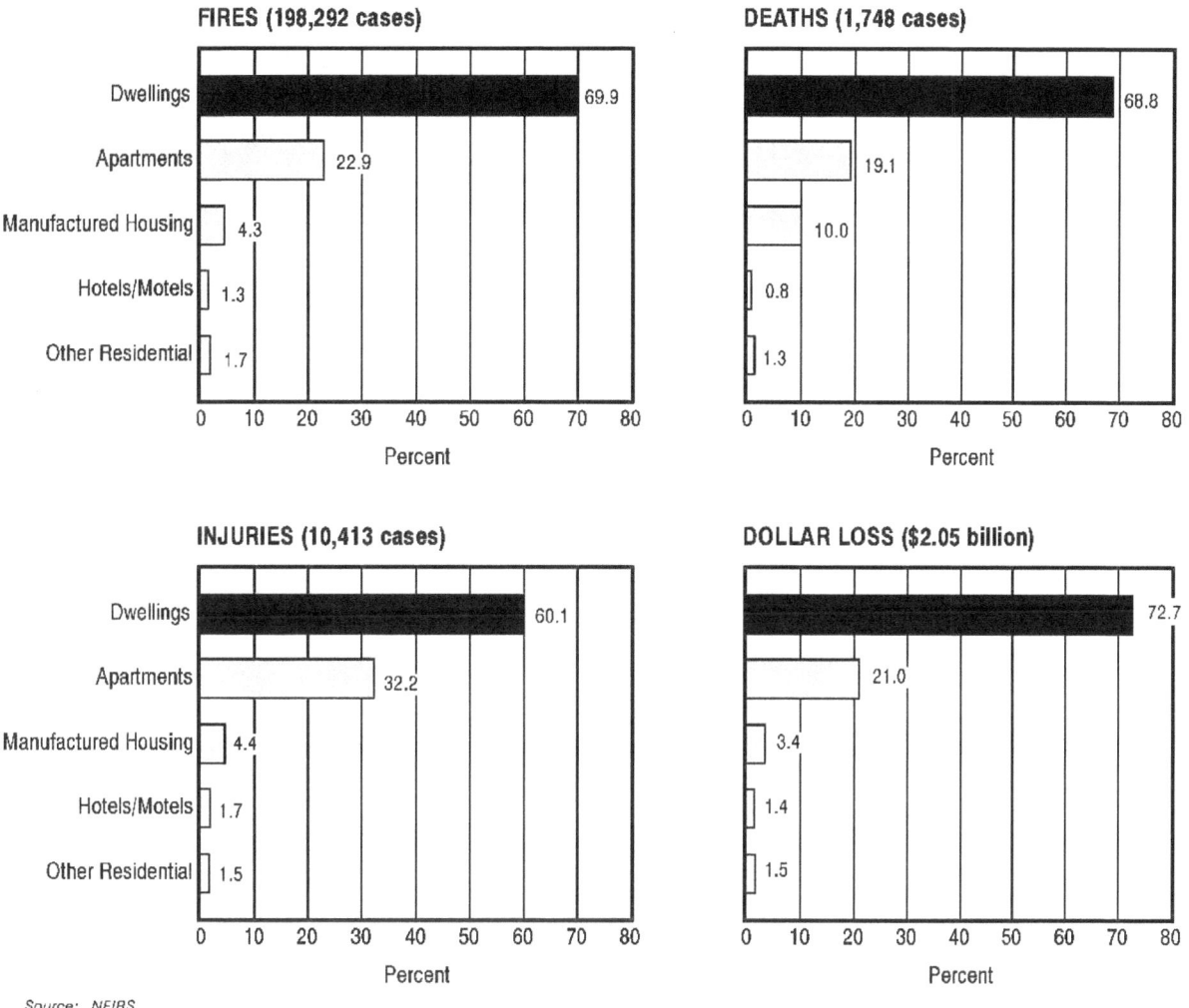

Figure 34. 1994 Residential Fires and Fire Losses by Property Type

Apartments account for 19 to 23 percent of these problems except for injuries, where they account for 32 percent. This higher incident of injuries may be because the total space and number of exits are significantly less in apartments than in dwellings.

Manufactured housing, separated from the dwelling category, have 2.5 times the number of deaths (10 percent) as number of fires (4 percent). Deaths per fire are approximately twice as high for manufactured housing as for other dwellings.

Hotels and motels, the target of legislation requiring sprinklers in the mid 1980s, account for just over 1 percent of the residential fire problem in the various measures.

Causes

Figure 35 shows the leading causes of fires, deaths, injuries, and dollar loss in 1994. They are dominated by the causes of one- and two-family dwellings, which account for the majority of residen-

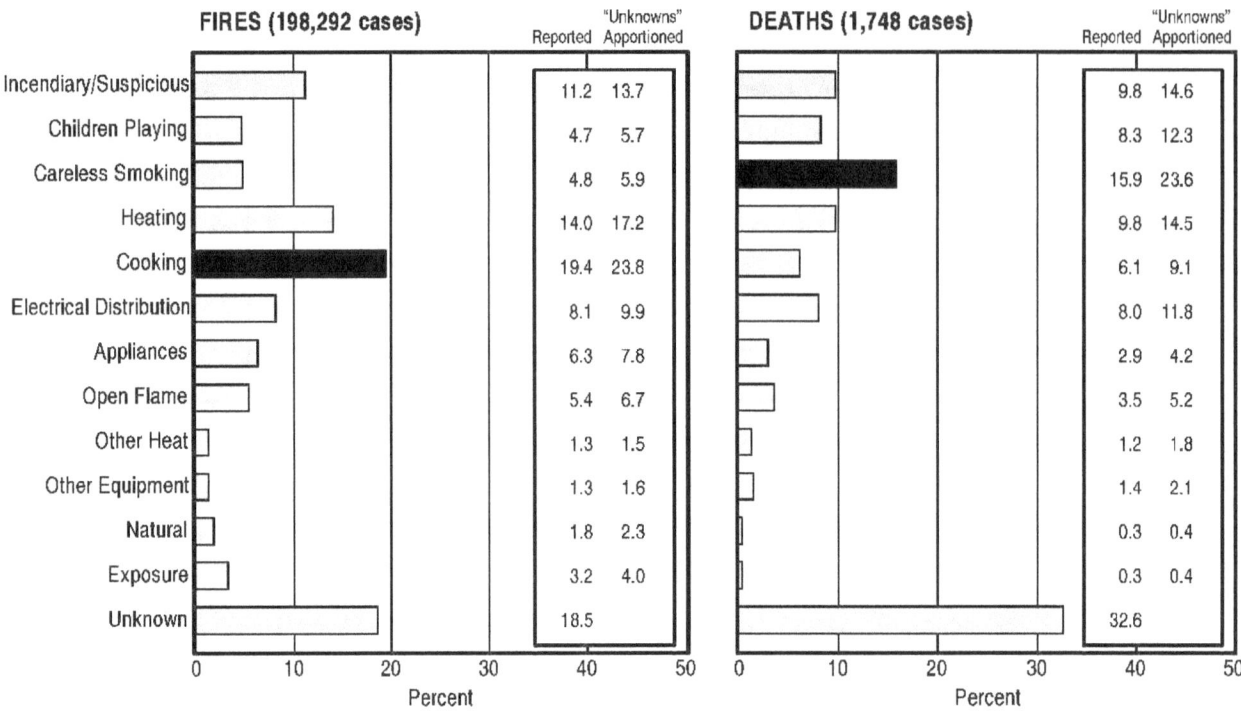

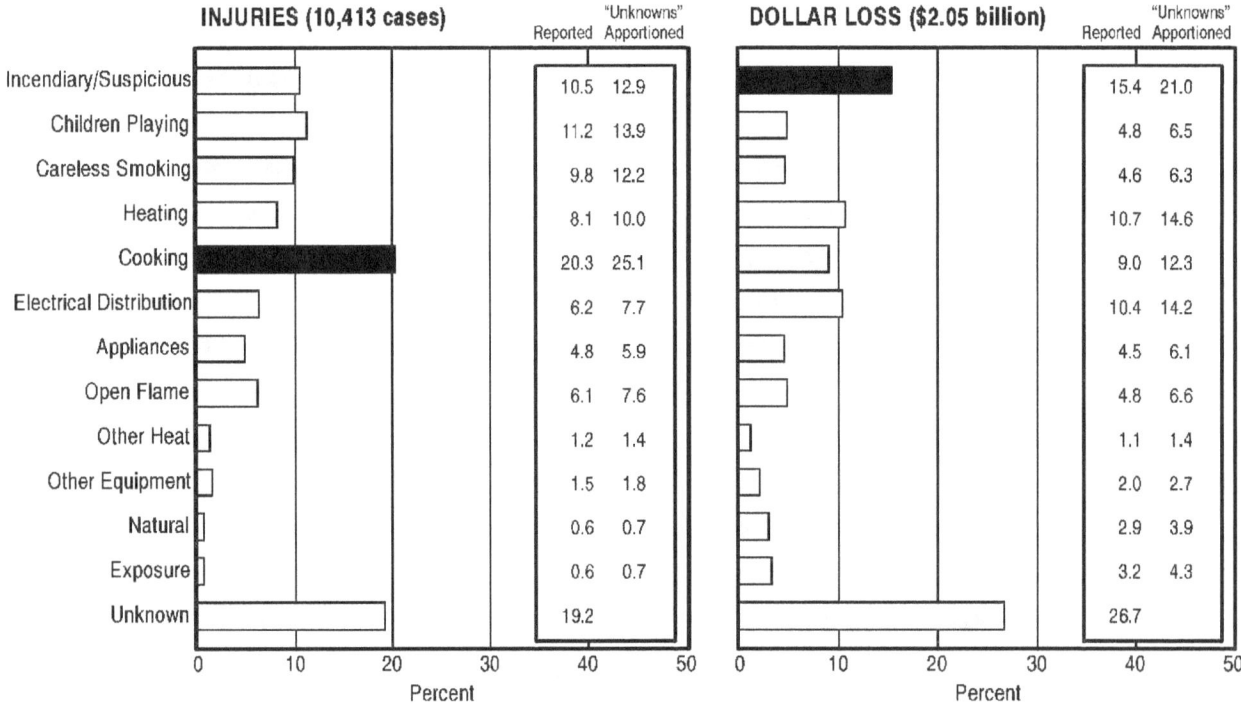

Figure 35. Causes of 1994 Residential Fires and Fire Losses

tial fires. The overall residential figures and those for one- and two-family dwellings discussed in the following section will seem to be quite similar. Larger differences from the overall residential causes will be found as one looks at the smaller subcategories of residences such as apartments and manufactured housing. Considering residential property types as a whole, the leading causes of fires are cooking, heating systems, and incendiary and suspicious.

Cooking, the leading cause of residential fires in 1994, has been the leading cause of residential fires most of the years since NFIRS inception. Heating passed cooking in the late 1970s when there was a surge in the use of alternative space heaters and wood heating. Cooking by far is the leading cause of fire injuries, nearly twice that of any other cause. Many cooking fires come from unattended cooking. These fires can be lessened by emphasizing the importance of vigilance while cooking. Also, the public should be better informed as to how to extinguish small cooking fires (e.g., a pot or pan lid, dousing it with baking soda). Wearing loose-fitting clothing such as bathrobes can be dangerous around cooking areas. Cooking, however, is only sixth leading cause of fire deaths.

Heating, the second leading cause of residential fires, includes those fires where the equipment involved in ignition is central heating, fireplaces, portable space heaters, fixed room heaters, wood stoves, and water heating. The central and water heating portions of the problem have remained relatively steady, while the portable space heater and wood burning stove portion of the problem, along with chimney fires, rose very sharply from the late 1970s to the early 1980s, but has since subsided somewhat. This last group seems to be the more volatile portion of this category of residential fires. Heating-related fires are also the second leading cause of dollar loss in residences and in fire deaths.

Incendiary and suspicious, which is called "arson" here even though that term has a narrower legal definition, is the leading cause of dollar loss, the second leading cause of fire deaths, and the third leading cause of fires and injuries in residences. That arson is so prominent a factor in the residential fire problem may be a surprise to many. There are a number of factors to residential arson fires—vandalism, revenge, fraud, and quarrels are common motives according to fire officials.[2] Because arson plays such a major role in the overall fire problem, this topic is examined in detail in Chapter 6.

It is important to note that the leading causes are different depending on what measure is used, as can be seen from Figure 35. The top three causes from each point of view are listed in Table 5.

Table 5. Leading Causes of 1994 Residential Fires and Fire Losses

[Numbers in parentheses reflect the 1990 ranking]

Rank	Fires	Deaths	Injuries	Dollar Loss
1	Cooking (1)	Careless Smoking (1)	Cooking (1)	Arson (1)
2	Heating (2)	Arson (2)	Children Playing (4)	Heating (2)
3	Arson (3)	Heating (3)	Arson (3)	Electrical (3)

Sources: NFIRS and Eighth Edition, Fire in the United States

---

[2] Motives are not reported to NFIRS, but are tabulated by some arson units.

In terms of residential fire deaths (and fire deaths overall), careless smoking is the leading cause, but in 1994 plays a less significant role than it did previously (down 2 percentage points from 1990). In terms of injuries and total fires, cooking is by far the leading cause. For dollar loss in residences, arson is the leading cause. The rank order of causes also varies among subcategories of residences, as discussed later.

Trends of Residential Causes

Figure 36 on the following four pages shows the trends in the causes of residential fires over the years 1985–1994.[3] All of these trends would appear lower if presented as per capita rather than in the absolute, because the population increased by an estimated 9 percent over the 10 years. Any upward change less than this population increase or any downward change at all represents doing better than expected over this period. One significant change in 1994 is that the hotel/motel category is now counted under "other" residences; therefore, certain data could not be collected for the hotel/ motel category in 1994.

In terms of numbers of fires, the number of heating fires has decreased more than 50 percent over 10 years and in 1990 dropped to second place, where it continues. Cooking has remained relatively constant and was the leading cause of residential fires in 1994. Incendiary and electrical distribution fires declined very slightly and are in third and fourth places, respectively.

For fire deaths, careless smoking remains the leading cause, but the number of deaths has dropped significantly in 10 years to its lowest level in 1994. This drop can be attributed to the overall decline in fire deaths and, in part, to the greater use of smoke detectors in the home. Heating peaked in the late 1970s, but has fallen sharply since. Although arson dropped to its lowest level in 1994, overall it is in second place, just above heating. Children playing as a cause of fire deaths rose until 1988 but has since declined.

For fire injuries, the leading cause, cooking, has risen steadily since 1985, although there was a significant drop in 1994. Children playing is the second leading cause of injuries. Incendiary injuries have risen steadily since 1985 and in 1994 became the third leading cause. Careless smoking injuries pinnacled in 1988 and have fallen steadily to the fourth leading cause of injuries in 1994.

For dollar loss, incendiary has dipped and risen, always maintaining its hold on first place. Heating has dropped significantly, but it continues in second place. Electrical distribution and cooking are in third and fourth places, respectively.[4]

---

[3] The data for each point on these figures may be found in Table B–1, Appendix B. Similar tables are presented in Appendix B for other graphs where data cannot conveniently be shown on the graph itself.

[4] When analyzing dollar loss trends, any precipitous increases must be checked to see if they might be due to errors in entering data for one or two fires. Often this happens when the data are entered on the incident report form left-adjusted instead right-adjusted. A $100 fire can be entered as a $100,000,000 fire.

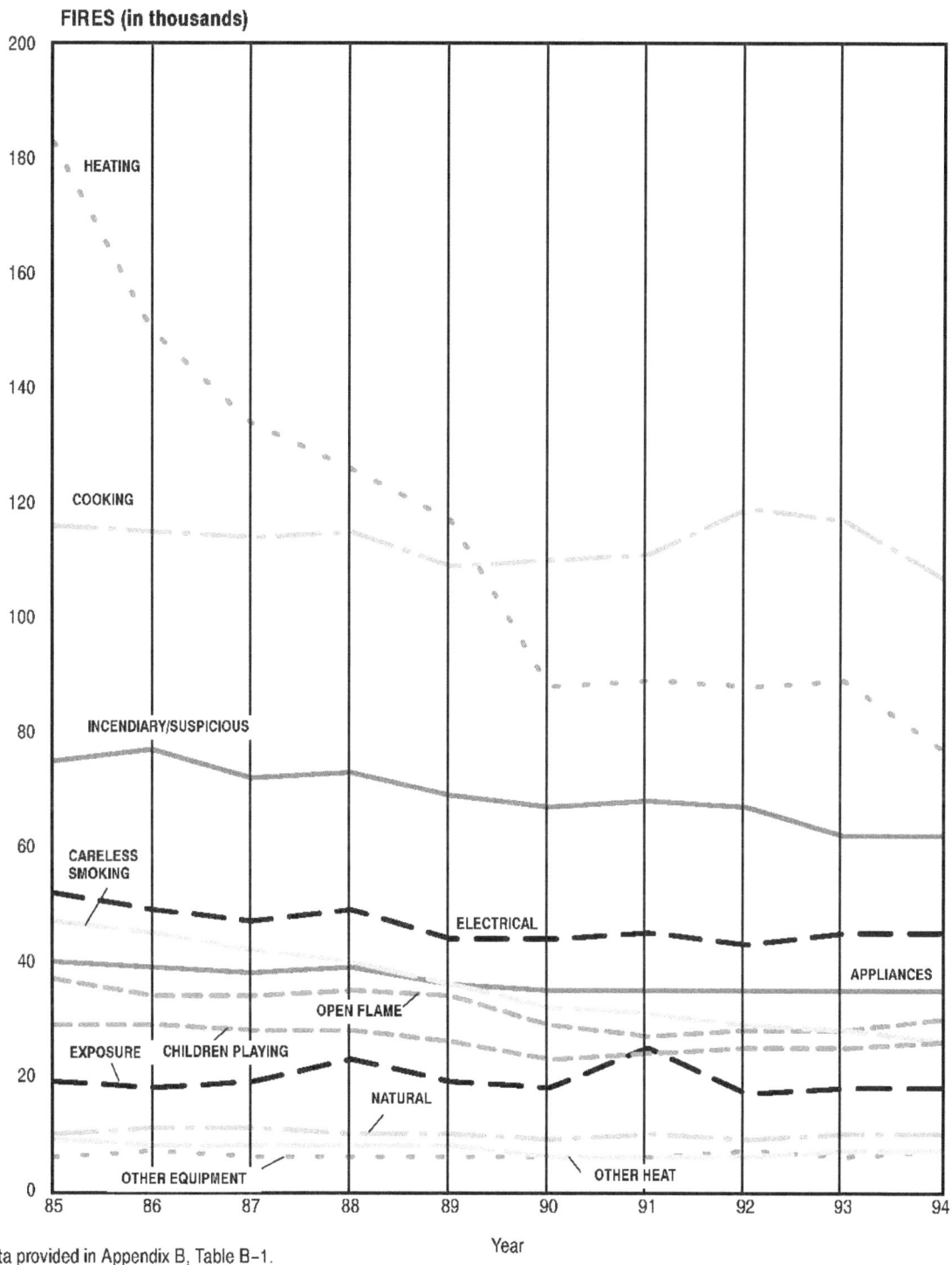

FIRES (in thousands)

Continued on next page

Note: Data provided in Appendix B, Table B-1.

Year

Figure 36. Trends in Causes of Residential Fires and Fire Losses

63

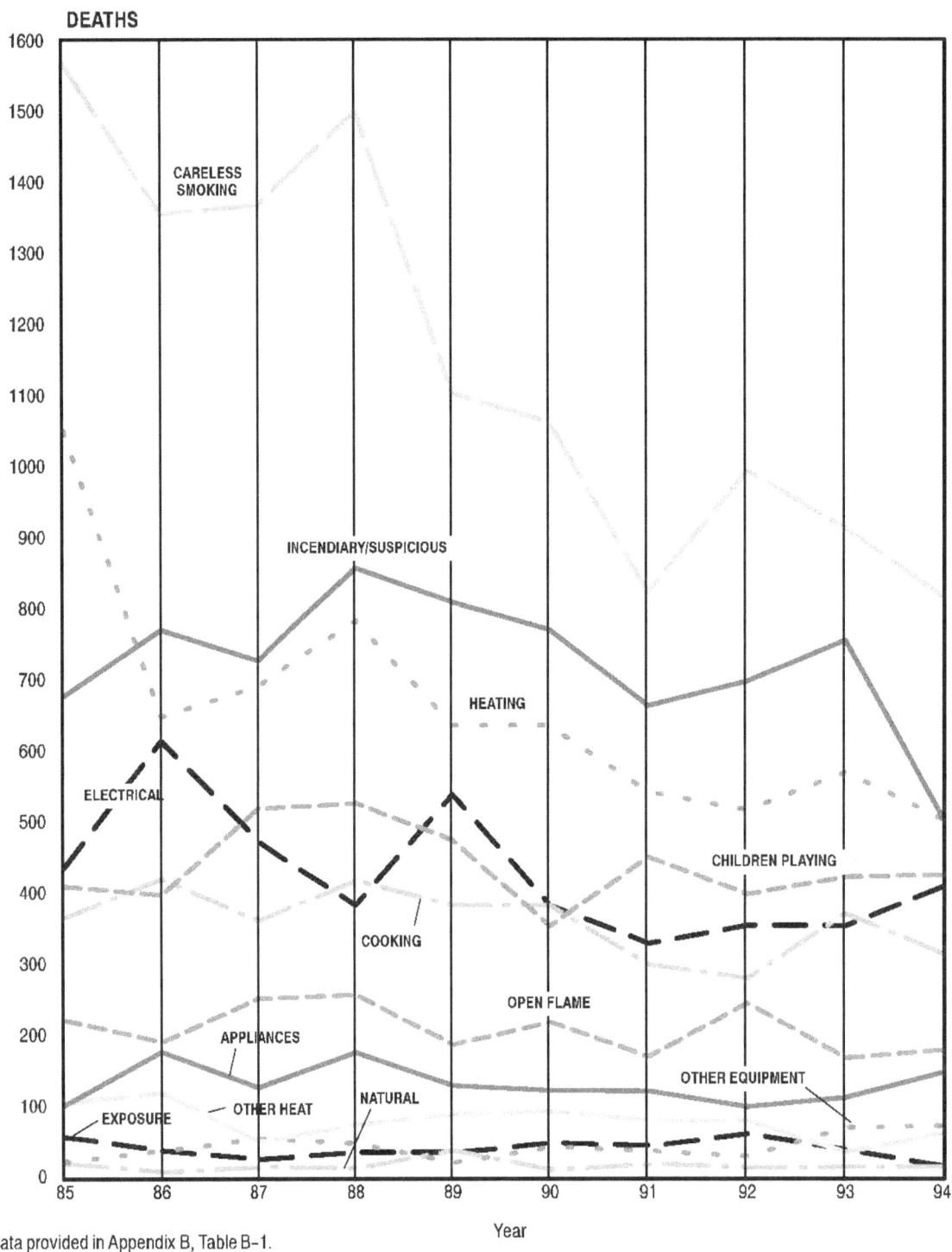

Figure 36. Trends in Causes of Residential Fires and Fire Losses (cont'd)

Note: Data provided in Appendix B, Table B-1.

Continued on next page

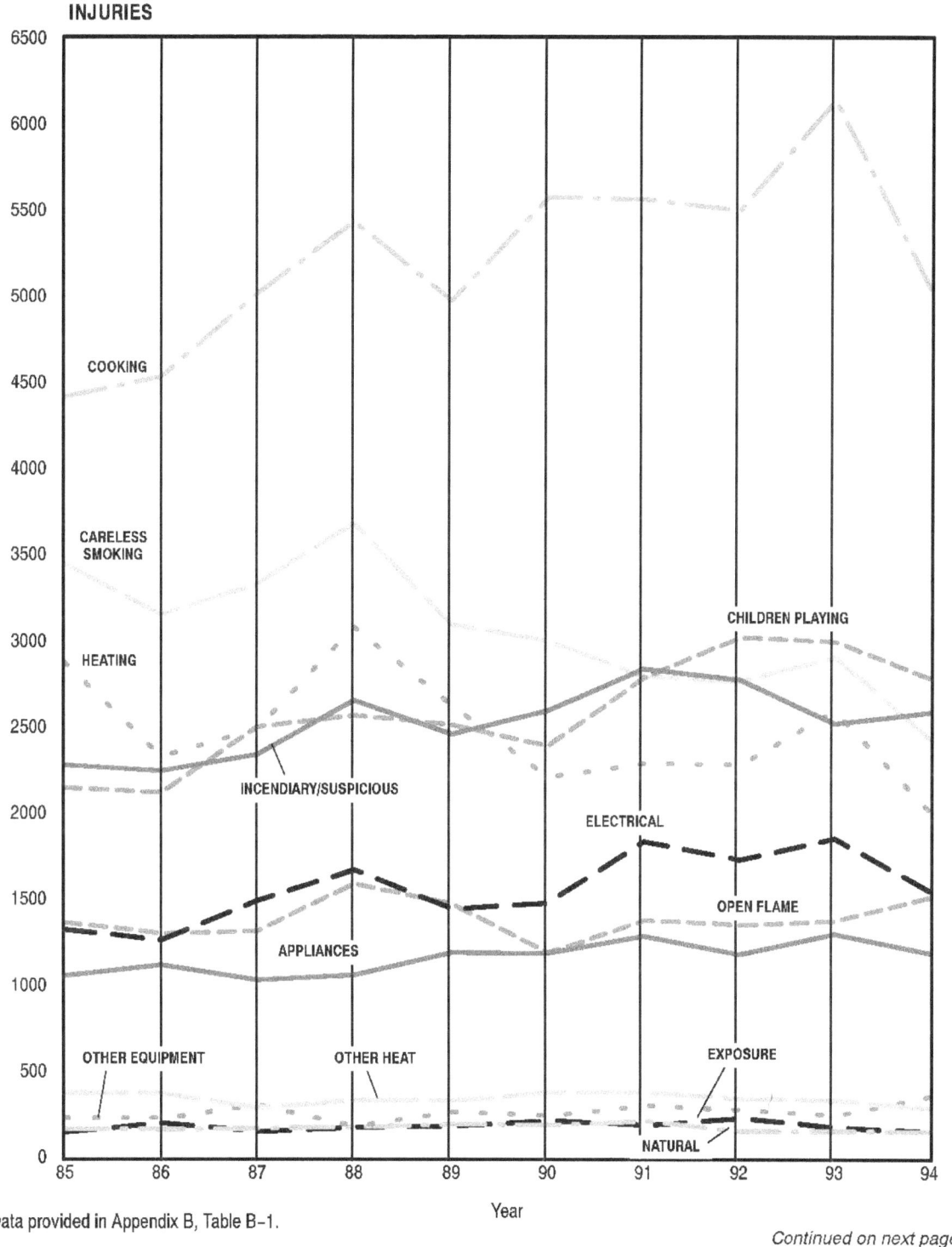

INJURIES

Note: Data provided in Appendix B, Table B-1.

Continued on next page

Figure 36. Trends in Causes of Residential Fires and Fire Losses (cont'd)

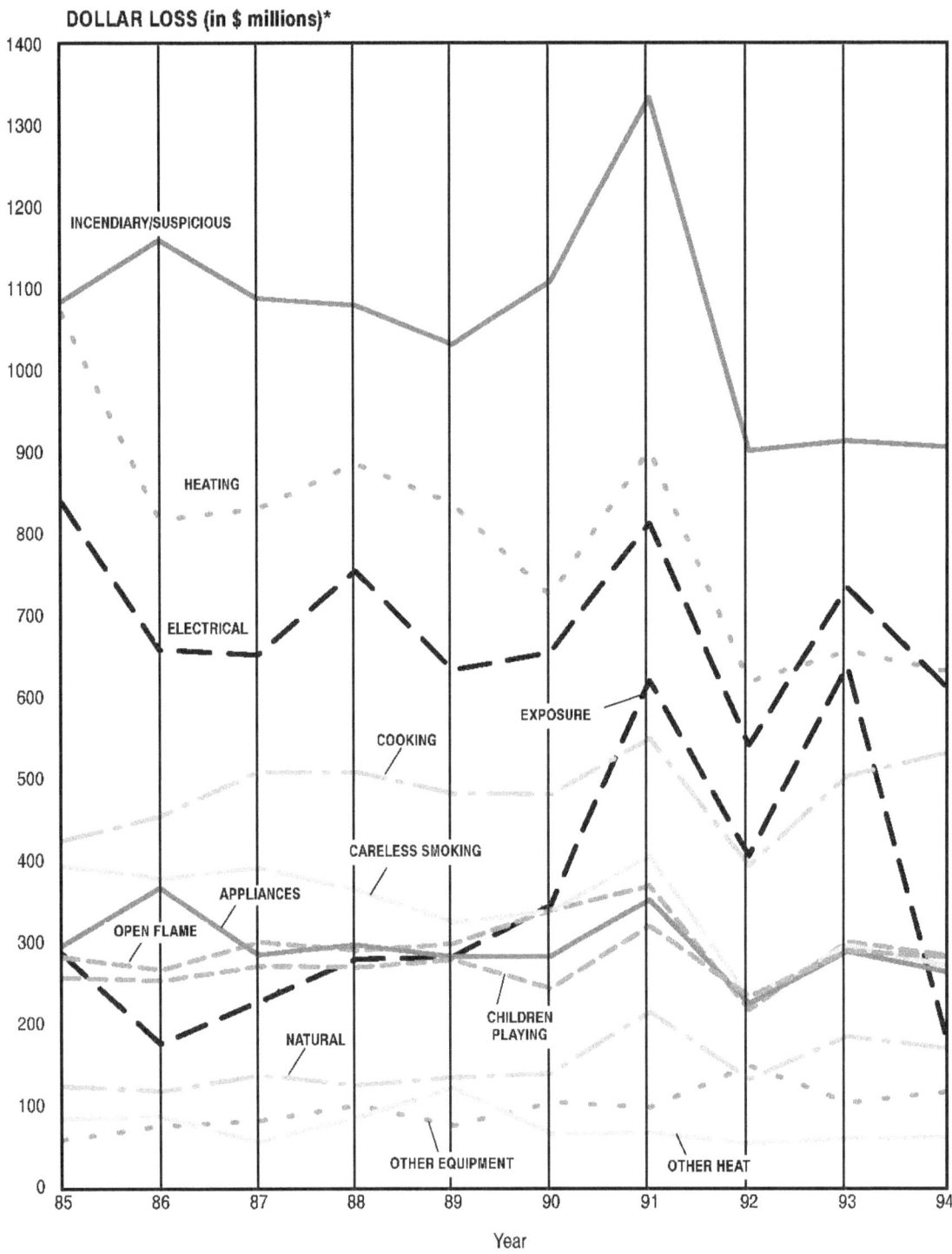

DOLLAR LOSS (in $ millions)*

Note: Data provided in Appendix B, Table B-1.

Sources: NFPA Annual Surveys and NFIRS

*Adjusted to 1994 dollars

Figure 36. Trends in Causes of Residential Fires and Fire Losses (cont'd)

Smoke Detector Performance

Smoke detectors are thought to account for a significant part of the decrease in reported fires and fire deaths since the mid 1970s. From previous surveys, we know that at least 88 percent of U.S. households have at least one smoke detector.[5] Only 39 (unadjusted) percent of households that had fires were reported to have detectors; considering only the incidents where smoke detector performance was reported, this percentage rises to 57 percent, still considerably less than the national average (Figure 37). Still, there has been a 33 percent rise since 1990. Households that have reported fires appear much less likely to have smoke detectors than others. Either people with detectors are more safety conscious or the detectors allow early detection and extinguishment so that the fires are not reported. Also, anecdotal information indicates that reported fires are more prevalent in older, less well cared for homes, and these are less likely to be equipped with a detector.

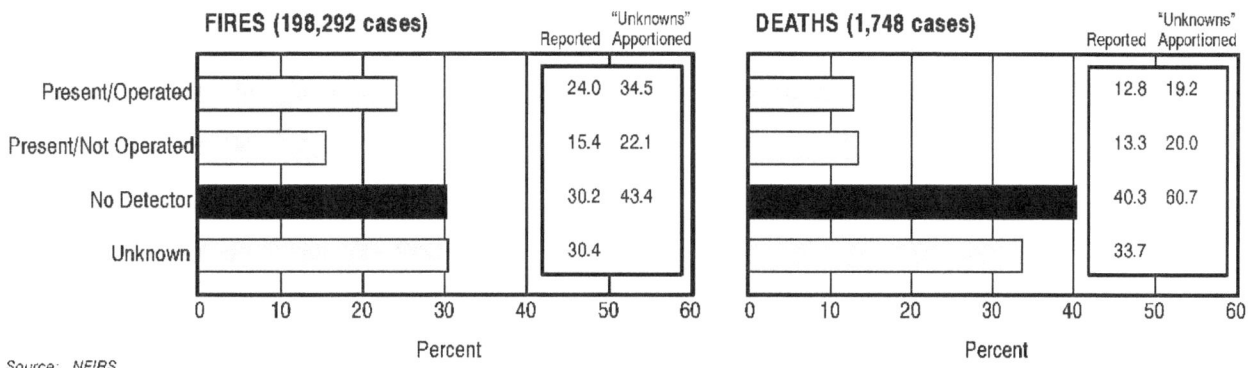

Figure 37. Smoke Detector Performance in 1994 Residential Fires and Fire Deaths

In only 35 percent of the residential fires where smoke detector performance was reported did a detector operate in the fire. That is, there was either no detector or the detector did not operate in at least 65 percent of the reported household fires. This figure is down from 71 percent in 1990.

As shown in Figure 37 for fire fatalities, detectors were not present in 61 percent of the fatalities in 1994, about 50 percent more than was the case for fires. In 19 percent of fire deaths, a detector did operate; this was the same percentage as in 1990. This is somewhat disturbing since there is a widespread belief that an operating detector will save lives. In some of these cases, the detector may have gone off too late to help the victim, the victim may have been too inebriated or feeble to react, or the fire may have been too close to the victim. Such cases merit further study.

When detectors were present in fires (57 percent), their performance based on where they were installed is shown in Table 6.

---

[5] *The Smoke Detector Operability Survey Report on Findings*, Consumer Product Safety Commission, Revised October 1994.

Table 6. Performance of Detectors When Present (Adjusted Percentages)

Detector Present	Present and Did Operate	Present but Did Not Operate	Total Present
In room of origin	20	7	27
Not in room of origin	15	11	26
In room, but fire too small	N/A	4	4
Total	35	22	57

Source: NFIRS

When households do have detectors and the fire is not small, they did not work in about one-third of the cases. The statistics are somewhat unclear because detector performance was not reported in 30 percent of the fires.

Figure 38 shows the trend of detector performance in fires and fire deaths. There has been an encouraging drop over 10 years in the percent of fires as well as the percent of fire deaths with no detector present—fires from 44 percent to 30 percent and fire deaths from 56 percent to 40 percent. Correspondingly, the percentage of fires where a detector operated has doubled from 12 percent to 24 percent. However, the percentage of fires where a detector was present but did not operate also increased. Public education programs need to focus on the proper maintenance of smoke detectors to reverse this trend.

Residential Sprinklers

Residential sprinklers are found in only a small fraction (2 percent) of residences other than hotels and newer apartment buildings today. Therefore, it is no surprise that they are reported to be present in only a small percentage of residential fires nationally, though they represent a great potential in the future.

Sprinkler data were reported in 3,890 residential fires out of the 198,292 cases reported to NFIRS in 1994 (Figure 39). They operated in 1,460 cases and did not operate in 2,430 fires, mostly because the fire was too small. In 21 percent of the total cases, sprinkler performance was not reported, a decrease of 50 percent since 1990.

The trend in use of residential sprinklers has been upward (Figure 40). However, the percentages are still minuscule because most homes are not equipped with sprinklers. They were reported as present in 1 percent of the residential fires in 1985 versus 2 percent in 1994. They operated 0.7 percent of fires in 1994, up from 0.3 percent in 1985.

When Fires Occur

TIME OF DAY. Fires do not occur uniformly throughout the day, as shown in Figure 41 (two pages). Fire incidents peak from 5:00 p.m. to 7:00 p.m., when cooking fires most occur. Fire incidents drop when people sleep.

FIRES

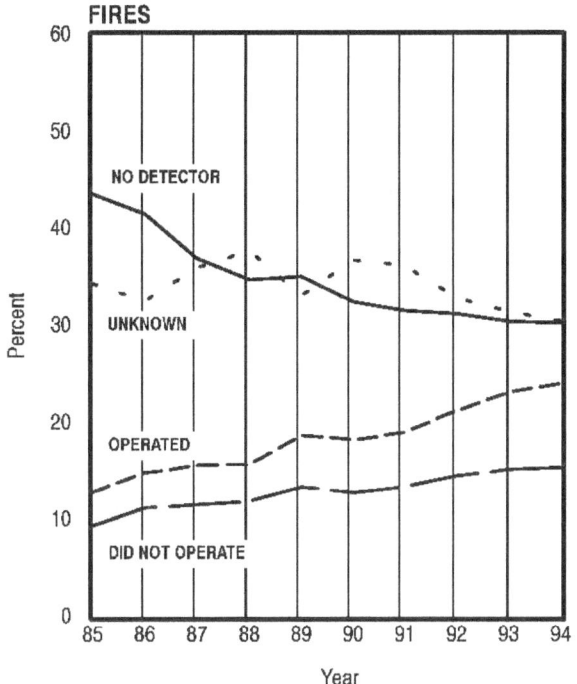

	No Detector	Operated	Did Not Operate	Unknown
1985	43.6%	12.8%	9.3%	34.3%
1986	41.5	14.8	11.2	32.5
1987	36.9	15.6	11.5	36.0
1988	34.7	15.7	11.9	37.7
1989	35.0	18.6	13.3	33.1
1990	32.4	18.2	12.7	36.7
1991	31.5	19.0	13.4	36.1
1992	31.2	21.2	14.5	32.8
1993	30.4	23.1	15.1	31.4
1994	30.2	24.0	15.4	30.4

DEATHS

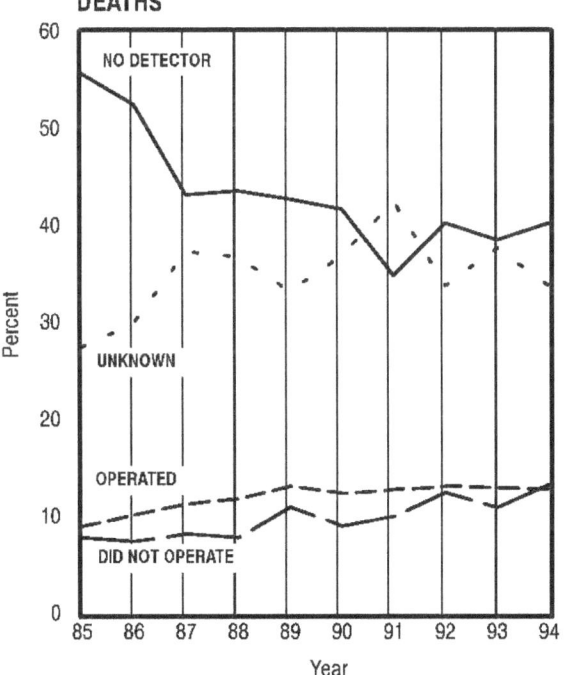

	No Detector	Operated	Did Not Operate	Unknown
1985	55.7%	9.0%	7.8%	27.5%
1986	52.5	10.2	7.4	29.9
1987	43.2	11.3	8.2	37.3
1988	43.6	11.9	7.8	36.7
1989	42.7	13.1	11.0	33.2
1990	41.7	12.4	9.0	36.9
1991	34.8	12.8	10.0	42.5
1992	40.3	13.1	12.5	33.8
1993	38.5	12.9	10.9	37.7
1994	40.3	12.8	13.3	33.7

Source: NFIRS

Figure 38. Trends in Smoke Detector Performance in Residential Fires and Fire Deaths

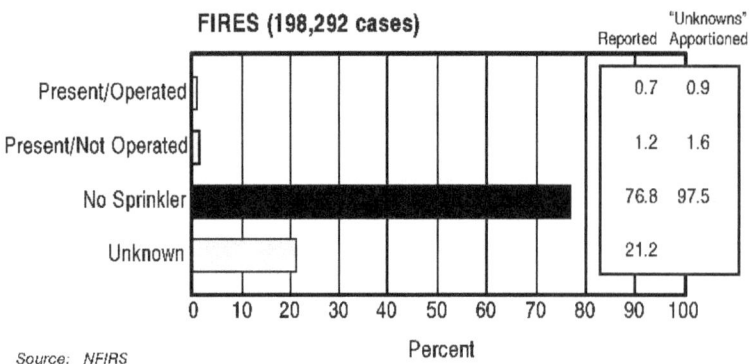

Figure 39. Sprinkler Performance in 1994 Residential Fires

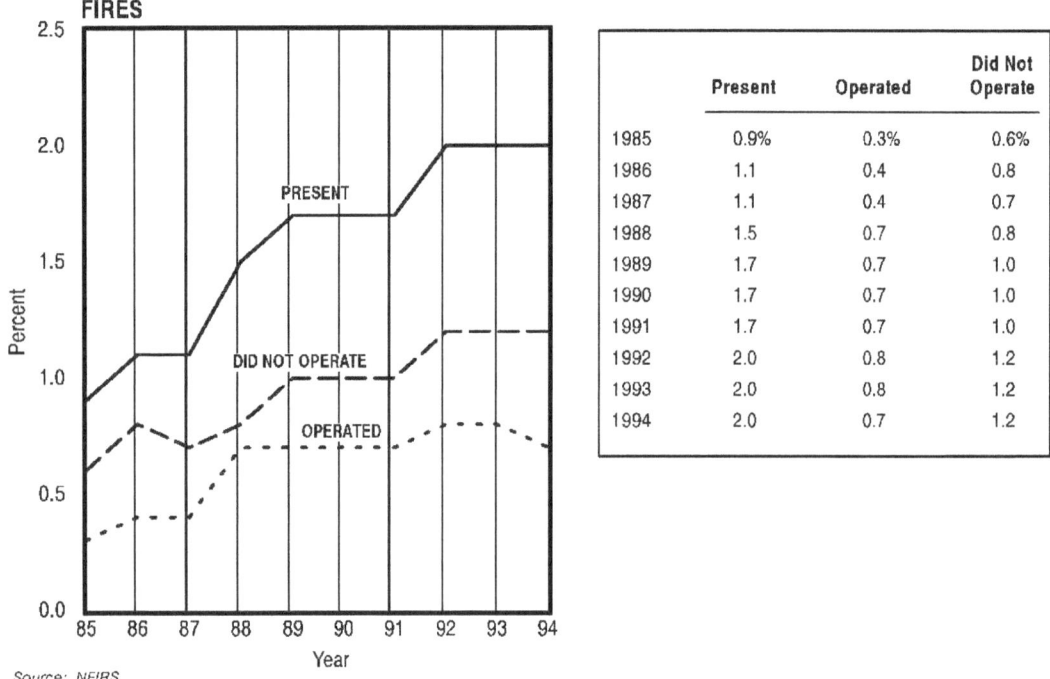

	Present	Operated	Did Not Operate
1985	0.9%	0.3%	0.6%
1986	1.1	0.4	0.8
1987	1.1	0.4	0.7
1988	1.5	0.7	0.8
1989	1.7	0.7	1.0
1990	1.7	0.7	1.0
1991	1.7	0.7	1.0
1992	2.0	0.8	1.2
1993	2.0	0.8	1.2
1994	2.0	0.7	1.2

Figure 40. Trends in Sprinkler Performance in Residential Fires

Fire deaths are usually associated with fires that start late at night and early morning. Nearly half of residential fire deaths occur in fires that start from 11:00 p.m. to 6:00 a.m. The peak night hours are from 3:00 to 5:00 a.m. when people are in deep (REM) sleep.

Fire injuries occur more uniformly throughout the day, peak slightly during dinner hours when people cook, and actually drop to their low point early in the morning hours.

The sharp peak in dollar loss in 1994 is between 7:00 p.m. and 8:00 p.m. Dollar loss is otherwise relatively constant, again with a drop in the early morning hours.

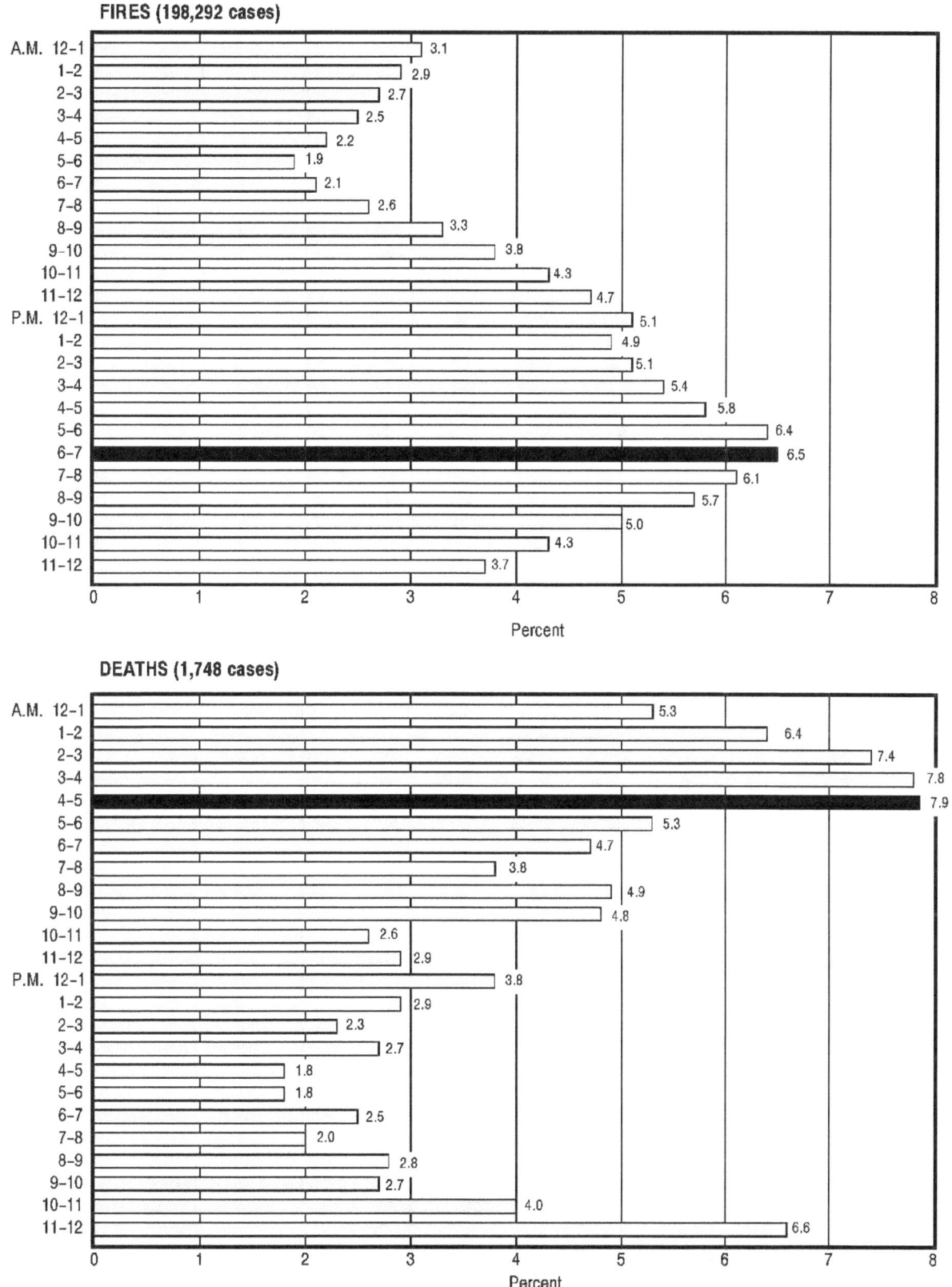

FIRES (198,292 cases)

DEATHS (1,748 cases)

Figure 41. Time of Day of 1994 Residential Fires and Fire Losses

Continued on next page

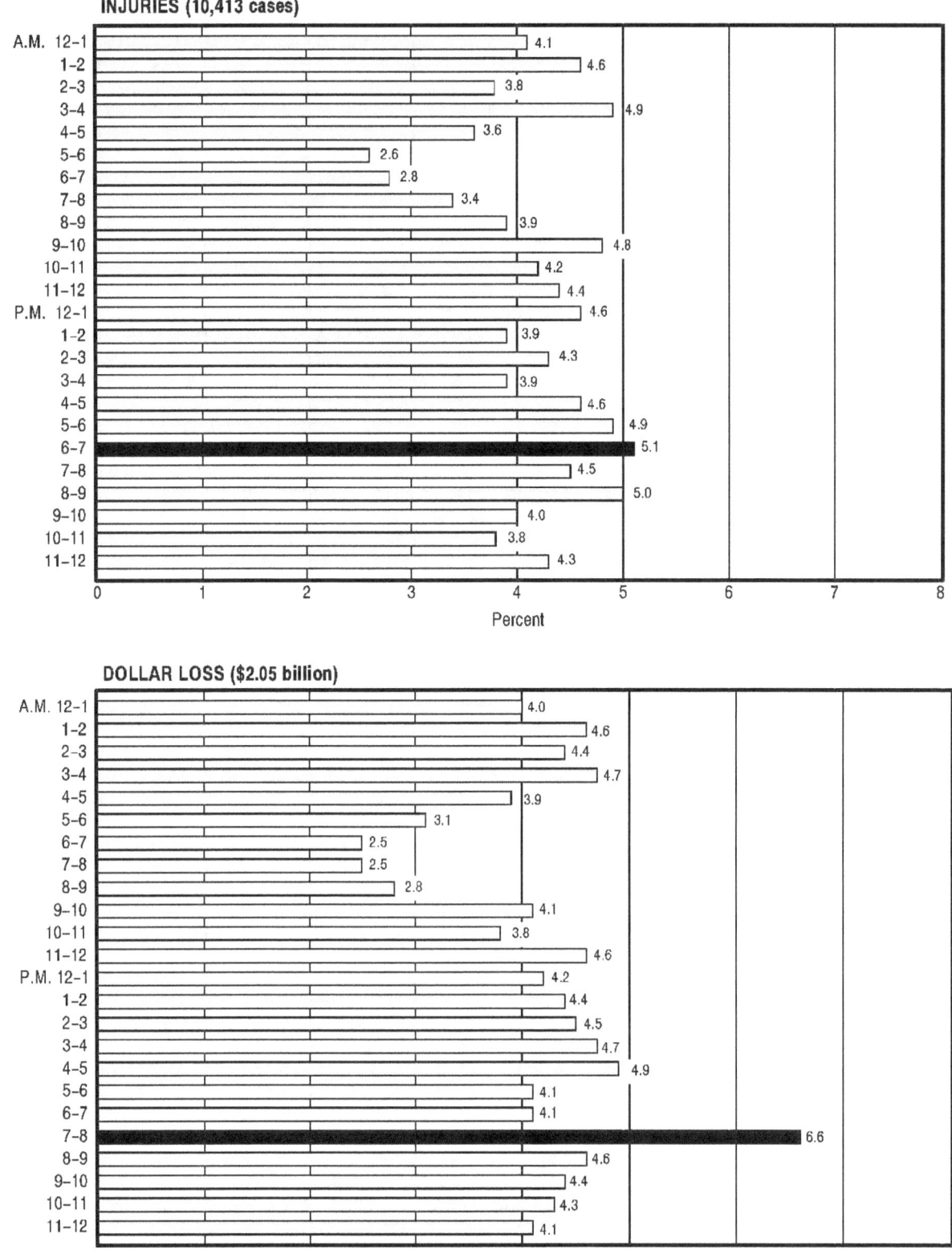

Figure 41. Time of Day of 1994 Residential Fires and Fire Losses (cont'd)

MONTH OF THE YEAR.   Residential fires and fire deaths are most frequent during winter months when heating systems play a dominant role. Forty percent of all deaths occur from December through February (Figure 42).

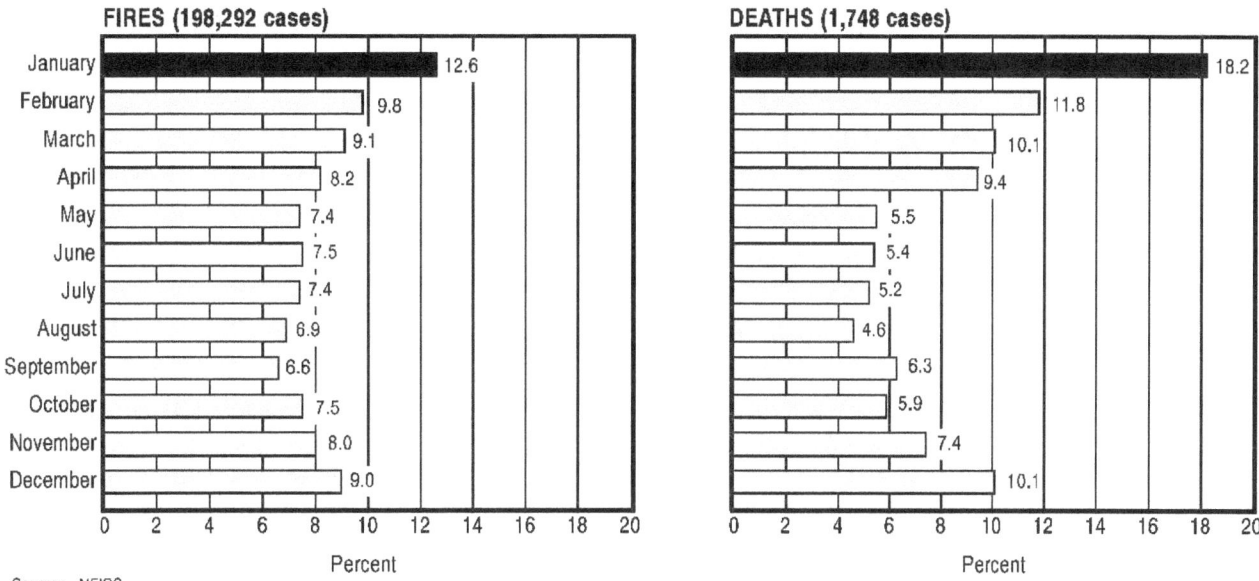

Figure 42. Month of Year of 1994 Residential Fires and Fire Deaths

DAY OF THE WEEK.   The incidence of residential fires is uniformly spread over the entire week, but one-third of all deaths occur on the weekend when a large portion of the populace is at home (Figure 43). The leading causes of residential fires—cooking and heating—are generally unaffected by the day of the week.

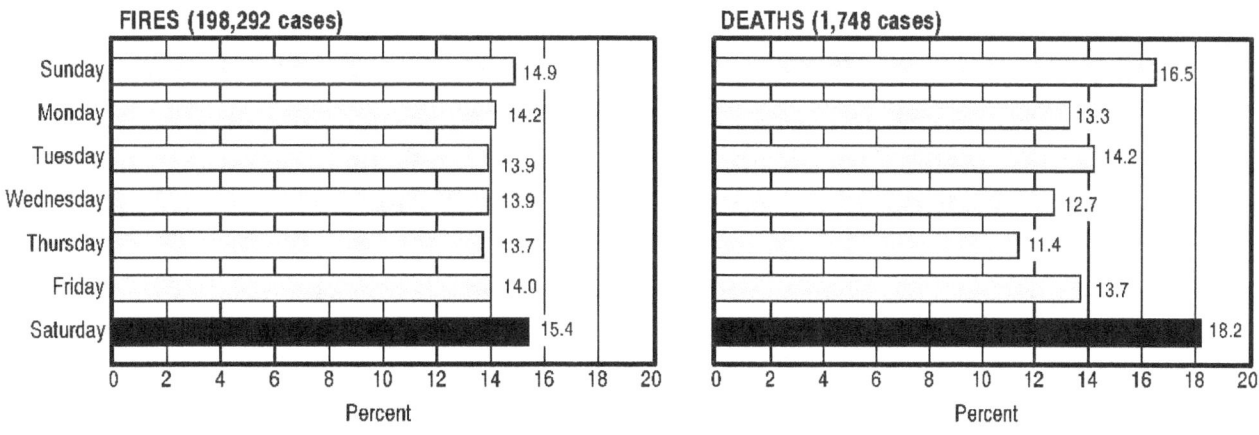

Figure 43. Day of Week of 1994 Residential Fires and Fire Deaths

ONE- AND TWO-FAMILY HOMES

One- and two-family homes are where two-thirds of the people in the United States reside. The fire profile, therefore, is similar to that for residences as a whole. Manufactured housing (mobile homes) is included in the profiles for one- and two-family homes unless otherwise noted. A separate discussion on manufactured housing is included toward the end of this section. Data on residential garage fires are also treated separately at the end of this section.

Overview of Trends

Figure 44 shows the numbers of fires, deaths, injuries, and adjusted dollar loss for single-family dwellings from 1985 to 1994. Fires and deaths steadily declined, with the 10-year trend at or above 32 percent. Dollar losses also declined, by 15 percent. Injuries have remained virtually level.

Because the number of fires has dropped faster than injuries or dollar loss, the statistics per fire are getting worse. One reason for this is that the ever-increasing number of smoke detectors detects fires in the early stages. Fires that are detected early are often extinguished before they are reported to the fire department, and so the number of reported fires decreases. When detectors are not present, the fire burns longer before detection and does more damage. This situation results in fires attended by fire departments being, on average, more serious.

When Fires Occur

TIME OF DAY. Figure 45 shows that fires and injuries in one- and two-family dwellings are highest between 5:00 and 7:00 p.m., when cooking fires sharply increase. Fire deaths, on the other hand, peak late at night and in the early morning hours. This result is often caused by careless smoking fires that smolder for several hours and then rapidly increase in smoke production and open flames. Also, the early morning hours are when people are in deep (REM) sleep so they do not awake in time to escape. Dollar loss is fairly uniform throughout the day with peaks at 1:00−2:00 a.m. and 2:00−3:00 p.m. and nadirs at 6:00−8:00 a.m.

MONTH OF THE YEAR. Fires and fire deaths in one- and two-family homes peak in mid winter, when heating fires add to the other types of year-round fires (Figure 46).

74

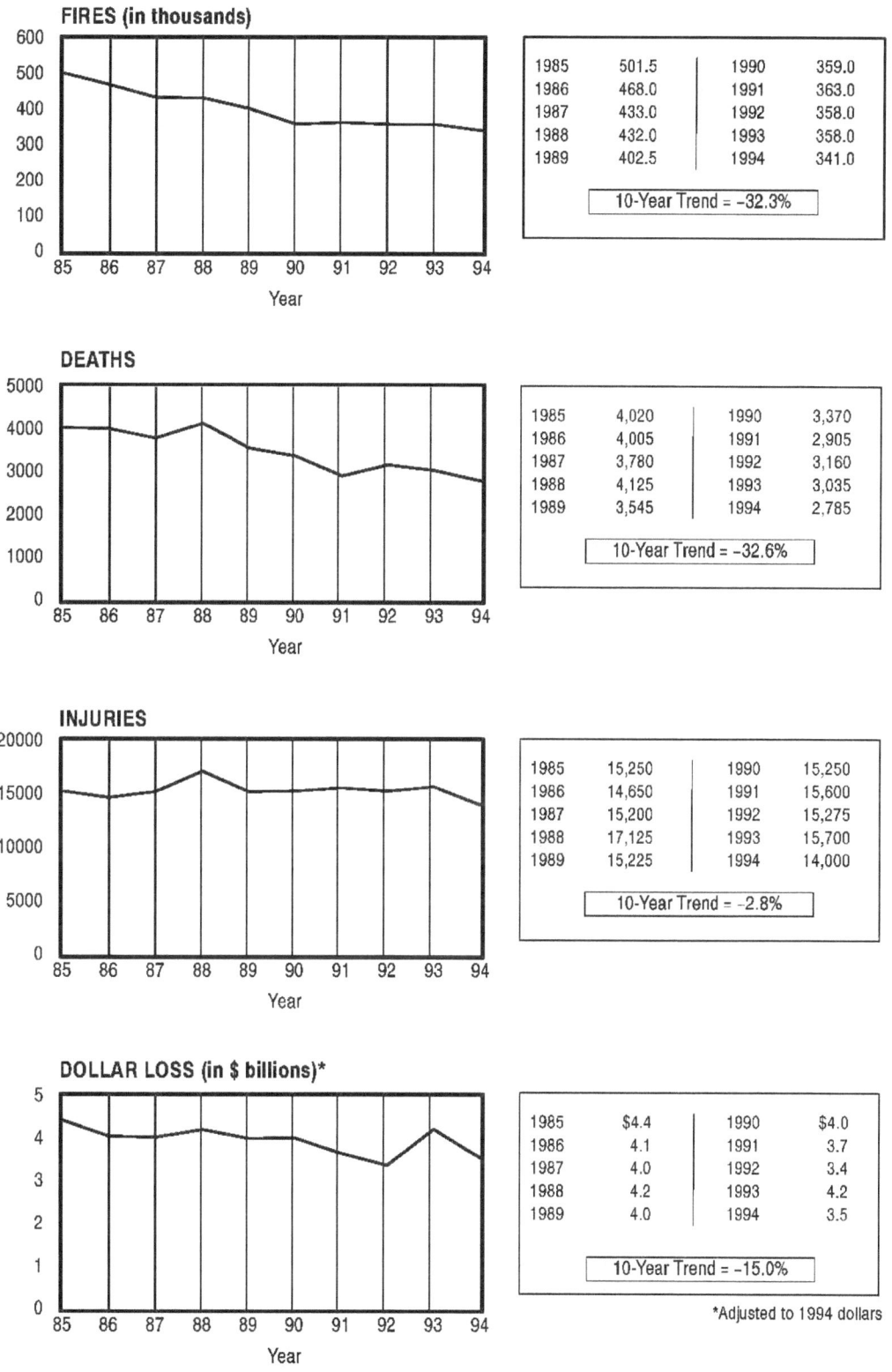

FIRES (in thousands)

1985	501.5	1990	359.0
1986	468.0	1991	363.0
1987	433.0	1992	358.0
1988	432.0	1993	358.0
1989	402.5	1994	341.0

10-Year Trend = −32.3%

DEATHS

1985	4,020	1990	3,370
1986	4,005	1991	2,905
1987	3,780	1992	3,160
1988	4,125	1993	3,035
1989	3,545	1994	2,785

10-Year Trend = −32.6%

INJURIES

1985	15,250	1990	15,250
1986	14,650	1991	15,600
1987	15,200	1992	15,275
1988	17,125	1993	15,700
1989	15,225	1994	14,000

10-Year Trend = −2.8%

DOLLAR LOSS (in $ billions)*

1985	$4.4	1990	$4.0
1986	4.1	1991	3.7
1987	4.0	1992	3.4
1988	4.2	1993	4.2
1989	4.0	1994	3.5

10-Year Trend = −15.0%

*Adjusted to 1994 dollars

Sources: NFPA Annual Surveys and Consumer Price Index

Figure 44. Trends in One- and Two-Family Dwelling Fires and Fire Losses

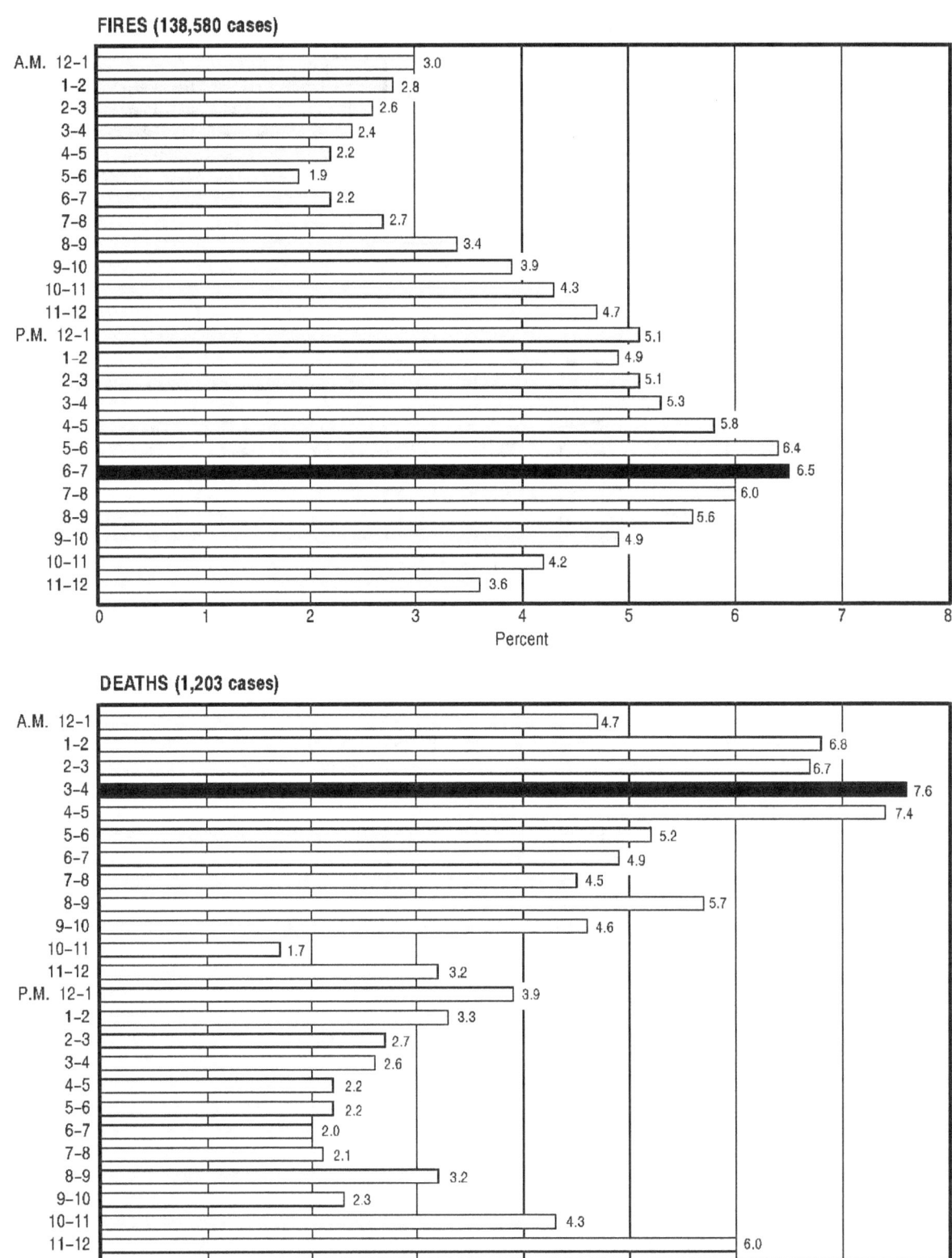

Figure 45. Time of Day of 1994 One- and Two-Family Dwelling Fires and Fire Losses

Continued on next page

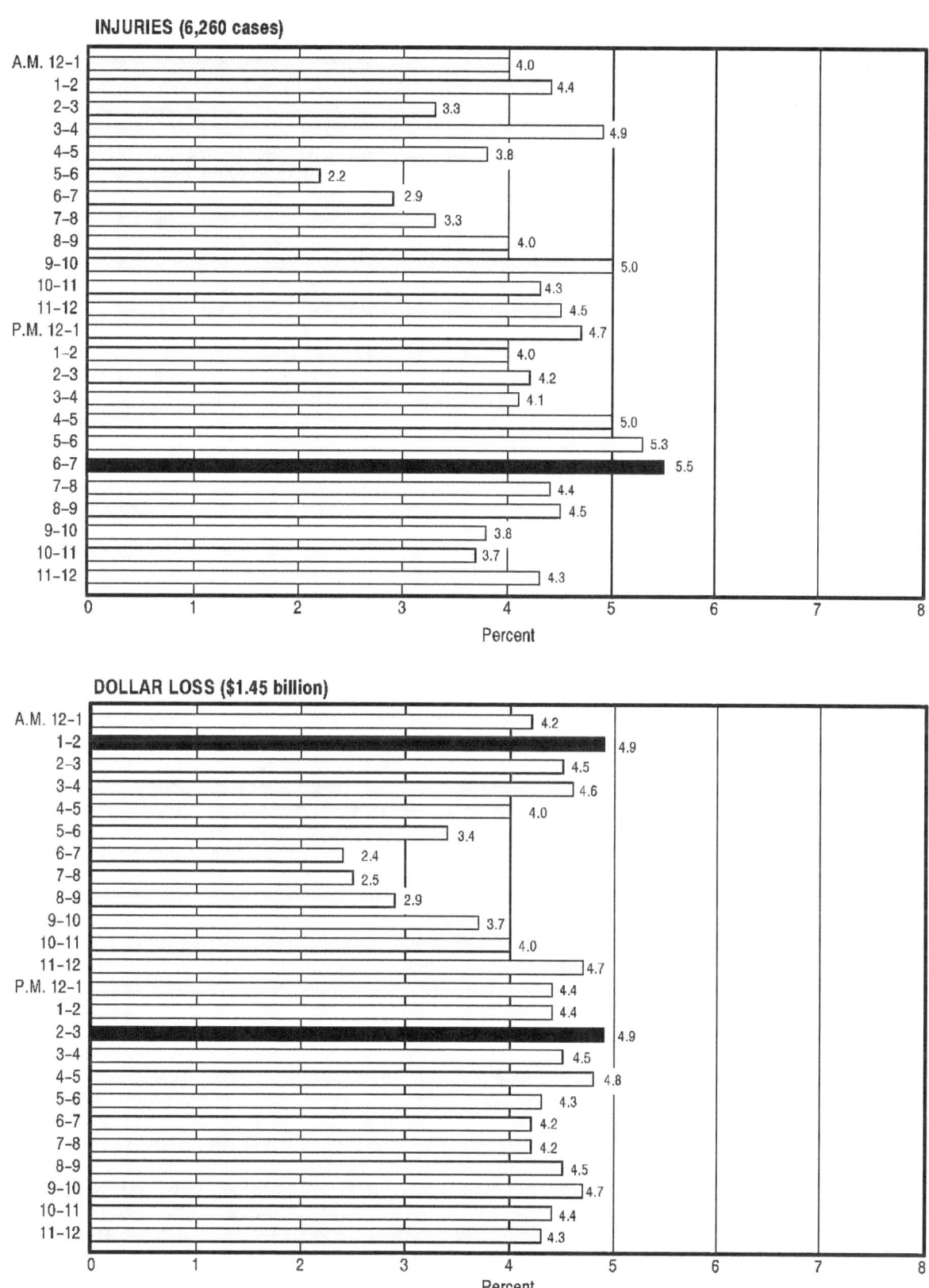

INJURIES (6,260 cases)

Time	Percent
A.M. 12–1	4.0
1–2	4.4
2–3	3.3
3–4	4.9
4–5	3.8
5–6	2.2
6–7	2.9
7–8	3.3
8–9	4.0
9–10	5.0
10–11	4.3
11–12	4.5
P.M. 12–1	4.7
1–2	4.0
2–3	4.2
3–4	4.1
4–5	5.0
5–6	5.3
6–7	5.5
7–8	4.4
8–9	4.5
9–10	3.8
10–11	3.7
11–12	4.3

Percent

DOLLAR LOSS ($1.45 billion)

Time	Percent
A.M. 12–1	4.2
1–2	4.9
2–3	4.5
3–4	4.6
4–5	4.0
5–6	3.4
6–7	2.4
7–8	2.5
8–9	2.9
9–10	3.7
10–11	4.0
11–12	4.7
P.M. 12–1	4.4
1–2	4.4
2–3	4.9
3–4	4.5
4–5	4.8
5–6	4.3
6–7	4.2
7–8	4.2
8–9	4.5
9–10	4.7
10–11	4.4
11–12	4.3

Percent

Source: NFIRS

Figure 45. Time of Day of 1994 One- and Two-Family Dwelling Fires and Fire Losses (cont'd)

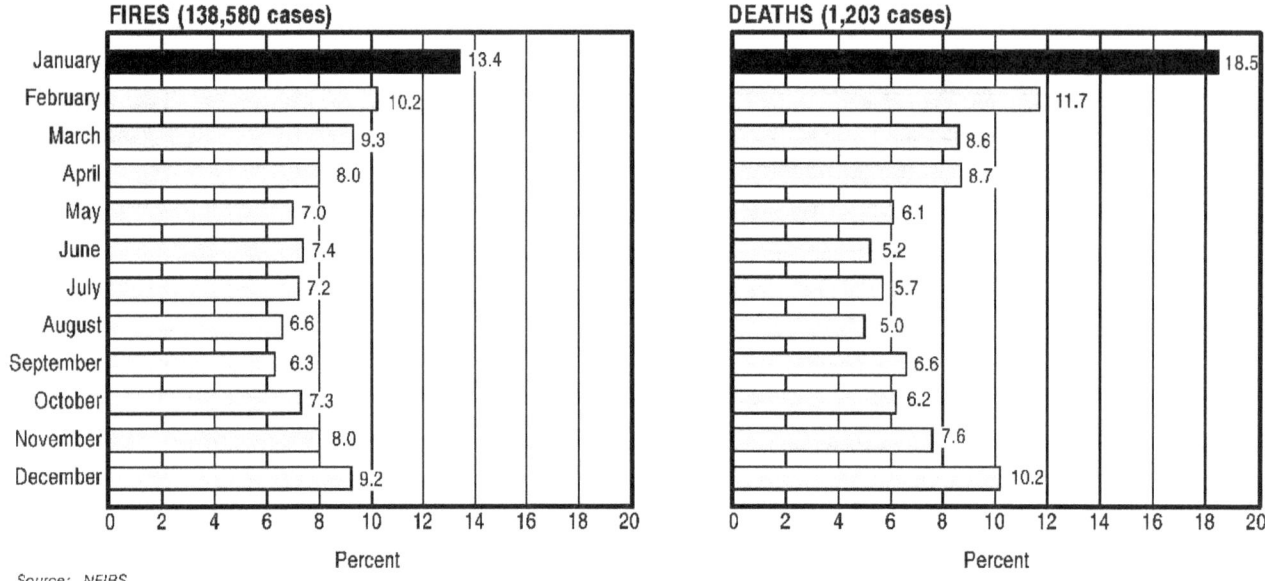

Source: NFIRS

Figure 46. Month of Year of 1994 One- and Two-Family Dwelling Fires and Fire Deaths

Causes

Although cooking is the overall leading cause of residential fires, heating is the leading cause of fires in one- and two-family dwelling, at 21 percent (Figure 47). Cooking at 19 percent and arson at 13 percent are the next two leading causes. That heating rather than cooking is the leading cause of fires is in part due to the role of the one-and two-family homeowner as chief maintenance officer. In the other residential property types, this role tends to be handled by professionals. The difference between heating and cooking fires has narrowed to less than 2 percentage points in 1994, most likely because the use of wood stoves and kerosene heaters has diminished over the past 10 years.

The leading cause of death is careless smoking, as in all NFIRS years, at 22 percent. This is down 24 percent from 1990 and 28 percent from 1985. Most of the careless smoking deaths come from cigarettes dropped on upholstered furniture or bedding, often by someone who has been drinking. Heating is the second leading cause of death at 16 percent and arson is third at 14 percent. These three causes account for over half of the total deaths.

For injuries, cooking is first at 24 percent. The most common cooking fires result from unattended cooking, when oil or grease catches fire, and from the ignition of loose clothing. Children playing is second at 14 percent and heating is third at 13 percent. These profiles are similar to the 1985 and 1990 causes.

Because heating is the leading cause of fires, Figure 48 examines the types of heating equipment involved in ignition of the heating fires. The largest group, nearly one-third of the total, are those from fixed or stationary area heating units, which includes baseboard units, wood heating stoves, and heating units mounted in walls. Second most common are chimney fires, especially those

78

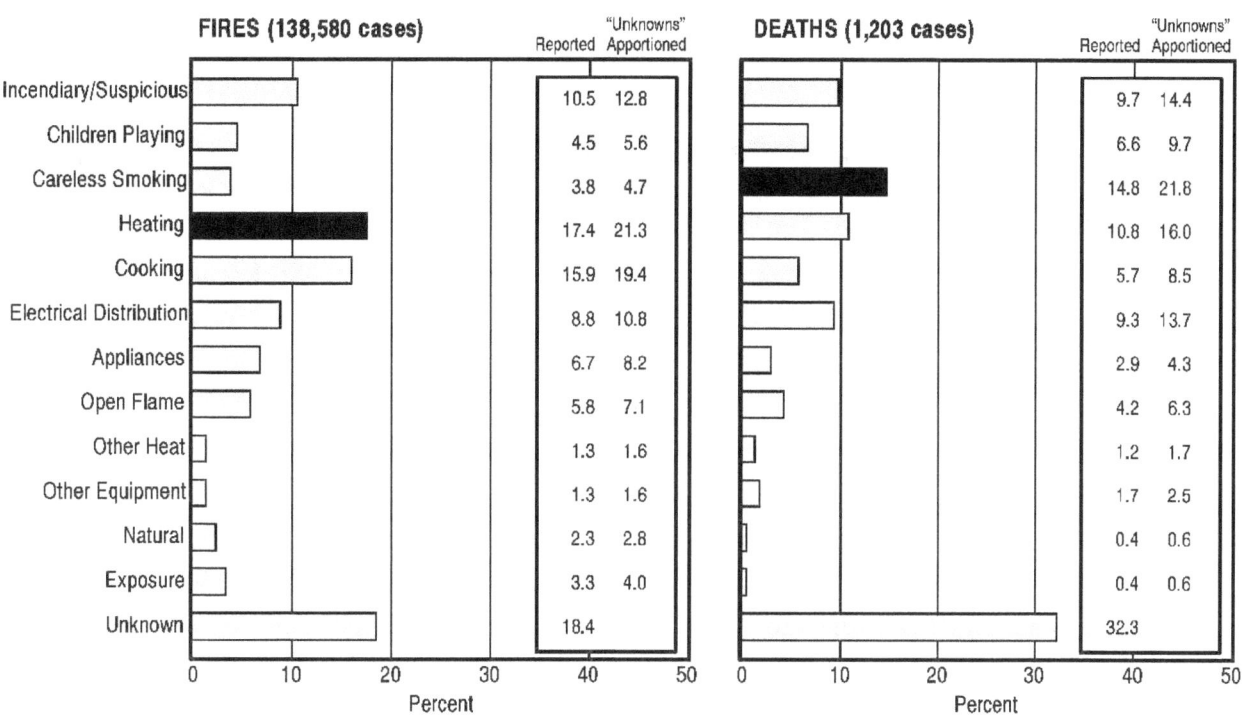

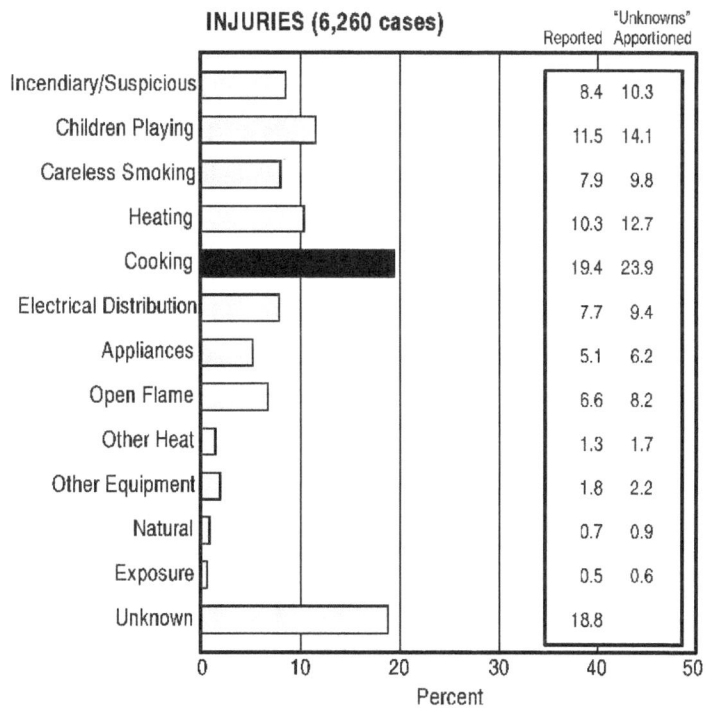

Figure 47. Causes of 1994 One- and Two-Family Dwelling Fires and Fire Casualties

Source: NFIRS

79

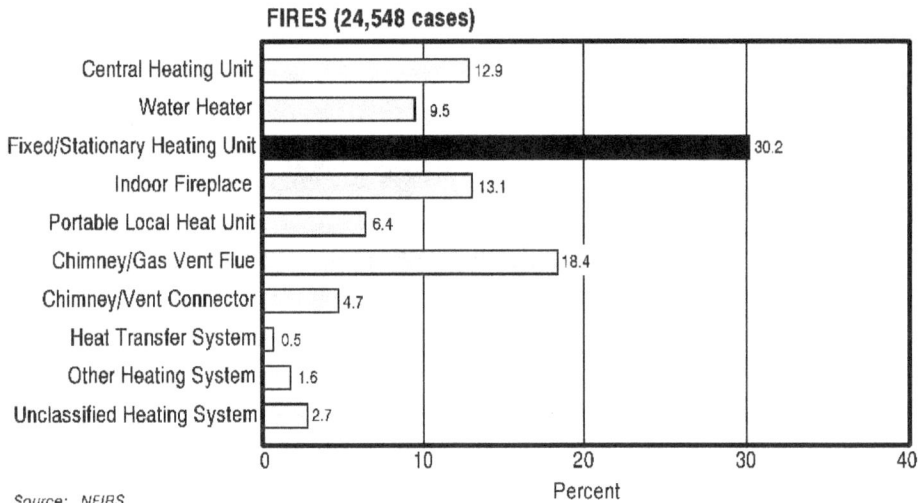

FIRES (24,548 cases)

Central Heating Unit — 12.9
Water Heater — 9.5
Fixed/Stationary Heating Unit — 30.2
Indoor Fireplace — 13.1
Portable Local Heat Unit — 6.4
Chimney/Gas Vent Flue — 18.4
Chimney/Vent Connector — 4.7
Heat Transfer System — 0.5
Other Heating System — 1.6
Unclassified Heating System — 2.7

Percent

Source: NFIRS

Figure 48. Types of 1994 One- and Two-Family Dwelling Heating Fires

from creosote accumulation. Third most frequent are fires associated with indoor fireplaces. Central heating units, water heaters, and portable heaters are also significant, but represent smaller parts of the problem in terms of number of fires.

Trends in Causes

The rank order of the leading five causes of one- and two-family dwelling fires has not changed over the 10-year period 1985–1994, as shown in Figure 49. The most dramatic change is the 62 percent drop in heating fires. Careless smoking fires dropped by 45 percent (not among the top five leading causes).

Smoking deaths, however, dropped 51 percent between 1985–1995, although it continues as the leading cause of fire deaths.[6] Heating deaths dropped 48 percent over this period, but were the second leading cause until 1994. Arson deaths dropped sharply from 1993 to 1994, but still remained at third. Electrical distribution peaked in 1986 and again in 1989 but dropped to 1985 levels in 1993.

Cooking has been the leading cause of fire injuries in all 10 years by a wide margin. There is little difference in 1994 cooking injuries from those in 1985. Heating and careless smoking injuries have dropped sharply, by 27 and 37 percent, respectively. Injuries caused by children playing fires are have increased 25 percent.

---

[6] In 1993, arson and smoking tied at 653 for the leading cause of dwelling fire deaths.

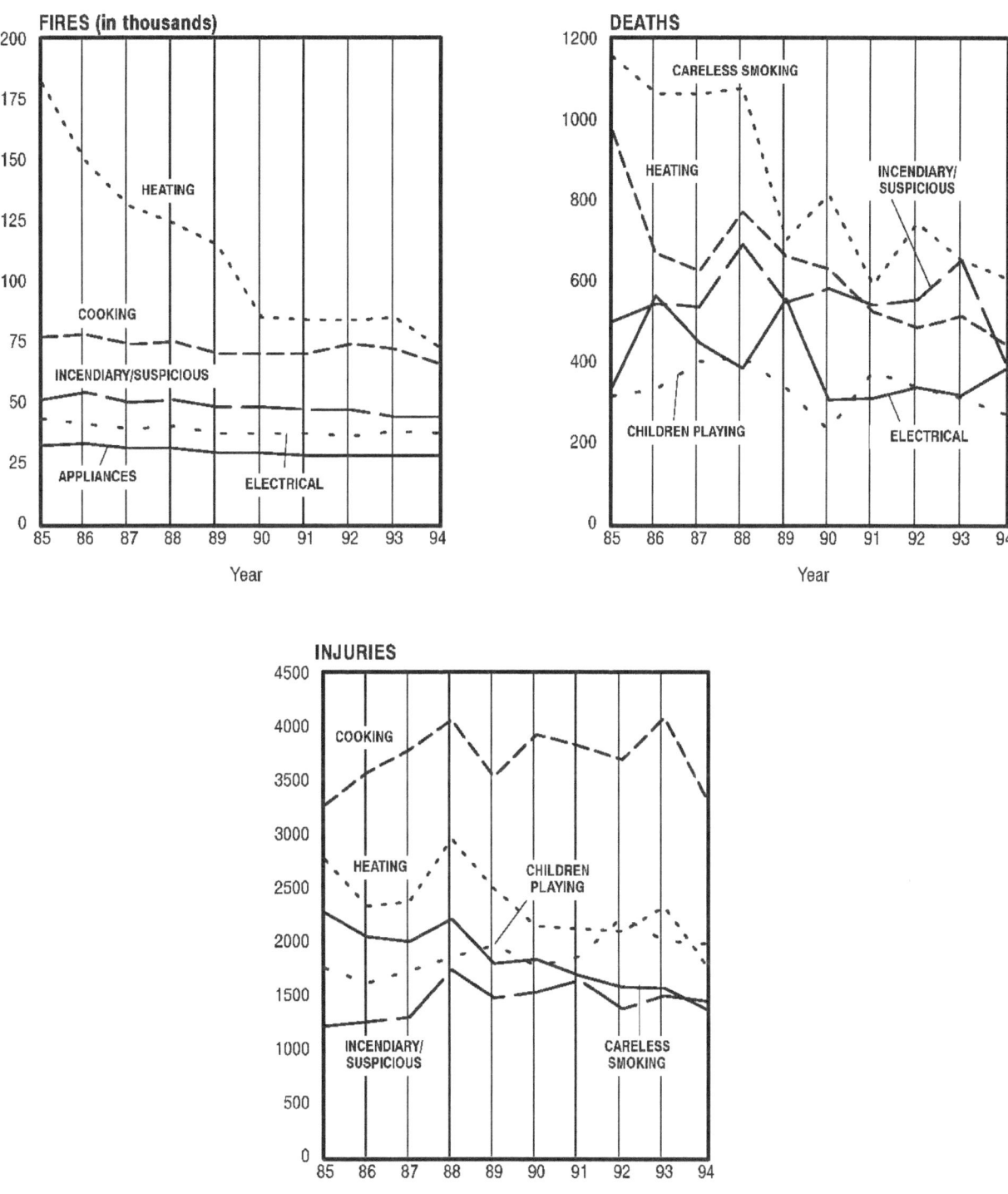

FIRES (in thousands)

HEATING

COOKING

INCENDIARY/SUSPICIOUS

APPLIANCES   ELECTRICAL

Year

DEATHS

CARELESS SMOKING

HEATING

INCENDIARY/
SUSPICIOUS

CHILDREN PLAYING   ELECTRICAL

Year

INJURIES

COOKING

HEATING

CHILDREN
PLAYING

INCENDIARY/
SUSPICIOUS

CARELESS
SMOKING

Year

Note : Data for all 12 causes are provided in Appendix B, Table B-2.

Sources: NFIRS and NFPA Annual Surveys

Figure 49. Trends in Leading Causes of One- and Two-Family Dwelling Fires and Fire Casualties

Area of Home

To help people visualize the fire problem more personally, it is useful to describe it in terms of where different types of fires occur in the home and what types of fires occur in each room. Figure 50 shows the rooms or areas of origin for fires, deaths, and injuries in one- and two-family homes in 1994. There is little change from 1990 in the overall rankings.

Twice as many fires occur in the kitchen than in any other area, most obviously those associated with cooking. The second most common location is the bedroom or sleeping area. Children playing, careless smoking, and intentionally set fires are the three most common fire causes here. Chimney fires, third place, often result from creosote buildup that ignites when the chimney has not been cleaned often enough or well enough. Lounge areas (living rooms and family rooms) are fourth, with heating fires and careless smoking as the primary causes.

Garage/storage areas are not shown as one of the leading areas, but they actually are more significant than implied. Because of a quirk in the NFPA 901 standard for reporting fires, many fires in residential garages are not included with residential fires but rather are in a separate category under storage properties.[7] If these storage/garage fires were counted, the total number of fires in dwellings would increase by about 5 percent. There were 17,100 such garage fires in 1994. This portion of the residential fire problem is sometimes overlooked. A fuller portrait of garage fires is presented later in this section.

In one- and two-family dwellings, more than half of all deaths occur in lounge areas and bedrooms, possibly because people fall asleep smoking on upholstered furniture. (More die from smoking on furniture than from smoking in bed.) For injuries, the kitchen is most common because of the large number of burns associated with cooking. Sleeping areas rank a close second.

Slicing the problem a slightly different way, Tables 7, 8, and 9 show the most common rooms for each of the leading causes of fires, fire deaths, and injuries in one- and two-family dwellings. For heating fires, the chimney accounts for 44 percent of the fires and heating equipment areas for another 13 percent. A number of cooking fire appeared on exterior balconies and porches, presumably from grills and barbecues. Arson fires in the home had the highest frequencies of occurrence in bedrooms and lounge areas. Electrical fires were most frequently reported in kitchens, but large numbers were also in ceilings, bedrooms, and laundry rooms.

Careless smoking deaths occur most often in lounge areas, often where the intoxicated victims fall asleep on upholstered furniture, and in bedrooms, where the cigarette ignites the bedding material. Together, these two rooms account for about 80 percent of smoking-related deaths. Heating fire deaths most often are from fires started in bedrooms and second most often in lounge areas. Portable

---

[7] In the version of the 1976 NFPA 901 reporting standard that is used in the National Fire Incident Reporting System, all residential garages were to be reported under storage properties. In later versions (e.g., NFPA 901, *Uniform Coding for Fire Protection*, 1981 and 1986), only detached garages are included in this category. Since not all reporting firefighters know that the old definition holds for NFIRS, and some never knew, there is some inconsistency in reporting these fires. The standard is under discussion for change in the future.

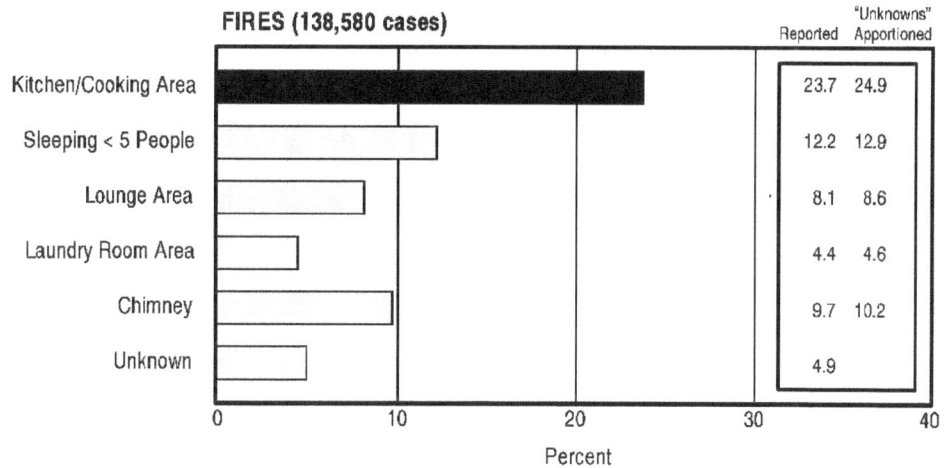

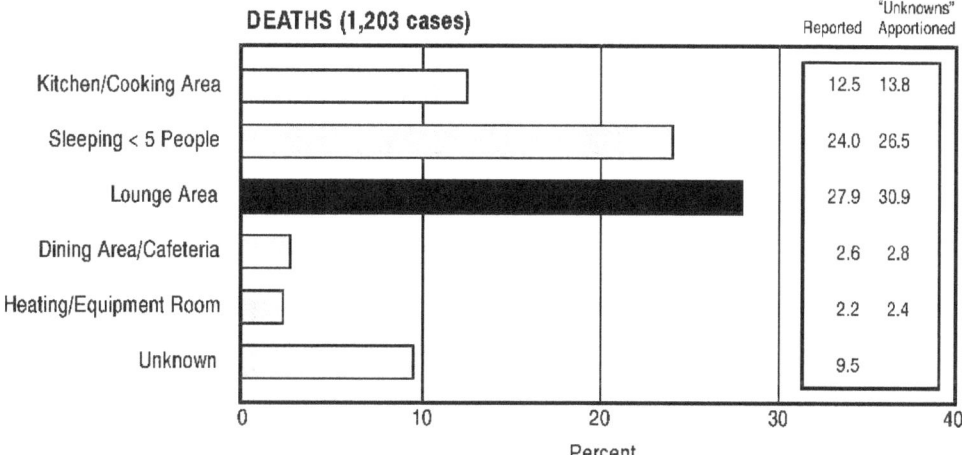

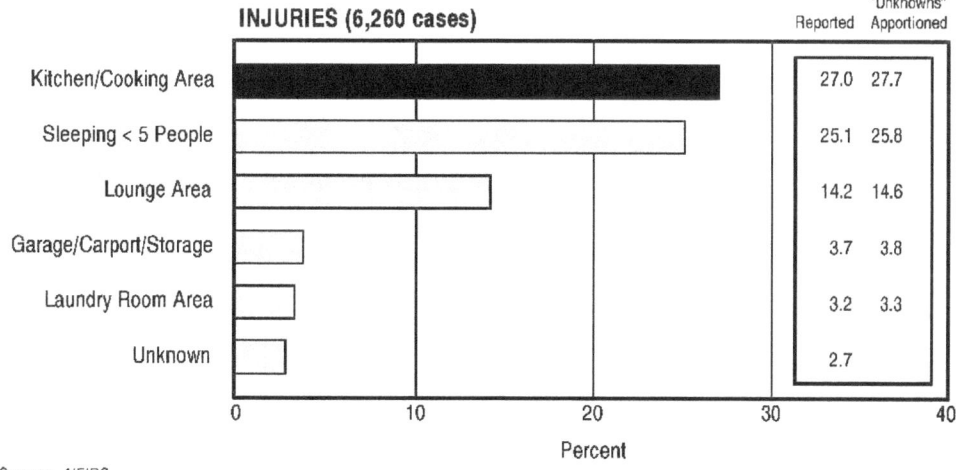

Source: NFIRS

**Figure 50. Leading Rooms of Origin of 1994 One- and Two-Family
Dwelling Fires and Fire Casualties**

Table 7. Leading Rooms of Origin by Cause for 1994 One- and Two-Family Dwelling Fires

Area of Home	Leading Causes				
	Heating	Cooking	Arson	Electrical Distribution	Appliance
Lounge	2,743 11.4%	45 0.2%	1,827 12.6%	1,373 11.2%	506 5.5%
Sleeping Under 5	937 3.9%	90 0.4%	2,541 17.5%	2,268 18.5%	1,040 11.3%
Kitchen/Cooking		20,877 95.0%	1,240 8.5%	1,147 9.4%	1,362 14.8%
Laundry Room					3,683 39.9%
Garage/Carport/Vehicle Storage		85 0.49%	679 4.7%		464 5.0%
Chimney	10,474 43.5%				
Heating Equipment Area	3,124 13.0%				
Exterior Balcony/Open Porch		231 1.1%	776 5.3%		
Crawl Space	760 3.2%				
Ceiling/Roof				1,109 8.3%	
Exterior Wall				693 5.7%	
All NFIRS Dwelling Fires	24,066	21,986	14,532	12,241	9,229

Note: For each cause, the five most common rooms or areas of origin reported are shown. Data here are NFIRS raw counts, *not* national estimates. Percentages shown are column percentages (e.g., percentages of heating or cooking fires, not percentages of lounge fires).

Source: NFIRS

and fixed space heaters play a big role here. Most arson fire deaths occur in lounge or sleeping areas. Electrical distribution fire deaths are most often in lounge areas. More than half of children playing fire deaths occur in bedrooms.

Cooking fire injuries are obviously almost all from fires in the kitchen. As with deaths (and not coincidentally with fires), over half of children playing fire injuries occurred in bedrooms. Parent and care providers need to be particularly vigilant in the supervision of young children. Heating fire injuries occur almost equally in heating equipment areas and lounge areas (which include most fireplaces). Arson fire injuries occur in a variety of places, but many of these fires occur in lounge or sleeping areas. Smoking fire injuries occur predominately in bedrooms.

Table 8. Leading Rooms of Origin by Cause for 1994 One- and Two-Family Dwelling Fire Deaths

Area of Home	Leading Causes				
	Smoking	Heating	Arson	Electrical Distribution	Children Playing
Hallway		10 7.7%			
Lounge	87 48.9%	24 18.5%	26 22.2%	24 21.4%	25 31.7%
Sleeping Under 5	53 29.8%	36 27.7%	26 22.2%	18 16.1%	40 50.6%
Sleeping Over 5					4 5.1%
Dining	6 3.4%		5 4.3%		4 5.1%
Kitchen/Cooking	13 7.3%	8 6.2%	5 4.3%	9 8.0%	
Garage/Carport/Vehicle Storage			5 4.3%		
Heating Equipment Area		18 13.8%			
Exterior Balcony/Open Porch			6 5.1%		
Crawl Space				6 5.4%	
Ceiling/Roof				3 2.7%	
Closet					2 2.5%
Unclassified Structural Area	6 3.4%				
All NFIRS Dwelling Deaths	178	130	117	112	79

Note: For each cause, the five most common rooms or areas of origin reported are shown. Data here are NFIRS raw counts, *not* national estimates. Percentages shown are column percentages (e.g., percentages of heating or cooking fires, not percentages of lounge fires).

Source: NFIRS

Table 9. Leading Rooms of Origin by Cause for 1994 One- and Two-Family Dwelling Fire Injuries

Area of Home	Leading Causes				
	Cooking	Children Playing	Heating	Arson	Smoking
Lounge		112 15.6%	112 17.3%	98 7.4%	161 32.4%
Sleeping Under 5		418 58.3%	91 14.1%	128 9.6%	238 47.9%
Dining	6 0.5%				
Kitchen/Cooking	1,138 93.9%	32 4.5%		49 3.7%	30 6.0%
Laundry Room			4.5 7.0%		
Garage/Carport/Vehicle Storage	5 0.4%	21 2.9%	40 6.2%	17 1.3%	7 1.4%
Heating Equipment Area			113 17.5%		
Crawl Space		8 1.1%			7 1.4%
Exterior Balcony	21 1.7%				
Court/Terrace/Patio	6 0.5%				
Multilocation Use				28 2.1%	
All NFIRS Dwelling Injuries	1,212	717	647	523	497

Note: For each cause, the five most common rooms or areas of origin reported are shown. Data here are NFIRS raw counts, *not* national estimates. Percentages shown are column percentages (e.g., percentages of heating or cooking fires, not percentages of lounge fires).

Source: NFIRS

Smoke Detector Performance

In 1994, detectors were present in 50 percent of homes that had fires and 31 percent of one- and two-family homes that had fire deaths (Figure 51). There are proportionally fewer detectors in homes than in apartments that have fires, primarily because detectors are often provided by landlords and more often required by law than in single-family houses. (See page 93 for more on apartments.)

Where present, the detectors operated about half the time for both fires and fire deaths, often because the fire was too small. Detectors when present operated less often in homes than apartments, perhaps because fires are more spread out relative to the detectors and perhaps because the homeowner is solely responsible for their maintenance (e.g., battery replacement).

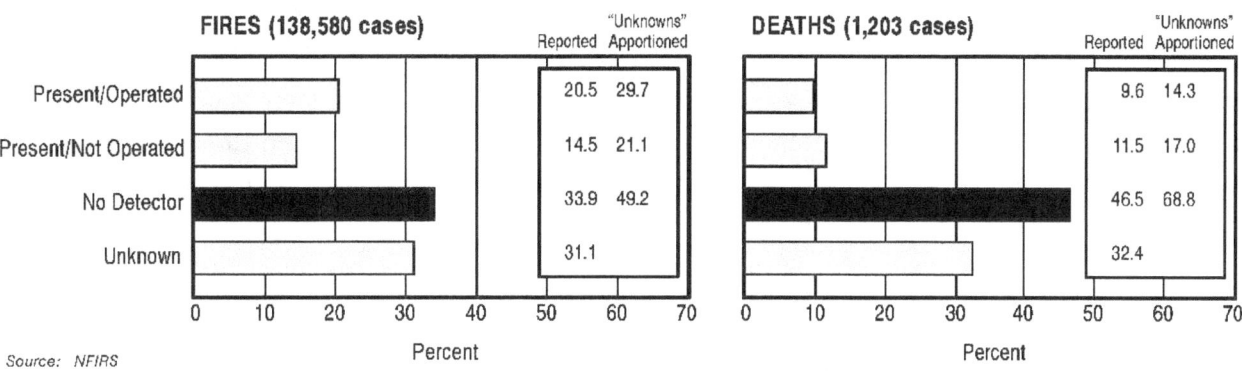

Figure 51. Smoke Detector Performance in 1994 One- and Two-Family Dwelling Fires and Fire Deaths

There was no operating detector in 49 percent of total fires and 69 percent of fire deaths. This is a modest improvement over 1990.

Sprinkler Performance

Sprinklers were present in less than 1 percent of fires or fire deaths in one- and two-family dwellings in 1994 (Figure 52). This is an insignificant amount. There were few deaths when sprinklers were present and operated, however.

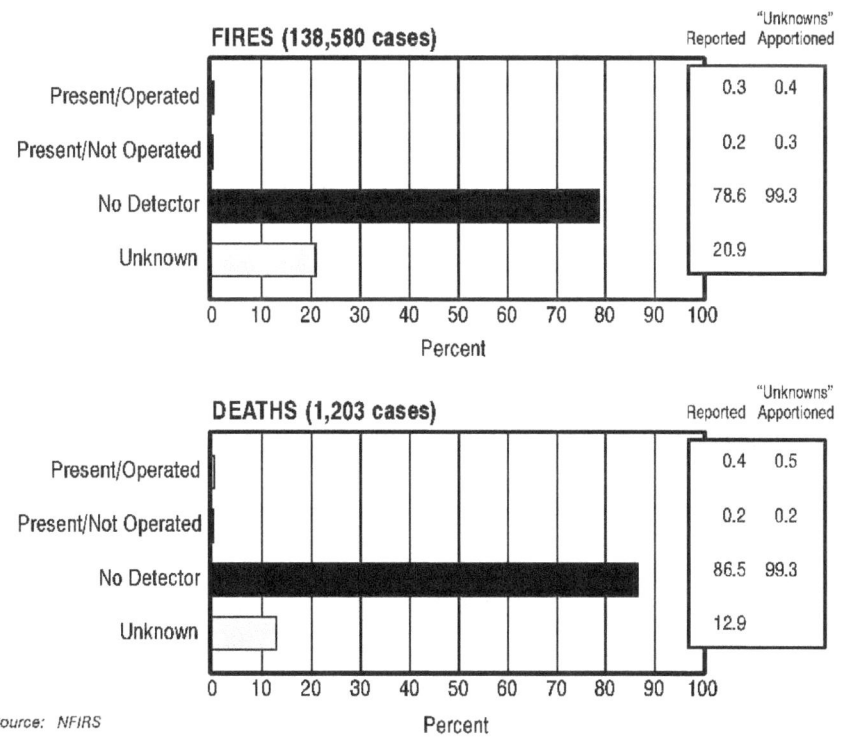

Figure 52. Sprinkler Performance in 1994 One- and Two-Family Dwelling Fires and Fire Deaths

87

Residential Garages

Not all residential garage fires are reported as they are supposed to be. A substantial number are reported as part of residential fires. The definition in the widely used early manual on NFPA 901 required that both attached and detached residential garages be included in the storage category. More recent versions of the 901 standard that are not in use by NFIRS require only detached garage fires to be included in the storage category. Undoubtedly, there is confusion in the field. To further complicate matters, the residential garage data often are overlooked or ignored altogether in discussions of residential fires.

Figure 53 shows the trends in fires, casualties, and losses for residential garages in fixed property use category #881, "residential parking garages." The number of fires for 1994 is rather large, about 17,100 per year or 5 percent of the total in dwellings; the 10-year trend is down 28 percent.

Because the numbers of deaths are small, there is considerable year-to-year variation. The range of from 11 to 25 deaths over the period 1985 to 1994 results in a 10-year downward trend of 15 percent. In every year, this was less than 1 percent of the fire deaths in one- and two-family dwellings.

Injuries in garage fires, which account for less than 1 percent of the injuries in dwelling fires, have trended downward sharply (36 percent). Dollar loss, about 2 percent of the total loss in dwelling fires, is down 4 percent.

Figure 54 shows that 40 percent of deaths in garage fires for 1994 is from arson. For injuries, the leading cause is heating, with arson and other equipment second and third, respectively. In terms of sheer numbers of fires, arson leads by approximately two to one over the next leading cause—exposures to house or outside fires.

The past confusion about coding garage fires has not distorted the residential fire profiles in any significant way, but it does lead to understating the fire problem by 1 to 5 percent, depending on the measure used.

Manufactured Housing

Manufactured housing is a special category of one- and two-family dwellings. Although only a small fraction of the U.S. population lives in manufactured housing, it has represented a severe problem in terms of fire fatalities in the past—double the fatality rate per fire compared to other homes. This caused the U.S. Department of Housing and Urban Development in 1976 to establish strict standards for improving the fire safety of such homes (often called "mobile homes"). The HUD standard clearly made an impact. However, the manufactured housing fire problem is still significant.

Figure 55 shows the magnitude and trends in manufactured housing fires. Despite an increase in the manufactured housing stock, fires have dropped 28 percent, deaths 29 percent, and dollar loss 37 percent over the last 10 years. Injuries over this period have remained relatively constant. These

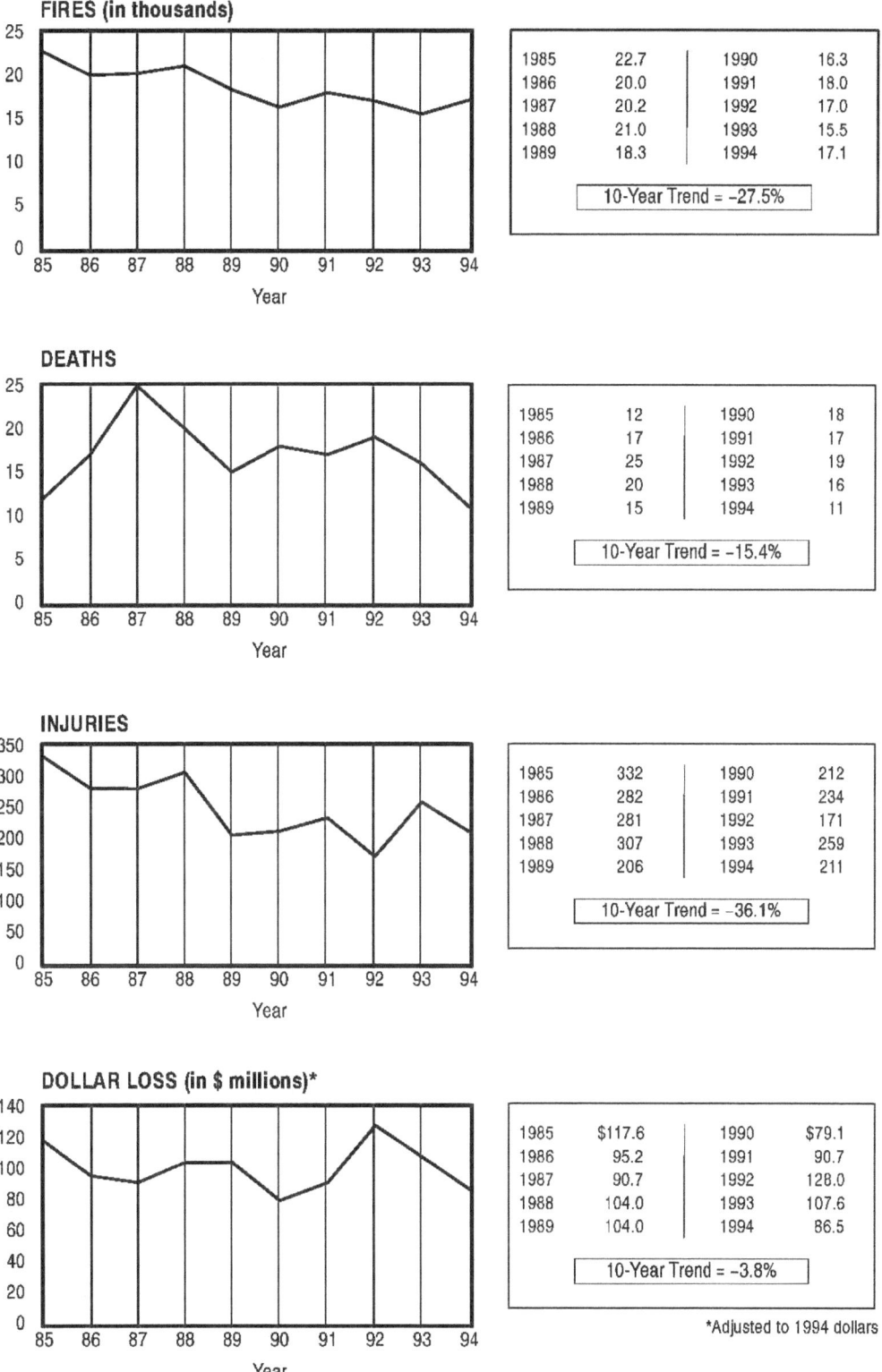

FIRES (in thousands)

1985	22.7	1990	16.3
1986	20.0	1991	18.0
1987	20.2	1992	17.0
1988	21.0	1993	15.5
1989	18.3	1994	17.1

10-Year Trend = –27.5%

DEATHS

1985	12	1990	18
1986	17	1991	17
1987	25	1992	19
1988	20	1993	16
1989	15	1994	11

10-Year Trend = –15.4%

INJURIES

1985	332	1990	212
1986	282	1991	234
1987	281	1992	171
1988	307	1993	259
1989	206	1994	211

10-Year Trend = –36.1%

DOLLAR LOSS (in $ millions)*

1985	$117.6	1990	$79.1
1986	95.2	1991	90.7
1987	90.7	1992	128.0
1988	104.0	1993	107.6
1989	104.0	1994	86.5

10-Year Trend = –3.8%

*Adjusted to 1994 dollars

Sources: NFIRS, NFPA Annual Surveys, and Consumer Price Index

Figure 53. Trends in Residential Garage Fires and Fire Losses

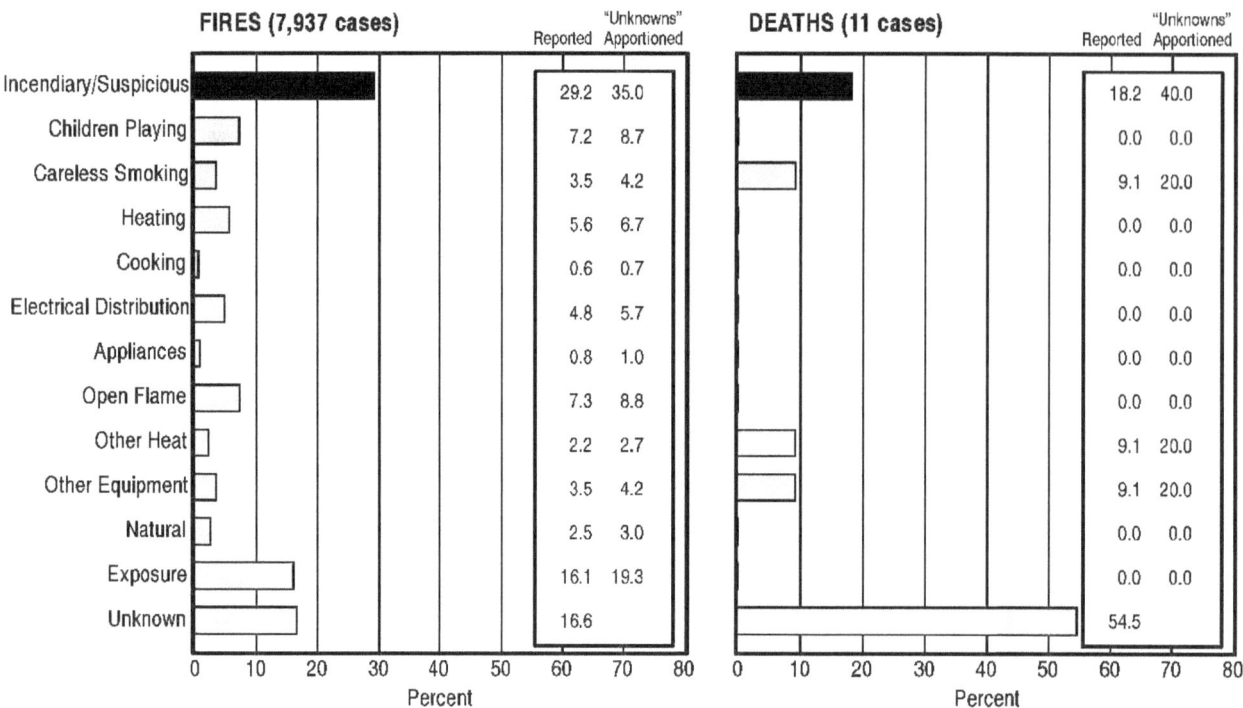

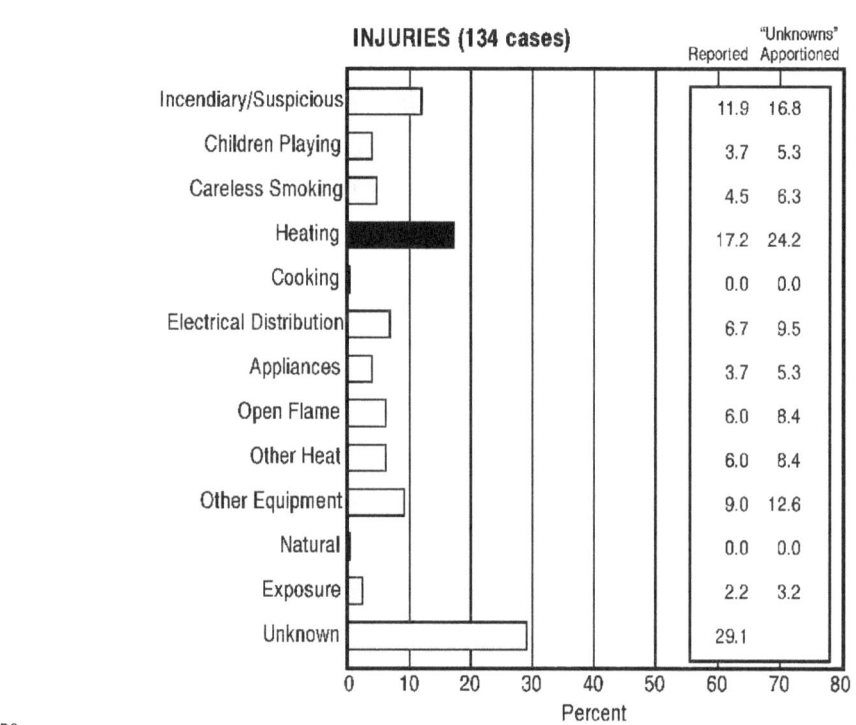

Source: NFIRS

Figure 54. Causes of 1994 Residential Garage Fires and Fire Casualties

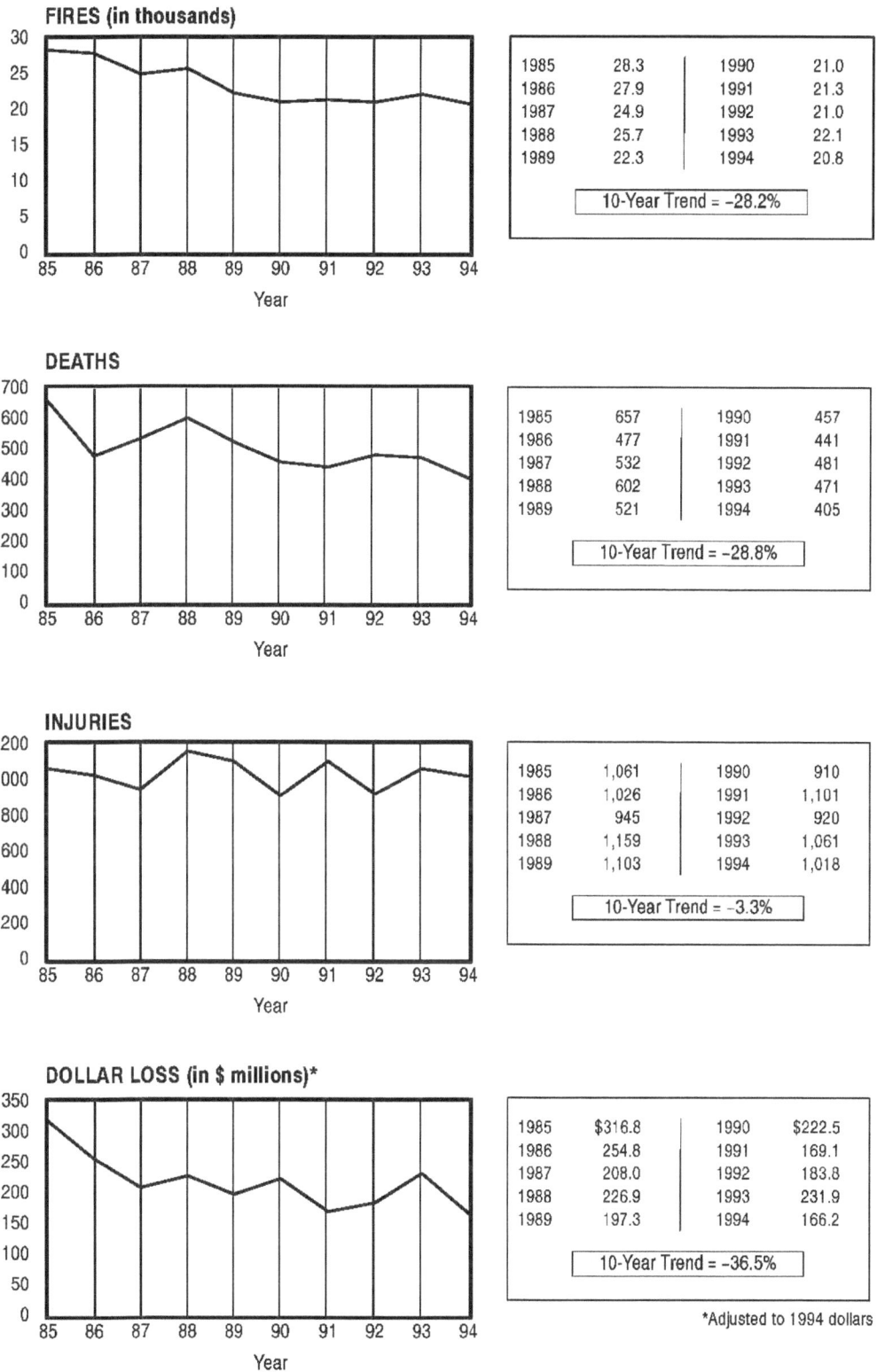

FIRES (in thousands)

1985	28.3	1990	21.0
1986	27.9	1991	21.3
1987	24.9	1992	21.0
1988	25.7	1993	22.1
1989	22.3	1994	20.8

10-Year Trend = –28.2%

DEATHS

1985	657	1990	457
1986	477	1991	441
1987	532	1992	481
1988	602	1993	471
1989	521	1994	405

10-Year Trend = –28.8%

INJURIES

1985	1,061	1990	910
1986	1,026	1991	1,101
1987	945	1992	920
1988	1,159	1993	1,061
1989	1,103	1994	1,018

10-Year Trend = –3.3%

DOLLAR LOSS (in $ millions)*

1985	$316.8	1990	$222.5
1986	254.8	1991	169.1
1987	208.0	1992	183.8
1988	226.9	1993	231.9
1989	197.3	1994	166.2

10-Year Trend = –36.5%

*Adjusted to 1994 dollars

Sources: NFPA Annual Surveys and Consumer Price Index

Figure 55. Trends in Manufactured Housing Fires and Fire Losses

are all quite similar to other single-family dwelling changes during that period, except for dollar loss, which had a much sharper decline for manufactured housing.

Figure 56 shows the manufactured housing trends based on 1,000 fires. Deaths remained almost constant with only a 1 percent decrease per 1,000 fires over the past 10 years. Injuries have increased significantly (31 percent), and adjusted dollar loss has decreased by about 10 percent.

The 1994 cause profile for manufactured housing fires is shown in Figure 57. Heating systems are the leading cause of fires in manufactured housing, with electrical distribution a close second. Cooking and arson were third and fourth, respectively. This cause profile is the same as that in 1990.

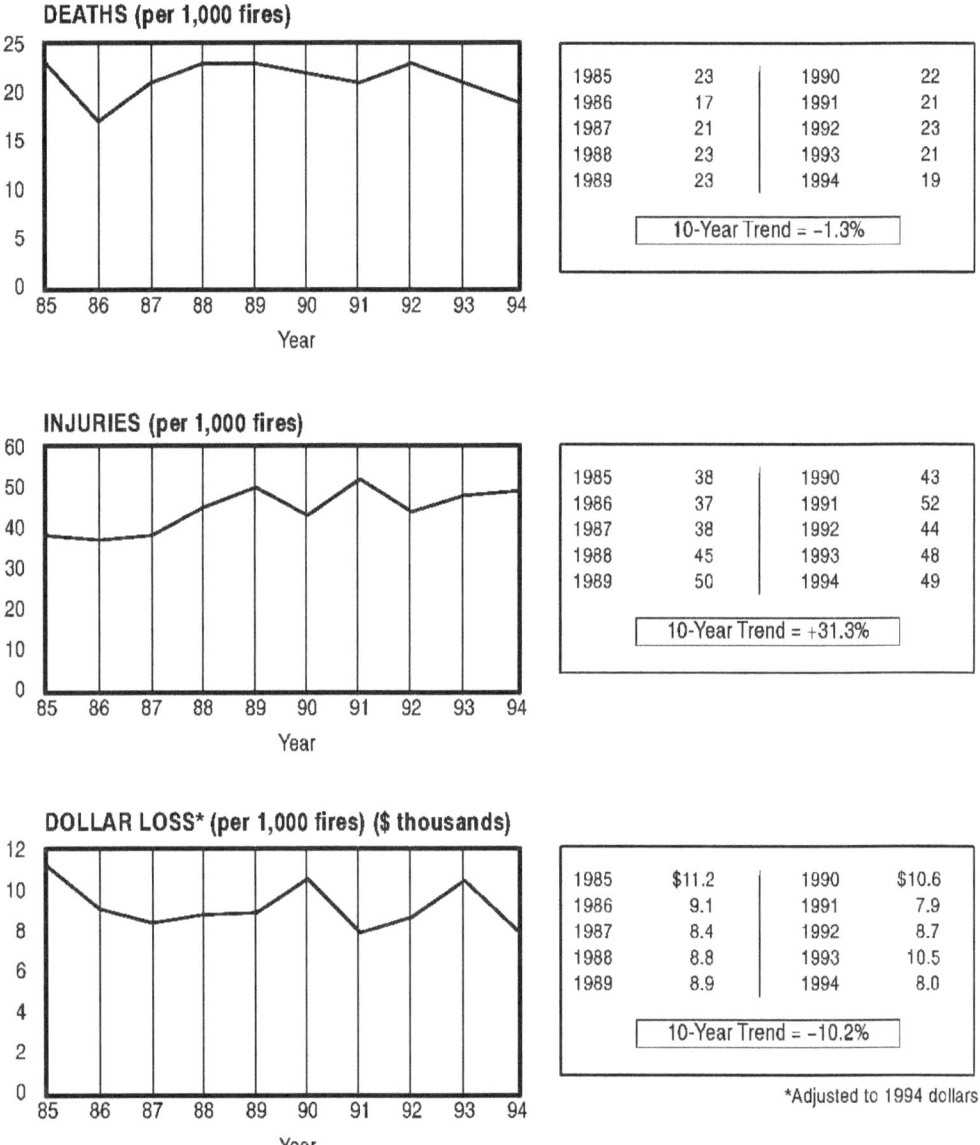

Sources: NFPA Annual Surveys and Consumer Price Index

Figure 56. Trends in Severity of Manufactured Housing Fire Losses

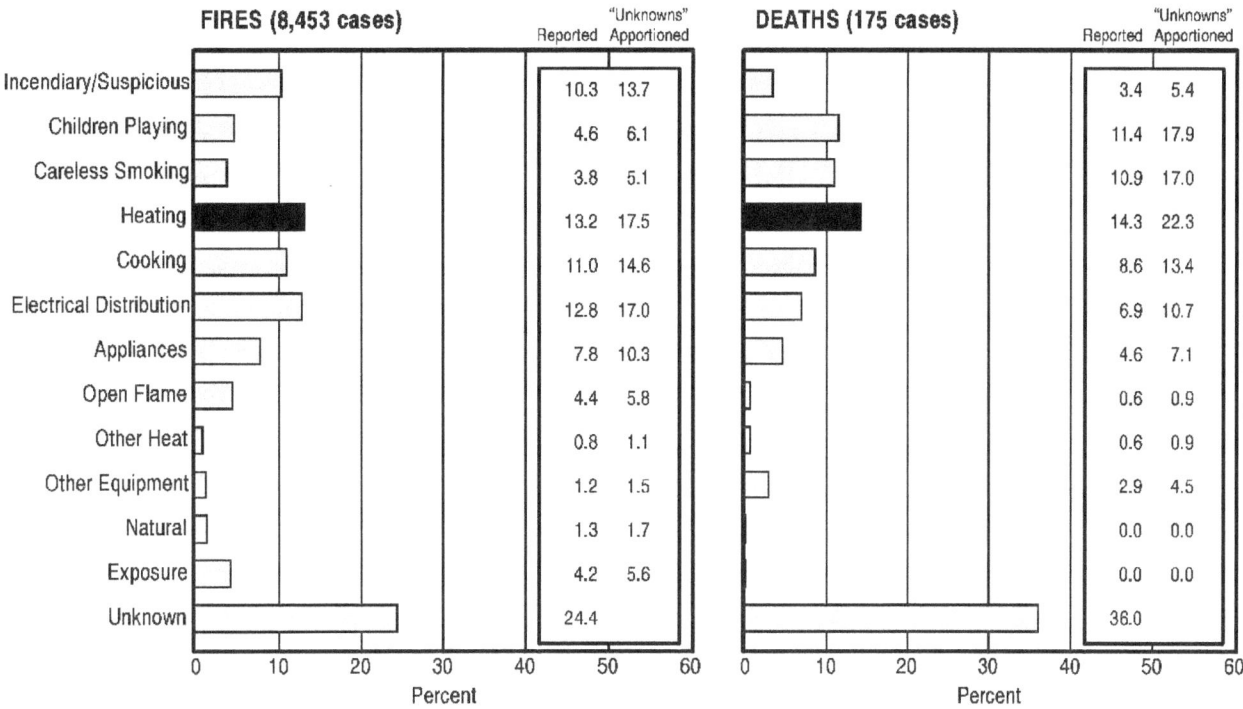

Source: NFIRS

Figure 57. Causes of 1994 Manufactured Housing Fires and Fire Deaths

For fire deaths, heating leads other causes of deaths at 22 percent. Children playing and careless smoking are second and third in manufactured housing, whereas heating and arson are so ranked in dwellings. It should be noted that deaths from smoking dropped 30 percent since 1990, from 25 to 17 percent.

APARTMENTS

The fire problem in multifamily dwellings, referred to as apartments in this report, is generally similar to that of one- and two-family dwellings with the exception of one major category: heating-related fires. Because apartments generally have central heating systems that are professionally maintained, heating-related fires from misuse and poor maintenance are significantly less than in single-family dwellings. This changes the proportions of the causes for apartments, with heating becoming less significant and the other causes moving up in importance.

Apartment buildings tend to be more regulated by building codes than single-family dwellings. Many apartments are rental properties, which may also fall under more stringent fire prevention statutes. In many communities, apartments have a significantly different socio-economic mix of residents compared to single-family dwellings. They may have more low-income families in housing projects or more high-income families in luxury high rises, or they may be centers of living for the elderly. In large cities, they may have all of these groups.

Because apartment buildings have large clusters of similar people, they can be given special attention with prevention programs based on the cause profiles of apartment buildings in different areas of the community.

Trends

Figure 58 shows the 10-year trend in apartment fire incidence, deaths, injuries, and dollar loss. Fire incidence dropped 4 percent in apartments, whereas dwelling fires dropped at nearly eight times this rate. The trend for deaths was down 26 percent, similar to that of one- and two-family dwellings. Injuries trended upwards a worrisome 51 percent, while they fell 3 percent for dwellings. Adjusted dollar losses were up by 1 percent, while dropping 15 percent in one- and two-family dwellings.

Compliance with stricter codes and the presence of detectors may be holding down the life loss in apartment fires. More detailed study of socio-economic and demographic changes over time might help explain some of the changes in fire incidence.

Causes

In terms of sheer numbers of reported fires, cooking in apartments is by far the most frequent cause, accounting for 40 percent of the fires. Arson is a distant second at 15 percent, and careless smoking third at 9 percent in 1994 (Figure 59).

The leading cause of deaths in apartments is careless smoking, with almost one-third of all fire deaths. Second was arson fires (21 percent), and third was children playing fires (20 percent). These three causes account for nearly three-quarters of all fire deaths in apartments; all other causes are relatively small, including heating. For fire injuries, cooking was first in 1994 at 30 percent, with arson second at 17 percent, careless smoking third at 16 percent, and children playing fourth at 12 percent. These four causes account for almost three-quarters of all injuries. There has been little change in the leading causes of fires, deaths, or injuries from the 1990 edition.

Figure 60 shows the trends in the top fire causes of apartment fires and casualties from 1985 to 1994. In terms of fire incidence, changes in occurrence of the leading causes were minimal. Arson fires were down 18 percent, although arson injuries increased 60 percent. Careless smoking fires were down 44 percent. The above data suggest that fire prevention programs aimed at apartment dwellers might emphasize the risk of careless smoking, the importance of supervising children, and the danger in leaving cooking unattended.

Careless smoking deaths dropped sharply, by 40 percent. Arson deaths reached a peak in 1989 but have experienced an overall 10-year downward trend. Children playing deaths have increased significantly since 1991. Overall, children playing deaths increased 20 percent over 1985. Some of this change may reflect more attention to defining fires as "children playing" rather than arson, but it is unlikely to account for all of the rise.

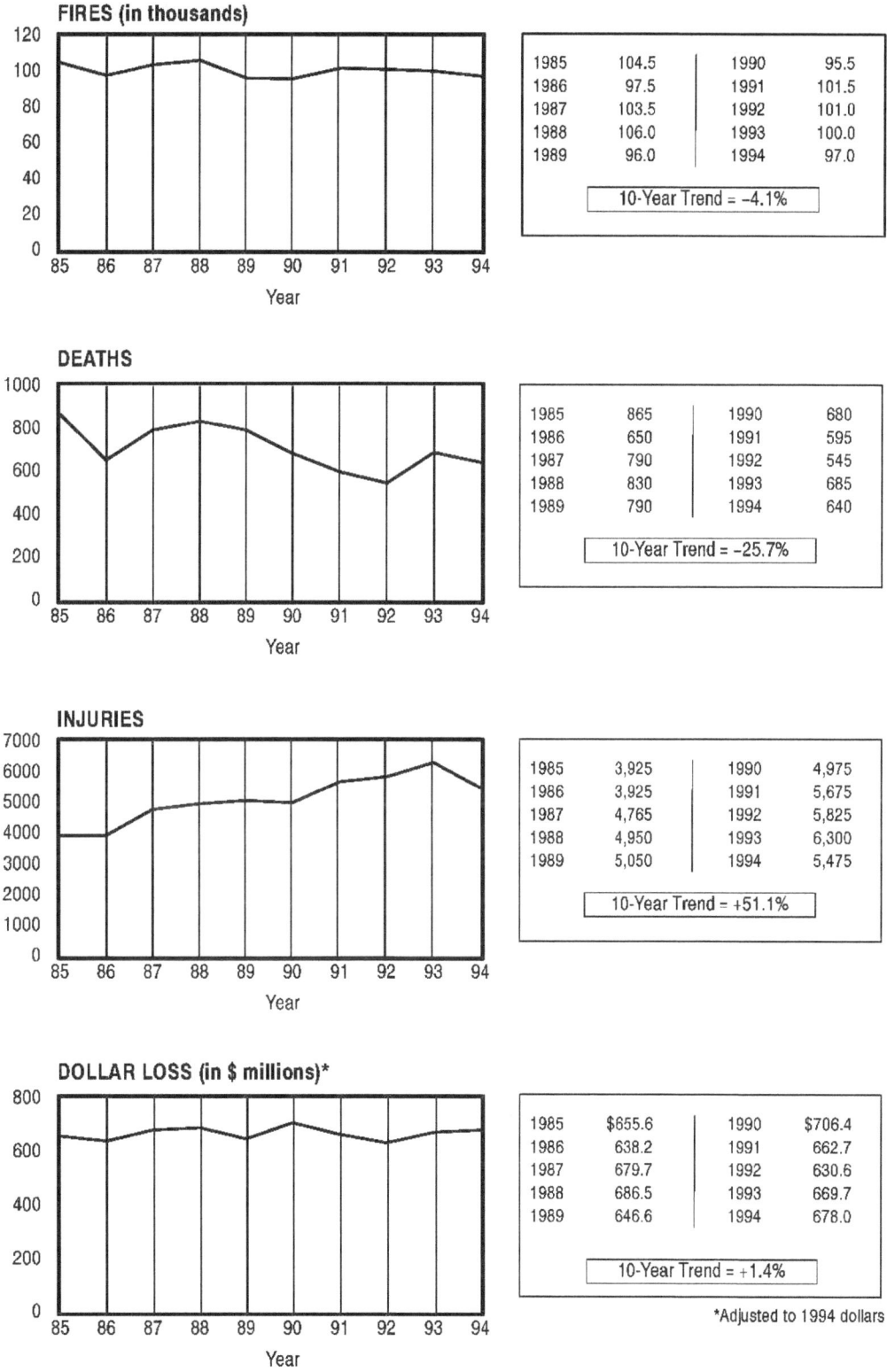

FIRES (in thousands)

1985	104.5	1990	95.5
1986	97.5	1991	101.5
1987	103.5	1992	101.0
1988	106.0	1993	100.0
1989	96.0	1994	97.0

10-Year Trend = –4.1%

DEATHS

1985	865	1990	680
1986	650	1991	595
1987	790	1992	545
1988	830	1993	685
1989	790	1994	640

10-Year Trend = –25.7%

INJURIES

1985	3,925	1990	4,975
1986	3,925	1991	5,675
1987	4,765	1992	5,825
1988	4,950	1993	6,300
1989	5,050	1994	5,475

10-Year Trend = +51.1%

DOLLAR LOSS (in $ millions)*

1985	$655.6	1990	$706.4
1986	638.2	1991	662.7
1987	679.7	1992	630.6
1988	686.5	1993	669.7
1989	646.6	1994	678.0

10-Year Trend = +1.4%

*Adjusted to 1994 dollars

Sources: NFPA Annual Surveys and Consumer Price Index

Figure 58. Trends in Apartment Fires and Fire Losses

95

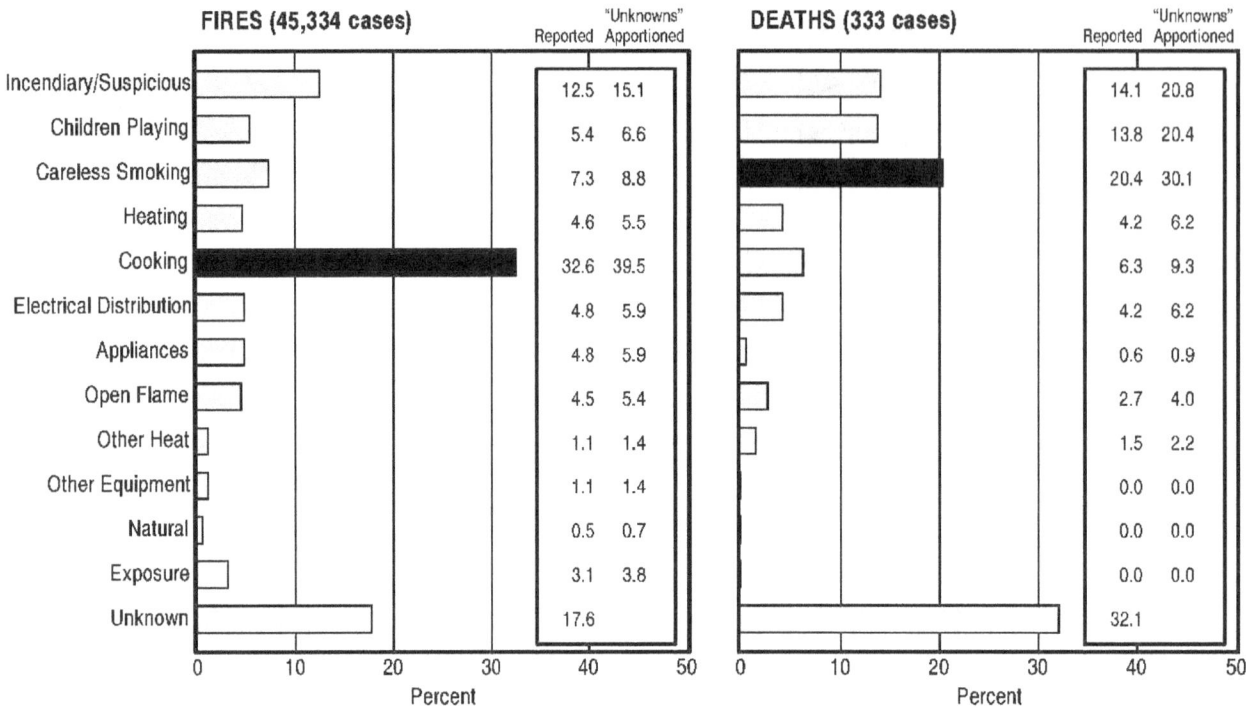

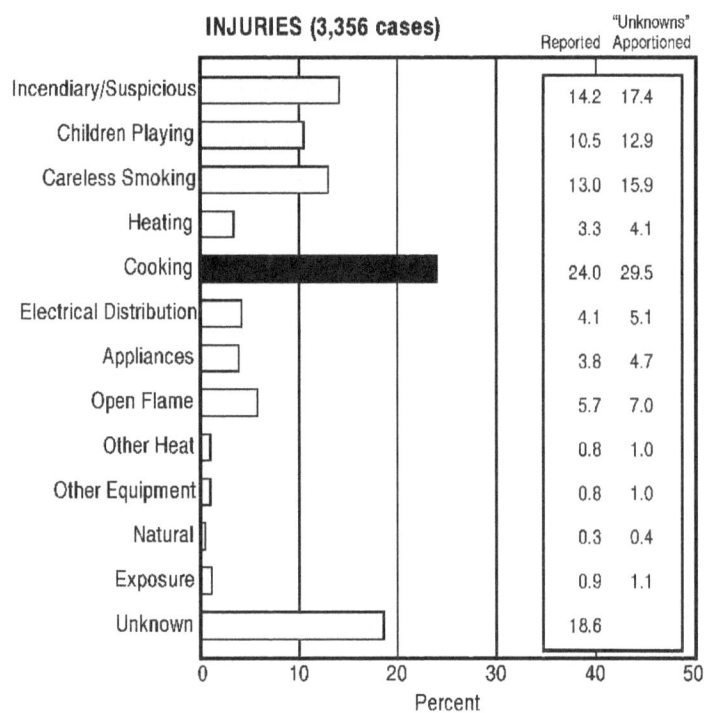

Figure 59. Causes of 1994 Apartment Fires and Fire Casualties

Source: NFIRS

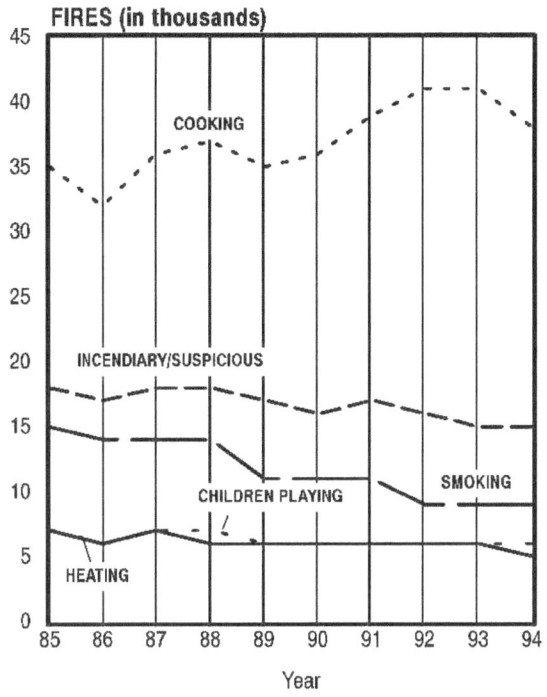

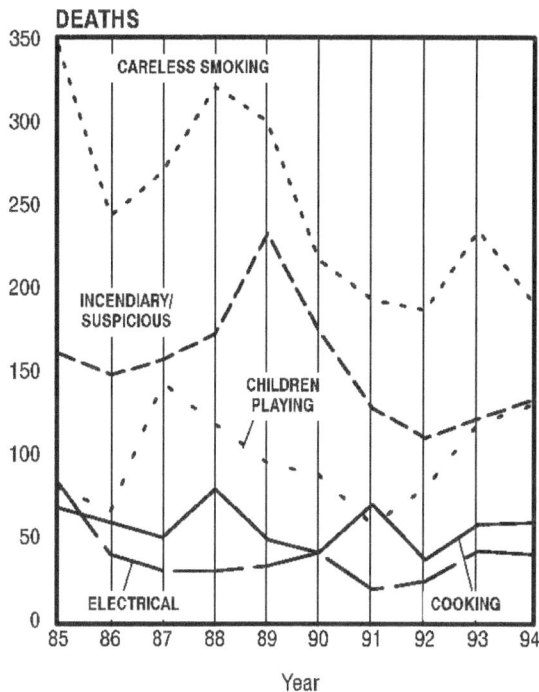

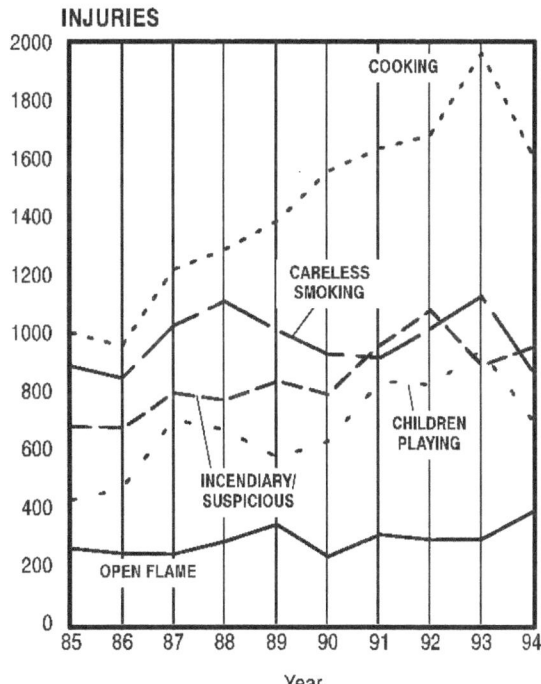

Note: Data for all 12 causes are provided in Appendix B, Table B–3.

Sources: NFIRS and NFPA Annual Surveys

Figure 60. Trends in Leading Causes of Apartment Fires and Fire Casualties

Injuries from cooking in apartments, by far the leading cause, nearly doubled over 10 years. Arson injuries have risen 47 percent. However, children playing accounted for a sharply higher number of injuries, rising by 78 percent over the 10-year period. Careless smoking and open flame injuries have changed very little.

Smoke Detector Performance

Figure 61 shows the performance of smoke detectors in apartments in 1994. Considering all fires, detectors were present 54 percent of the time (74 percent adjusted). Detectors are more likely to be installed in apartments, where they are provided by landlords, than in dwellings, where the occupants/owners provide and maintain them.

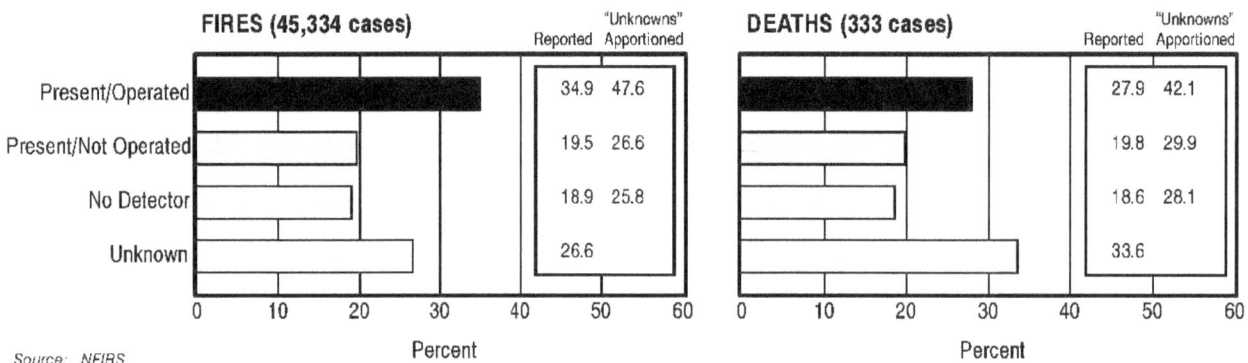

Figure 61. Smoke Detector Performance in 1994 Apartment Fires and Fire Deaths

The usage of detectors where fires occur in apartments, as well as dwellings, is considerably below the national average of over 88 percent of households having detectors according to recent national surveys. Apartment dwellers who have fires are probably less likely than average to be safety conscious and to provide and maintain detectors.

Smoke detectors in 1994 were present and operating in 28 percent (unadjusted) of fire deaths in apartments, and as high as 42 percent if only fire deaths with known detector performance are counted. Why detectors worked and people still died is a subject for further study. One possibility is that hallway detectors or detectors in other apartments operated after the victims were overcome. Also, apartments have fewer ways to escape, especially apartments on higher floors; at night, escaping from an apartment can be particularly confusing when people are awakened suddenly. This situation suggests the need to provide sprinklers in apartments and to emphasize fire prevention.

Detectors were present and did not operate in 20 percent of deaths (30 percent adjusted). This is 50 percent higher than the rate of nonworking detectors in dwellings. These statistics are unexpected as apartment detectors are more likely to be hardwired into the electrical system and professionally maintained than detectors in dwellings.

Since 1990, the presence of detectors in apartments with fire deaths has increased significantly from 39 to 48 percent. The percentage of detectors that were not working has increased (from 15 to 20 percent unadjusted).

Sprinkler Performance

There are few fire deaths being reported in apartment buildings where sprinklers were present and operating (Figure 62), but it is not known from NFIRS whether the sprinklers were in the apartment of origin. Overall, there are more sprinklers present in apartment fires than in dwelling fires, but the percentage is still small in 1994—just over 6 percent adjusted, which is only a 5 percent increase since 1990.

When Fires Occur

TIME OF DAY. Figure 63 shows the alarm times for fires, deaths, injuries, and dollar loss in apartment fires. The profiles are not as smooth as those for one- and two-family dwellings due to the smaller numbers of incidents involved.

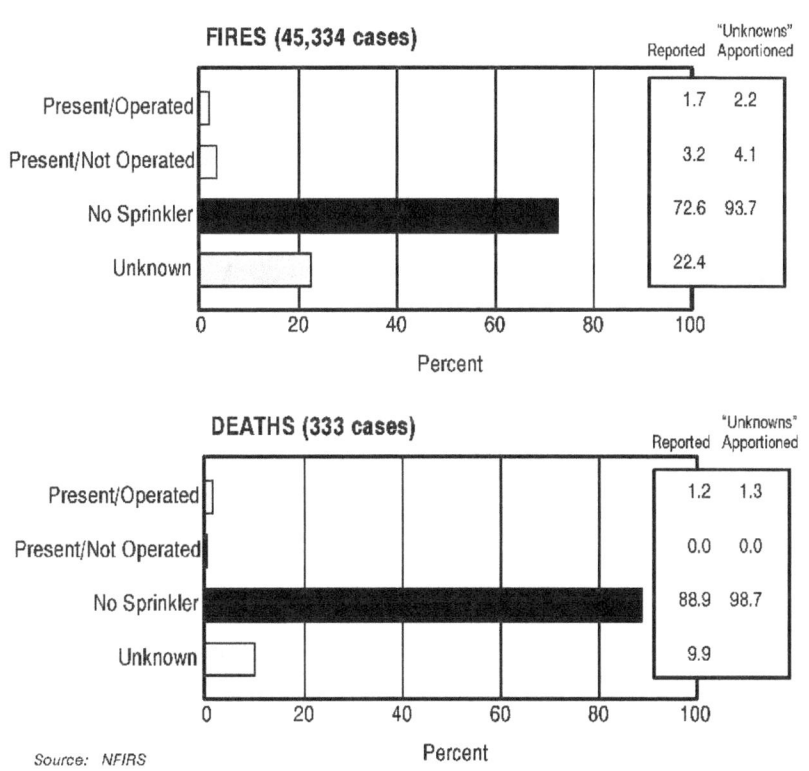

Figure 62. Sprinkler Performance in 1994 Apartment Fires and Fire Deaths

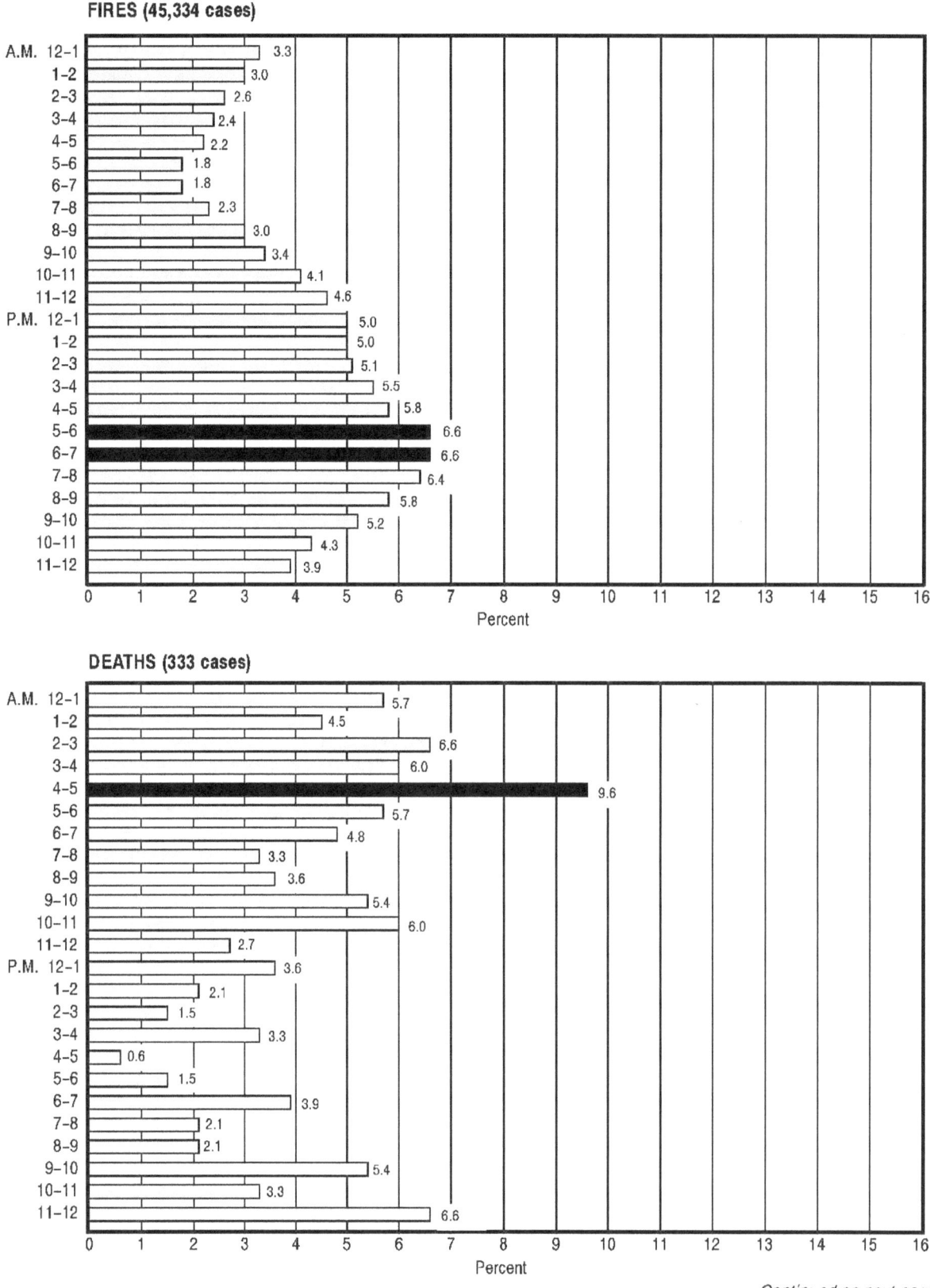

FIRES (45,334 cases)

Time	Percent
A.M. 12–1	3.3
1–2	3.0
2–3	2.6
3–4	2.4
4–5	2.2
5–6	1.8
6–7	1.8
7–8	2.3
8–9	3.0
9–10	3.4
10–11	4.1
11–12	4.6
P.M. 12–1	5.0
1–2	5.0
2–3	5.1
3–4	5.5
4–5	5.8
5–6	6.6
6–7	6.6
7–8	6.4
8–9	5.8
9–10	5.2
10–11	4.3
11–12	3.9

Percent

DEATHS (333 cases)

Time	Percent
A.M. 12–1	5.7
1–2	4.5
2–3	6.6
3–4	6.0
4–5	9.6
5–6	5.7
6–7	4.8
7–8	3.3
8–9	3.6
9–10	5.4
10–11	6.0
11–12	2.7
P.M. 12–1	3.6
1–2	2.1
2–3	1.5
3–4	3.3
4–5	0.6
5–6	1.5
6–7	3.9
7–8	2.1
8–9	2.1
9–10	5.4
10–11	3.3
11–12	6.6

Percent

Continued on next page

Figure 63. Time of Day of 1994 Apartment Fires and Fire Losses

100

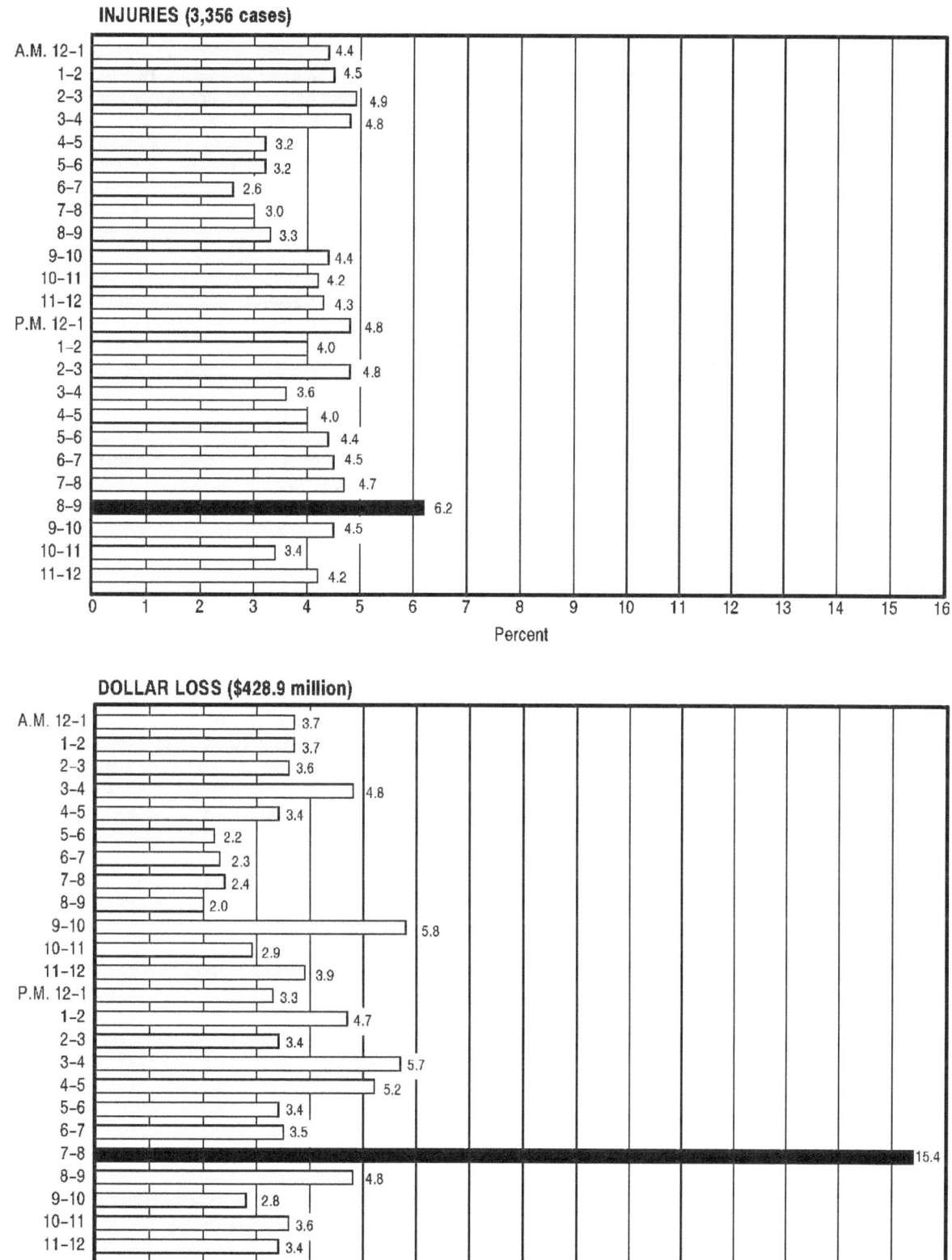

INJURIES (3,356 cases)

Time	Percent
A.M. 12–1	4.4
1–2	4.5
2–3	4.9
3–4	4.8
4–5	3.2
5–6	3.2
6–7	2.6
7–8	3.0
8–9	3.3
9–10	4.4
10–11	4.2
11–12	4.3
P.M. 12–1	4.8
1–2	4.0
2–3	4.8
3–4	3.6
4–5	4.0
5–6	4.4
6–7	4.5
7–8	4.7
8–9	6.2
9–10	4.5
10–11	3.4
11–12	4.2

Percent

DOLLAR LOSS ($428.9 million)

Time	Percent
A.M. 12–1	3.7
1–2	3.7
2–3	3.6
3–4	4.8
4–5	3.4
5–6	2.2
6–7	2.3
7–8	2.4
8–9	2.0
9–10	5.8
10–11	2.9
11–12	3.9
P.M. 12–1	3.3
1–2	4.7
2–3	3.4
3–4	5.7
4–5	5.2
5–6	3.4
6–7	3.5
7–8	15.4
8–9	4.8
9–10	2.8
10–11	3.6
11–12	3.4

Percent

Source: NFIRS

Figure 63. Time of Day of 1994 Apartment Fires and Fire Losses (cont'd)

Apartment fires peak from 5:00 to 7:00 p.m.—the cooking period—and are at a low from 5:00 to 7:00 a.m. As is the case in one- and two-family dwellings, the late night hours are most common for fire deaths, especially those associated with latent smoldering fires from careless smoking in the 11:00 p.m. to 7:00 a.m. period. Nearly 50 percent of all deaths occur in this 8-hour period.

Injuries have less pronounced peaks, with one from 8:00 to 9:00 p.m. in 1994. Because cooking is the leading cause of injuries, this probably accounts for the evening peak. Except for the huge peak between 7:00 and 8:00 p.m. ($64 million combined loss), dollar loss is fairly uniform.

MONTH OF YEAR. Fires in apartments are more uniform throughout the year than for dwellings because of the reduced role that heating plays (Figure 64). Still, they are somewhat more common in winter than in summer, perhaps because of heating fire problems in low-income apartments and increased indoor activity such as children playing.

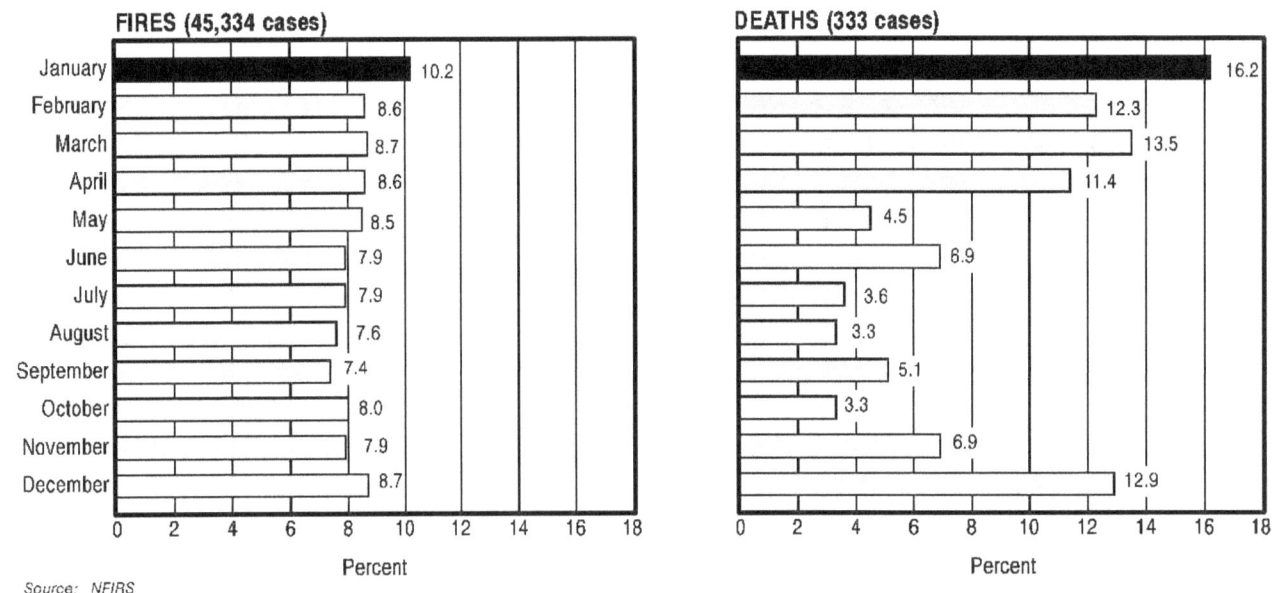

Figure 64. Month of Year of 1994 Apartment Fires and Fire Deaths

Apartment fire deaths are much more common in the winter than in the summer, as they are for one-and two-family dwellings, even though heating is a relatively minor cause for apartment fire deaths. Clearly there are other seasonal factors in addition to heating—perhaps a greater propensity to stay at home.

Room of Origin

Figure 65 shows the leading rooms of origin of fires, deaths, and injuries in apartments. As in every year for the past 10, the kitchen is the most common place for a fire and injury because of cooking. As in dwellings, the lounge area and bedrooms are the most common place for a fatal fire to start because of smoking on upholstered furniture.

102

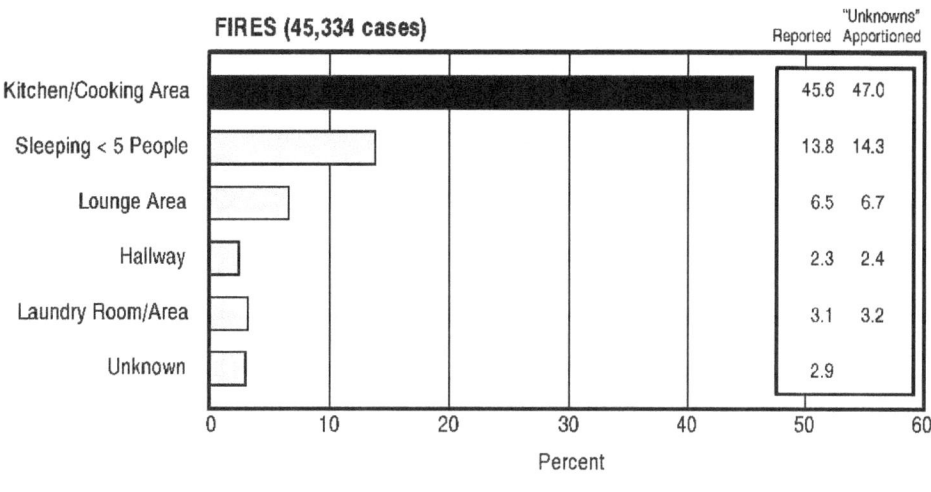

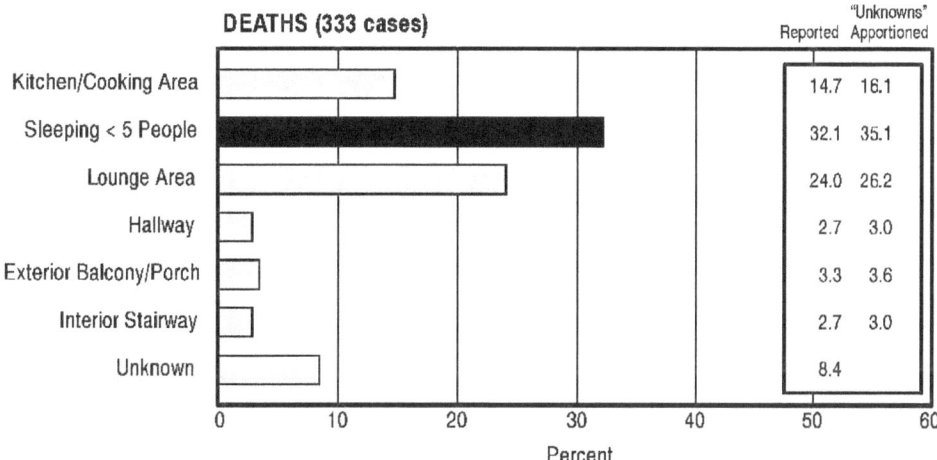

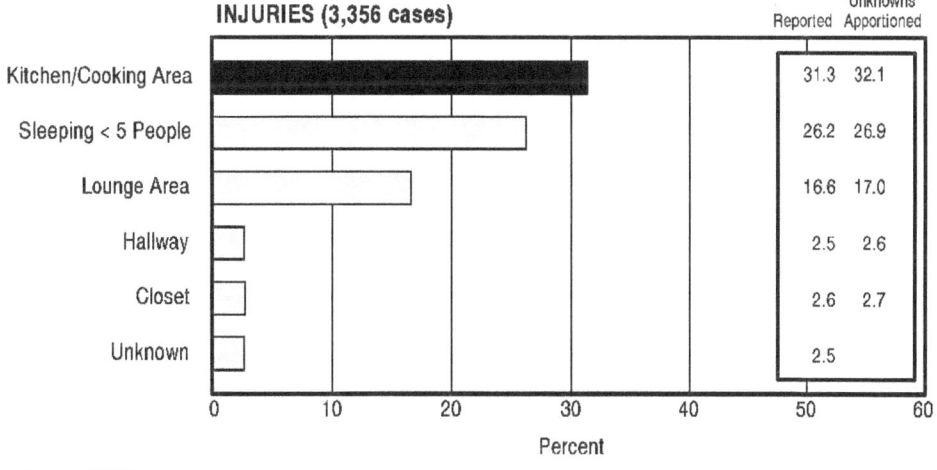

Source: NFIRS

Figure 65. Leading Rooms of Origin of 1994 Apartment Fires and Fire Casualties

103

Tables 10, 11, and 12 shows the leading rooms for each major cause of apartment fires, deaths, and injuries. Although the leading causes differ somewhat in apartments versus dwellings, the rooms for each cause are generally similar.

Table 10. Leading Rooms of Origin by Cause for 1994 Apartment Fires

Area of Home	Leading Causes				
	Cooking	Arson	Smoking	Children Playing	Electrical Distribution
Hallway		571 10.1%			
Interior Stairway		233 4.1%			
Lounge Area	20 0.14%	542 9.6%	744 22.5%	204 8.3%	253 11.5%
Sleeping Under 5	37 0.25%	967 17.1%	1,138 34.5%	1,464 59.4%	605 27.6%
Dining	26 0.18%				
Kitchen/Cooking	14,350 97.2%	536 9.5%	246 7.4%	149 6.0%	259 11.8%
Lavatory			119 3.6%	60 2.4%	138 6.3%
Closet				201 8.2%	83 3.8%
Trash Area/Container			189 5.7%		
Court/Terrace/Patio	30 0.20%				
All NFIRS Apartment Fires	14,759	5,661	3,303	2,466	2,196

Note: For each cause, the five most common rooms or areas of origin reported are shown. Data here are NFIRS raw counts, *not* national estimates. Percentages shown are column percentages (e.g., percentages of heating or cooking fires, not percentages of lounge fires).

Source: NFIRS

104

Table 11. Leading Rooms of Origin by Cause for 1994 Apartment Fire Deaths

Area of Home	Leading Causes				
	Smoking	Arson	Children Playing	Cooking	Heating
Hallway	1 1.5%	2 4.3%			2 14.3%
Interior Stairway		9 19.1%			
Laundry Room			4 8.7%		
Lounge	30 44.1%	7 14.9%	7 15.2%		7 50.0%
Sleeping Under 5	29 42.6%	5 10.6%	26 56.5%		1 7.1%
Kitchen/Cooking	3 4.4%		4 8.7%	21 100.0%	1 7.1%
Closet	1 1.5%		1 2.2%		
Supply Storage Room	1 1.5%		1 2.2%		
General Storage Area					3 21.4%
Exterior Balcony/Open Porch		8 17.0%			
All NFIRS Apartment Deaths	68	47	46	21	14

Note: For each cause, the five most common rooms or areas of origin reported are shown. Data here are NFIRS raw counts, not national estimates. Percentages shown are column percentages (e.g., percentages of heating or cooking fires, *not* percentages of lounge fires).

Source: NFIRS

105

Table 12. Leading Rooms of Origin by Cause for 1994 Apartment Fire Injuries

Area of Home	Leading Causes				
	Cooking	Arson	Smoking	Children Playing	Open Flame
Hallway		59 12.4%			
Interior Stairway		29 6.1%			
Lounge	5 0.6%	81 17.1%	186 42.8%	47 13.4%	51 26.6%
Sleeping Under 5		110 23.2%	157 36.1%	235 67.0%	76 39.6%
Dining	5 0.6%				
Kitchen/Cooking	784 97.3%	24 5.1%	15 3.4%	20 5.7%	15 7.8%
Wall Assembly					6 3.1%
Closet	2 0.2%		5 1.1%	17 4.8%	5 2.6%
Product Storage			6 1.4%		
Exterior Balcony	3 0.4%				5 2.6%
Incinerator Room				18 5.1%	
All NFIRS Apartment Injuries	806	475	435	351	192

Note: For each cause, the five most common rooms or areas of origin reported are shown. Data here are NFIRS raw counts, *not* national estimates. Percentages shown are column percentages (e.g., percentages of heating or cooking fires, not percentage of lounge fires).

Source: NFIRS

HOTELS AND MOTELS

Fires, deaths, injuries, and dollar loss in hotels and motels continued their decline from 1985 to 1993 (Figure 66). Because of the small sample sizes, the 9-year trends appear to be quite dramatic. Because of the notable improvement in the number of hotel and motel fires and fire losses—which in turn resulted in small sample sizes—NFPA no longer tabulates this residential category separately. As of 1994, hotel and motel fires are included in the other residential property category. NFIRS, however, still tabulates these fires separately.

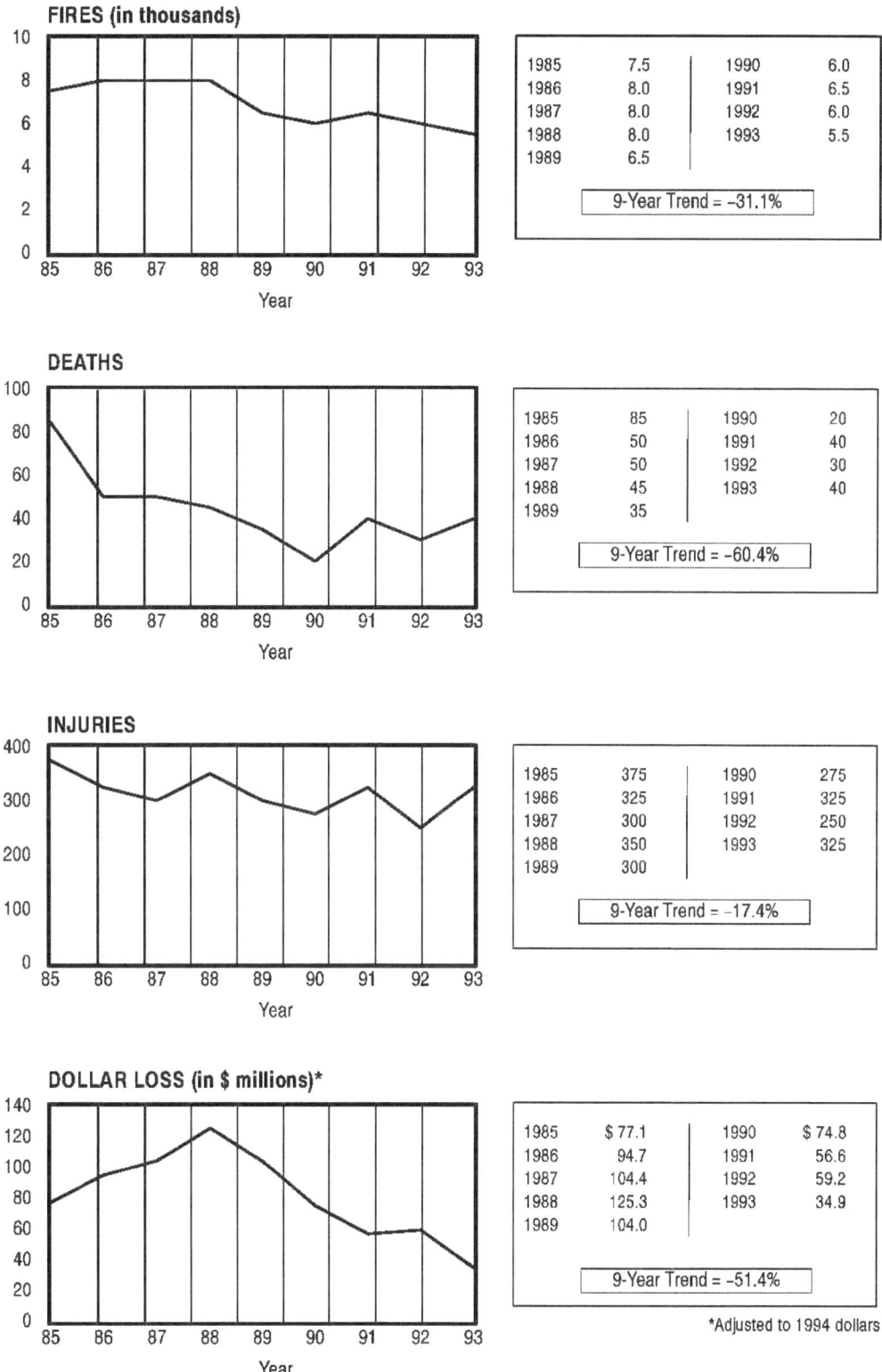

FIRES (in thousands)

1985	7.5	1990	6.0
1986	8.0	1991	6.5
1987	8.0	1992	6.0
1988	8.0	1993	5.5
1989	6.5		

9-Year Trend = –31.1%

DEATHS

1985	85	1990	20
1986	50	1991	40
1987	50	1992	30
1988	45	1993	40
1989	35		

9-Year Trend = –60.4%

INJURIES

1985	375	1990	275
1986	325	1991	325
1987	300	1992	250
1988	350	1993	325
1989	300		

9-Year Trend = –17.4%

DOLLAR LOSS (in $ millions)*

1985	$ 77.1	1990	$ 74.8
1986	94.7	1991	56.6
1987	104.4	1992	59.2
1988	125.3	1993	34.9
1989	104.0		

9-Year Trend = –51.4%

*Adjusted to 1994 dollars

Note: Hotel/motel fires in 1994 are treated as other residential property fires (see next section).

Sources: NFPA Annual Surveys and Consumer Price Index

Figure 66. Trends in Hotel/Motel Fires and Fire Losses

107

Causes

Most fires in hotels start in the guest rooms. Since heating is central and professionally maintained, the leading causes tend to be careless acts that guests can commit in hotel rooms, especially intentional acts (arson) by employees or guests. Cooking fires are second, but these usually originate in the hotel's centralized restaurant, not in the rooms. Appliances and careless smoking rate third and fourth, respectively.

The overwhelming leading cause of hotel fire deaths reported in 1994 is careless smoking at 67 percent (Figure 67). Cooking, electrical distribution, and other heat all tied for second at 11 percent. For fire injuries, careless smoking is first, followed by arson. These two causes have switched position since 1990.

Trends

Trends in causes of hotel/motel fire injuries are erratic; Table 13 compares the top three causes of fire injuries in 1985 and in 1994.

**Table 13. Trends in Leading Causes of
Hotel/Motel Fire Injuries**

Rank	1985	1994
1	Arson	Careless Smoking
2	Careless Smoking	Arson
3	Cooking	Cooking

Note: Data provided in Appendix B, Table B–4.

Source: NFIRS

Figure 68 shows the top five causes of fires. The trend in fires was downward for each, but less so than their drops for deaths and injuries. A survey conducted for the American Hotel and Motel Association found that virtually all hotel rooms in the United States have detectors and about half have sprinklers. This may be why the casualties are down sharply.

Trends in causes of deaths were erratic because of the small numbers in each cause category and are not shown here; however, there were dramatic drops in careless smoking deaths and arson deaths, which together caused total hotel fire deaths to be cut in half during the 9-year period examined. The hotel industry has instituted major changes—built-in fire protection systems and employee fire-awareness training are but two—that have been instrumental in this drop in fire deaths. The data on all 12 causes of hotel/motel deaths are shown in Appendix B, Table B–4.

OTHER RESIDENTIAL PROPERTIES

Other residential properties include rooming houses, dormitories, home hotels, halfway houses, and miscellaneous and unclassified properties reported as residences (and coded as NFIRS

108

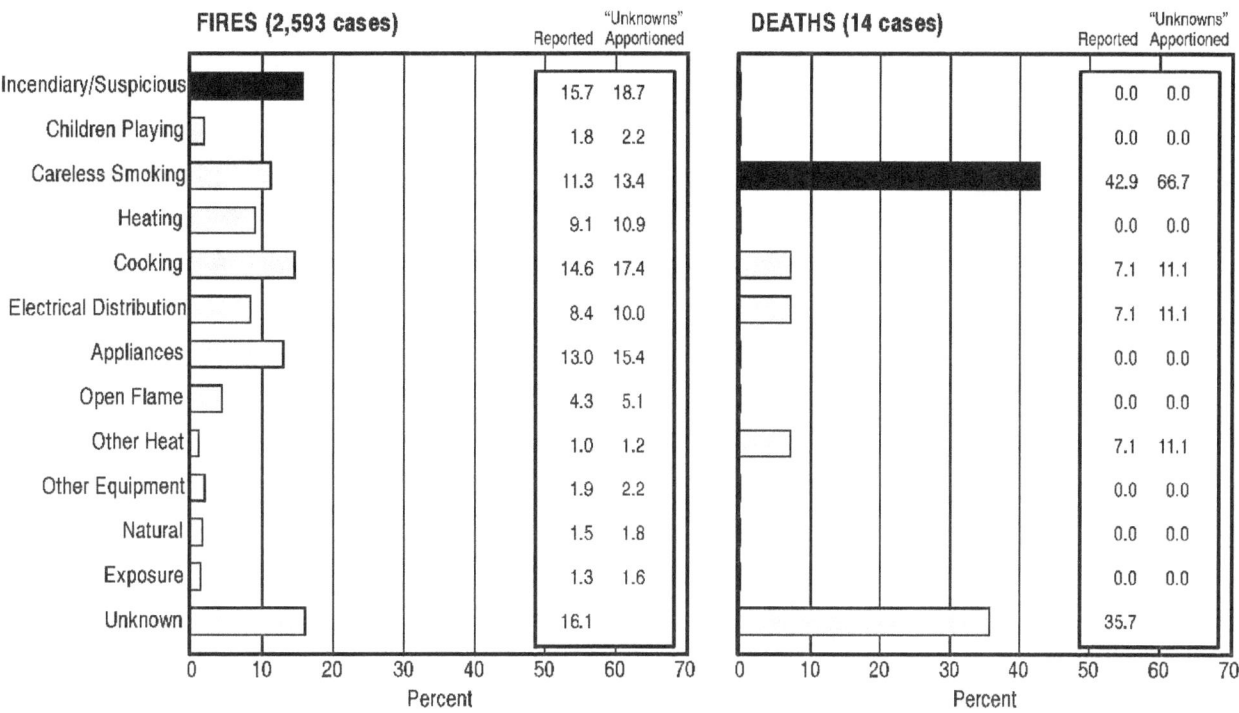

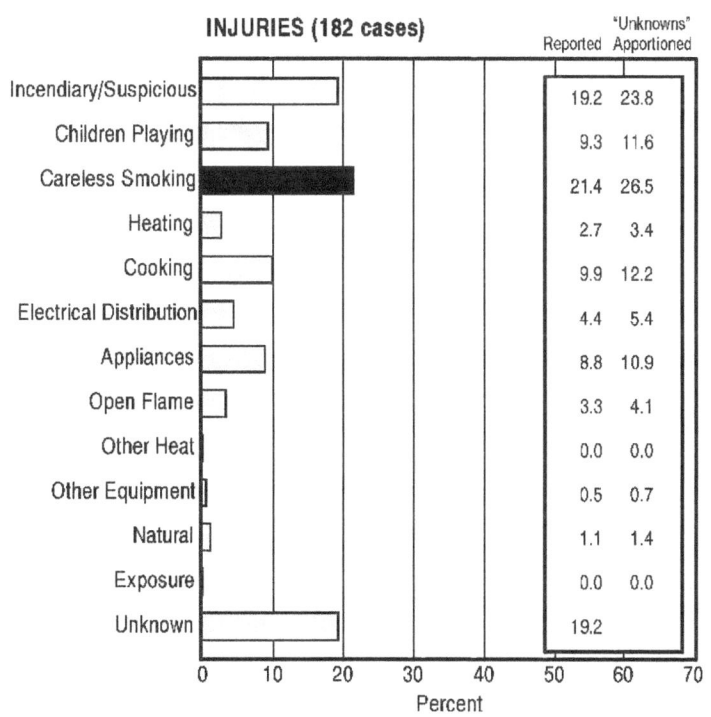

Figure 67. Causes of 1994 Hotel/Motel Fires and Fire Casualties

Source: NFIRS

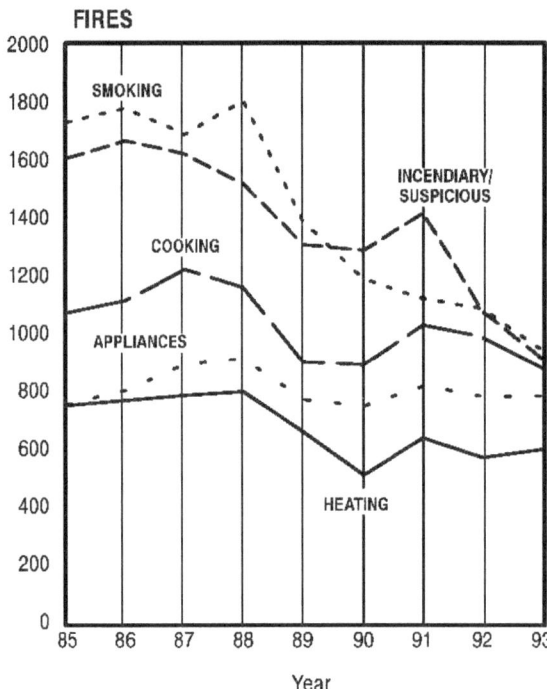

FIRES

Note: Data for all 12 causes provided in Appendix B, Table B–4.

Sources: NFIRS and NFPA Annual Surveys

Figure 68. Trends in Leading Causes of Hotel/Motel Fires

fixed property use #4). In 1994, hotels/motels are also in this category. The other residential properties category does not include homes for the elderly, prisons, orphanages, or other "institutions"; these have their own NFIRS categories under fixed property use #3 and are addressed in Chapter 4.

Trends

Figure 69 shows that the number of fires in the other residential category rose and fell over the 10-year period, spiking in 1994 due to the addition of hotels and motels to this category. Trends are not shown for this reason. Through 1993, the overall trend, however, is downward. Fire deaths ranged from 65 to 30 a year. Injuries ranged from 125 to 550 and adjusted dollar loss from $26 to $102 million. As with fires, the trends are downward through 1993. Other residential properties are a smaller portion of the residential fire problem than are detached garage fires, except in fire deaths.

Property Types

Figure 70 shows that hotels and motels in 1994 accounted for more fires, injuries, and dollar loss than all of the miscellaneous other residential categories combined, but far less than one- and two-family dwellings or apartments. Rooming house fires account for only slightly fewer fire deaths than hotels and motels, but receive much less attention. Some of these rooming houses are halfway

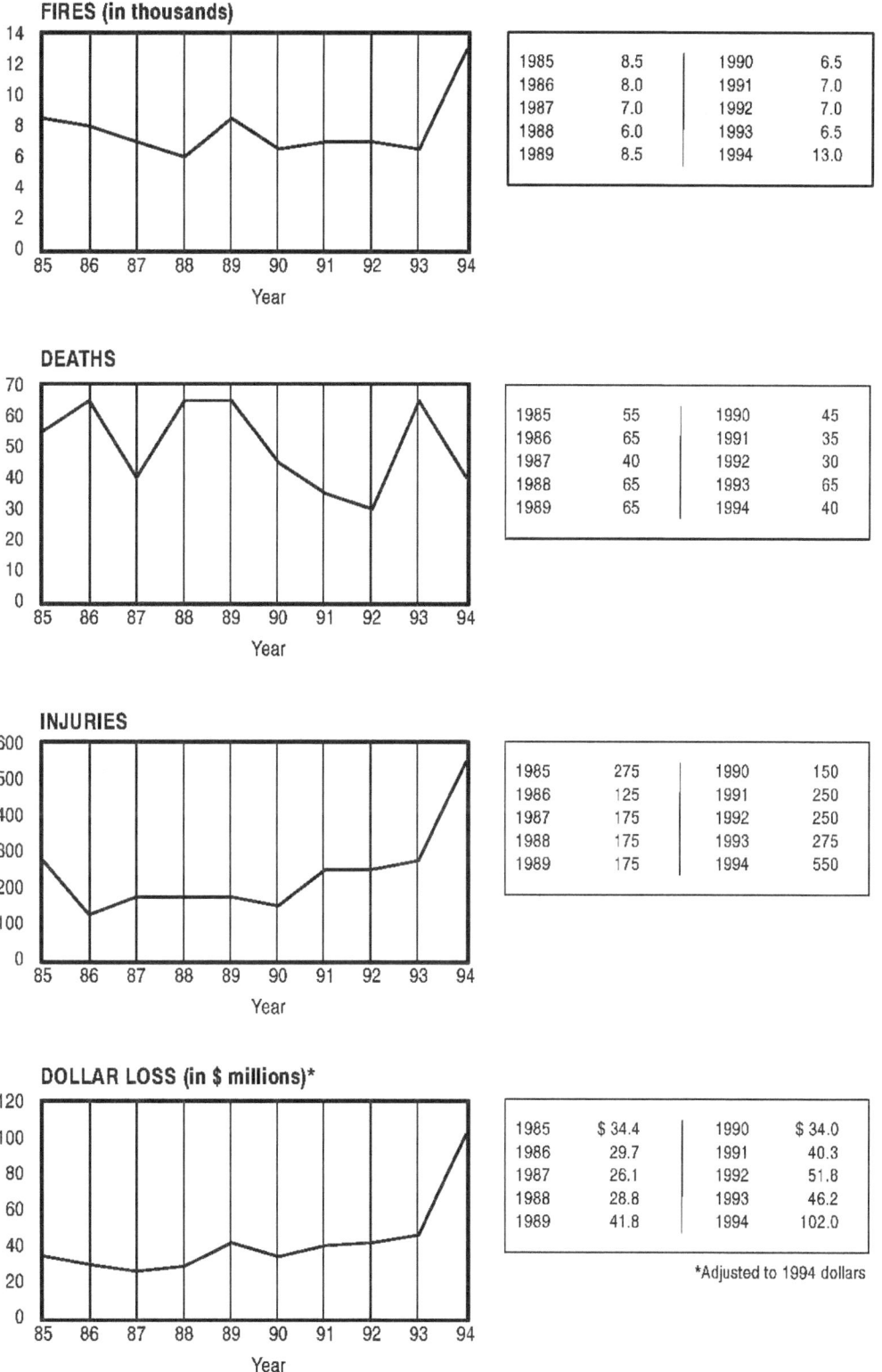

FIRES (in thousands)

1985	8.5	1990	6.5
1986	8.0	1991	7.0
1987	7.0	1992	7.0
1988	6.0	1993	6.5
1989	8.5	1994	13.0

DEATHS

1985	55	1990	45
1986	65	1991	35
1987	40	1992	30
1988	65	1993	65
1989	65	1994	40

INJURIES

1985	275	1990	150
1986	125	1991	250
1987	175	1992	250
1988	175	1993	275
1989	175	1994	550

DOLLAR LOSS (in $ millions)*

1985	$ 34.4	1990	$ 34.0
1986	29.7	1991	40.3
1987	26.1	1992	51.8
1988	28.8	1993	46.2
1989	41.8	1994	102.0

*Adjusted to 1994 dollars

Note: Hotels/motels were included in 1994.

Sources: NFPA Annual Surveys and Consumer Price Index

Figure 69. Trends in Other Residential Property Fires and Fire Losses

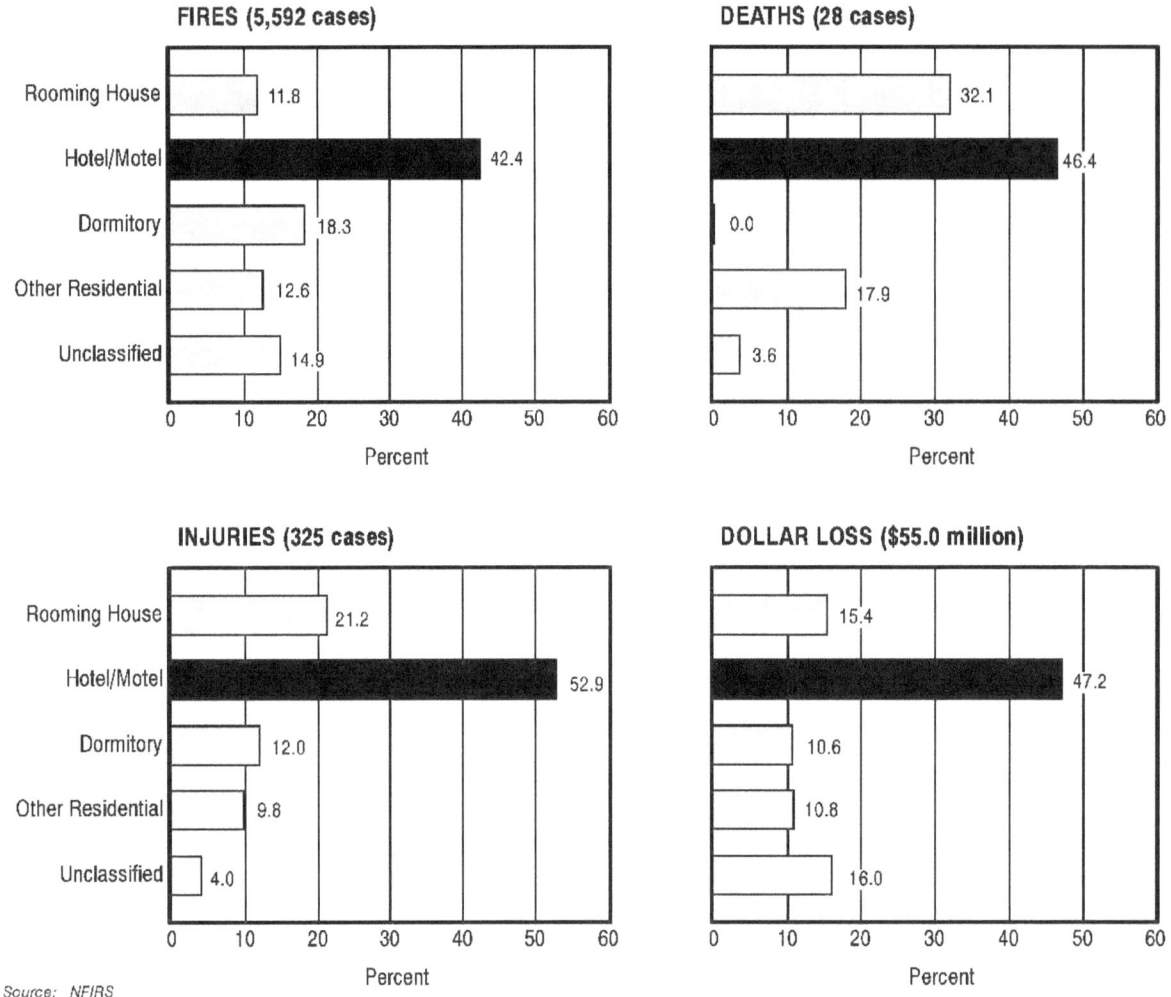

Figure 70. 1994 Other Residential Property Fires and Fire Losses by Property Type

houses for the physically or mentally handicapped. Dormitory fires do cause deaths, of course, despite their being none in the NFIRS sample in 1994.

Causes

As in 1990, arson was the leading causes of fires and injuries in other residential occupancies, whereas careless smoking was the leading cause for fire deaths (Figure 71).

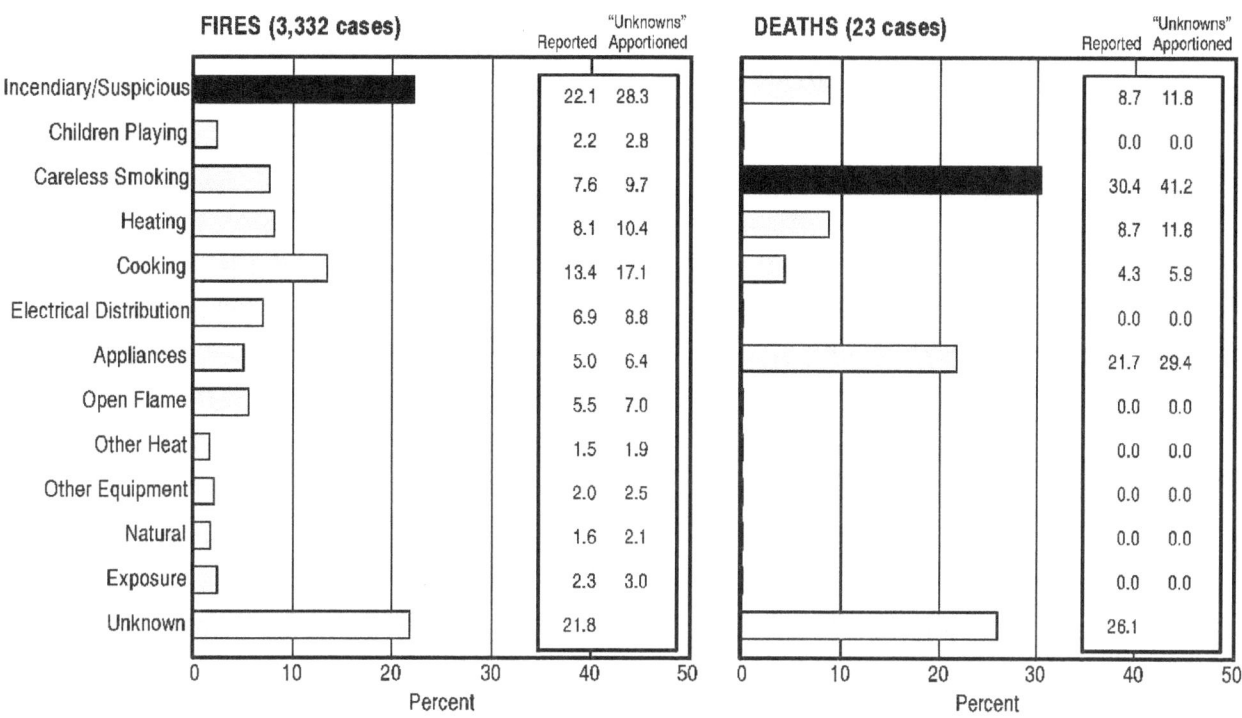

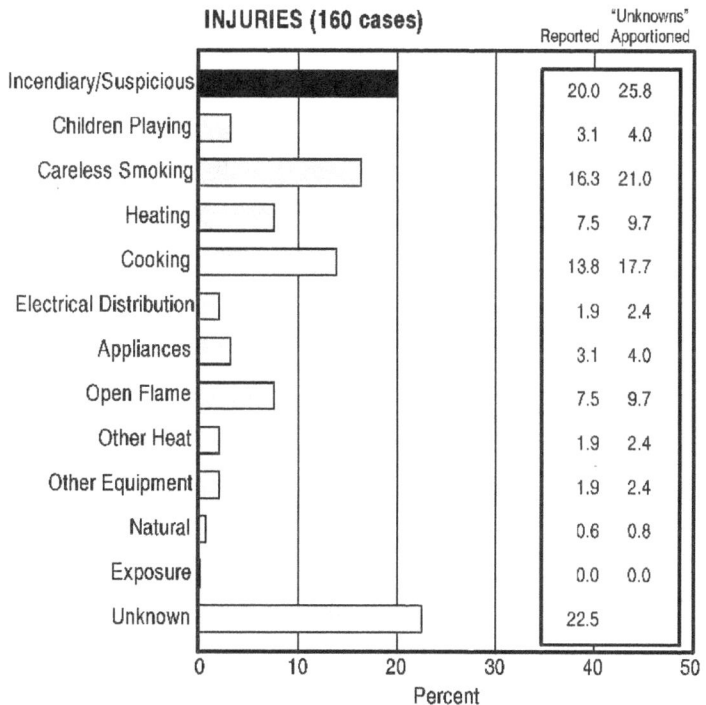

Figure 71. Causes of 1994 Other Residential Property Fires and Fire Casualties

Source: NFIRS

USFA RESOURCES ON FIRES IN RESIDENCES

The vast majority of civilian fire deaths and injuries continue to occur in residences. Residential occupancies also account for the largest annual dollar loss, and more firefighter injuries occur fighting fires in residences than in other types of occupancies. For these reasons, the U.S. Fire Administration has a variety of initiatives that focus on reducing residential fires and the deaths and injuries that they cause.

Public fire education is a cornerstone of USFA's fire prevention programs. USFA continues to attack barriers to public fire education programs at the state and local levels by developing public education tools, public awareness campaigns, and technical materials. USFA also promotes school system acceptance of fire safety education in K−12 and encourages private sector commitment and support for community fire prevention.

Many of the following topics are also addressed at the USFA's Web site at *http://www.usfa.fema.gov*.

Publications

To support and encourage public fire education, USFA has developed a series of public awareness campaign kits, which are described below. Each campaign kit has a variety of high-quality, ready-to-use materials for use by educators, community organizations, fire departments, and the private sector. Most campaigns promote home fire safety, primarily in one- and two-family dwellings, where about 70 percent of residential fires, deaths, and dollar loss occur.

USFA also produces a number of materials designed to improve the quantity and quality of public fire education efforts throughout the country. *Leadership in Public Fire Safety Education−2000* (#FA−135) presents the findings of a national public education conference held at the National Emergency Training Center. *Public Fire Education Today* (#FA−98) profiles hundreds of quality fire safety education programs being conducted across the United States. *Directory of National Community Volunteer Fire Prevention Programs* (#FA−92) is a catalog of local fire safety education programs addressing such issues as fire and burn prevention in the home, eliminating hazards, fire survival and escape, smoke detections and fire extinguishers, and the proper use of home heating devices. The *Short Guide to Evaluating Local Public Fire Education Programs* (#101) is a tool public fire educators can use to evaluate the effectiveness of a local fire prevention program.

These USFA publications are available by writing (include publication number if given in parentheses):

U.S. Fire Administration
Federal Emergency Management Agency
Publications Center, Room N310
16825 South Seton Avenue
Emmitsburg, MD 21727

Documents may also be ordered via the World Wide Web: *http://www.usfa.fema.gov*. USFA publications are free.

Video Training

FEMA's Emergency Education NETwork (EENET) provides video training and education via satellite for the fire service and emergency management community. EENET programs are satellite-distributed videoconferences broadcast over the "C" band and allow for audience interaction when originally broadcast. Each program is designed as a standalone training activity of 4-1/2 hours, and student materials are provided for each workshop.

Past programs focusing on fire prevention and public fire education topics include the following:

- *The Last Great Challenge: Fire Prevention*, September 27, 1989

- *Corporate Commitments to Fire Prevention*, October 17, 1991

- *Fire Administration Fire Prevention/Education Showcase*, October 23, 1991

- *How Effective Is Your Fire Prevention Program?*, May 6, 1992

- *Leadership in Public Fire Education: Trends and Issues*, July 8, 1992

Tapes of broadcasts from 1989 to the present are available for a modest cost through:

The National Audiovisual Center
National Technical Information Service (NTIS)
5285 Port Royal Road
Springfield, VA 22161
(703) 487–4650
http://www.ntis.gov

In addition, all broadcasts prior to 1989 and all current broadcasts may be borrowed from your State Emergency Management Office or from your FEMA regional office.

For further information on EENET, or if you would like to get on the EENET mailing list, contact:

Emergency Education NETwork
National Emergency Training Center
16825 South Seton Avenue
Emmitsburg, MD 21727
(301) 447–1068; (800) 527–4893; Fax: (301) 447–1363

National Fire Academy Courses

USFA's National Fire Academy (NFA) works to enhance the ability of the fire service and allied professions to deal more effectively with fire and related emergencies. Courses are delivered on campus at the resident facility in Emmitsburg, Maryland, and off campus throughout the nation in cooperation with state and local fire training officials and local colleges and universities. An initiative begun in 1992 offers NFA resident courses on a regional basis.

While issues related to fires in residences and fire prevention are addressed in numerous academy courses, several offerings include these issues as a major thrust. NFA's course on *Strategic Analysis of Fire Prevention Programs* (Resident Course #R309) is a 2-week course that helps senior fire executives identify strategies for improving their fire departments' fire prevention activities. *Management of Fire Prevention Programs* (Resident Course #R225) uses management concepts as a basis for the efficient operation of a fire prevention organization. *Fire Prevention Principles* (Resident Course #R220) provides the student with the fundamental knowledge, skills, and attitudes to conduct basic fire safety inspections. *Principles of Fire Protection: Structures and Systems* (Resident Course #R222) is a 2-week course designed to meet the professional development needs of the advanced fire prevention officer. *Code Management: A Systems Approach* (Resident Course #R101) addresses the management of code development, evaluation, and enforcement processes. Other offerings include *Fire Service Instructional Methodology* (Resident Course #R113) and *Developing Fire and Life Safety Strategies* (Resident Course #R352).

For information about course offerings, eligibility, and application procedures, write to:

The National Fire Academy
U.S. Fire Academy
16825 South Seton Avenue
Emmitsburg, MD 21727

Information on National Fire Academy course offerings can also be obtained on the USFA Web site at *http://www.usfa.fema.gov.*

The National Fire Academy off-campus materials for the *Public Fire Education Planning* (1985) and *Conducting Basic Fire Prevention Inspections* (1985) courses are available for purchase for locally sponsored delivery from the National Audiovisual Center. Current academy off-campus courses, consisting of an instructor guide, student manual, and supporting audiovisual aids are also available. For information on how to order courses, contact the National Audiovisual Center at the address and phone number listed earlier.

Campaign Materials

USFA has developed a series of public awareness campaign kits with a variety of high-quality materials for use by educators, community organizations, fire departments, and the private sector. *It's a Real Protector, It's a Smoke Detector* (English and Spanish) promotes the use of smoke detectors. It includes radio and print public service announcements (PSAs), sample letters to the editor, a fill-in-the-blank press release, and a resource guide. *Make the Right Call EMS (Emergency Medical Services)* is a campaign that uses the media to educate Americans about EMS and its proper use through new coverage, feature and entertainment articles, and PSAs.

ONE- AND TWO-FAMILY DWELLINGS. Most campaigns promote home fire safety, primarily in one- and two-family houses, where most residential fires, deaths, and dollar loss occur.

The most recent campaign, *Home Fire Safety: Act On It*, was developed in cooperation with the Sleep Products Safety Council, the National Association of Broadcasters, the National Board of Realtors, and the "Just Say No" campaign. It contains a variety of materials on general home fire safety themes. *Working Together for Home Fire Safety—A USFA Public Private Partnership* was designed for corporate managers to disseminate home fire safety tips to their employees. Camera-ready materials focus on careless use of smoking materials, unsafe use of alternate heaters, facts about the basic myths of fires, and the importance of practicing a home escape plan.

Other campaign kits focus on specific topics of home fire safety. *Let's Retire Fire* (#5−0129) provides fire safety tips for older Americans. *Curious Kids Start Fires* (#5−0101 in English or #5−0132 in Spanish) focuses on curiosity fireplay. *This Is Fire* (#5−0123) includes hard-hitting messages about the fatal characteristics of fire—that it is fast, dark, hot, and deadly. *Check Your Hot Spots!* (#5−0101) addresses safe use of alternate heaters, especially in rural homes. *Children and Fire: The Experience of Children and Fire in the United States* presents a detailed look at fires involving children, including the number of children killed in fires in 1991, U.S. counties with the highest rates of child fire deaths, death rates among different ethnicity groups, characteristics of fires that kill and injure children, importance of smoke detectors in preventing child deaths and injuries, and characteristics of children playing fires.

USFA also conducts special studies to address specific problems and current issues facing the nation's fire and rescue service. The technical reports produced under the Major Fires Investigations series analyze major or unusual fires with emphasis on sharing lessons learned. They are directed primarily to chief fire officers, training officers, fire marshals, and investigators as a resource for training and prevention.

A number of Major Fires Investigations reports focus on residential fires in one- and two-family homes: *Seven-Fatality Christmas Tree Fire, Canton, Michigan, December 1994* (#046); *Power Off to Hard-Wired Detector in Nine-Fatality House Fire, Peoria, Illinois, April 1989* (#031); *Eight Children and Two Adults Die in Rural House Fire, Remer, Minnesota, January 1989* (#028); *Four House Fires That Killed 28 Children, September−December 1987* (#020); *Eight-Fatality Row-House Fire, Chester, Pennsylvania, December 1992* (#067); and *Children Left Home Alone: Eleven Die in Two Fires, Detroit, Michigan−February 1993* (#070).

APARTMENTS. *An Information Backgrounder on Fire Resistant Construction and Building Codes* (#FA−70) is a three-page booklet explaining the importance of fire-resistant construction and how construction features contribute to fire control in multifamily dwellings.

USFA also has recent Major Fires Investigation reports that address major fires in apartment buildings: *Apartment Complex Fire—66 Units Destroyed, Seattle, Washington, September 21, 1991* (#059); *New York—Schomberg Plaza Fire, Harlem, 1987* (#004); *Apartment Building Fire, East 50th Street, New York City, January 1988* (#019); *Sixteen-Fatality Fire in High-Rise for the Elderly, Johnson City, Tennessee, December 1989* (#030); *Nine Elderly Fire Victims in Residential Hotel, Miami Beach, Florida, April 1994* (#041); *Fire, Police, and EMS Coordination at Apartment Building Explosion, New York City, November 1992* (#068); and *Nine-Fatality Apartment House Fire, Luddington, Michigan, February 1993* (#072).

HOTELS AND MOTELS. USFA has worked very diligently in the implementation of PL 101–391, The Hotel/Motel Fire Safety Act of 1994. By working closely with the American Hotel and Motel Association and the National Association of State Fire Marshals, USFA provided a variety of support services to states to help them identify facilities that meet the fire safety requirements of the Act.

Major Fires Investigations reports studying fires in hotels and motels include the following: *Seven-Fatality Hotel Fire, Grand Carais, Minnesota, July 1991* (#055); *Indiana—Ramada Inn Air Crash and Fire, Wayne Township, October 1987* (#014); *Louisiana—Doubletree Hotel Fire, New Orleans, July 1989*; *Texas—La Posada Hotel Fire, McAllen, February 1987* (#001); and *National Guard Plane Crash at Hotel Site, Evansville, Indiana, February 1992* (#064).

OTHER RESIDENTIAL PROPERTIES. Several Major Fires Investigations reports share lessons learned in major fires in other types of residential properties, including *Nine Elderly Fire Victims in Intermediate Care Facility, Colorado Springs, Colorado, March 1991* (#050); *Virginia—Twelve-Fatality Nursing Home Fire, Norfolk, November 1989*; *Virginia—Shenandoah Retirement Home Fire, Roanoke County, December 14, 1989*; *Virginia—Success Story at Retirement Home Fire, Sterling, December 16, 1989*; *Nine Elderly Fire Victims in Residential Hotel, Miami Beach, Florida, April 1994* (#041); *Delaware and Virginia—College Dormitory Fires in Dover and Farmville, April 1987*; and *North Carolina—Nine-Fatality Manufactured Housing Fire, Maxton, November 1989*.

RESIDENTIAL SPRINKLERS. USFA has completed extensive research to develop installation and application standards for quick-acting residential sprinklers and has conducted a variety of demonstrations of the quick-response sprinkler technology to demonstrate the practicality of these systems. USFA's report *Residential Fire Sprinklers Retrofit Demonstration Project Final Report and Case Studies* (#FA–89, #FA–90, #FA–96, #FA–97) describes a multiple-stage demonstration project in multifamily residences undergoing rehabilitation where quick-response residential fire sprinklers were installed.

USFA also worked with Factory Mutual and Underwriters Laboratories to complete design and testing of new limited-water-supply fire sprinkler systems for manufactured housing.

There are a number of other publications on sprinklers available from USFA. An *Information Backgrounder on Fire-Resistant Construction and Building Codes* (#FA–70) is a three-page booklet explaining the importance of fire-resistant construction and how construction features contribute to fire control in multifamily dwellings. *An Ounce of Prevention* (#FA—76) is an 18-page booklet for homeowners, insurance underwriters, building designers and developers, legislators, and building officials. The booklet provides a comprehensive discussion of why and how the combination of automatic sprinklers and early warning systems in all types of buildings can have a major impact on fire-related deaths, injuries, and property loss.

Home Fire Protection—Quick-Response Fire Sprinkler Systems (#FA—43) is a five-page pamphlet for the general public explaining the merits of home sprinklers and the financial and insurance benefits.

4

NON-RESIDENTIAL PROPERTIES

The non-residential property category includes industrial and commercial properties, institutions (such as hospitals, nursing homes, prisons), educational establishments (from preschool through university), mobile properties, and properties that are vacant or under construction. Each of these categories corresponds to one of the major divisions of the NFPA 901 coding system used by NFIRS. Each is quite different, and their cause profiles and magnitudes need to be examined separately.

NON-RESIDENTIAL STRUCTURES

Magnitude and Trends

Much of the effort in fire prevention, both public and private, has gone into protecting non-residential structures, and the results have been highly effective in the main, especially relative to the residential fire problem. Even with all the diverse properties included here, non-residential structures consistently accounted for only 5−6 percent of the fire deaths annually, as was shown in Figure 28, Chapter 2. They also accounted for 13−14 percent of all fire injuries, 32−46 percent of total fire dollar loss, and 9−10 percent of all fires. These proportions tended to be similar over the 10-year period.

In absolute numbers, non-residential fire deaths from 1985 to 1994 dropped from 240 to 125, or 37 percent (Figure 72). In 1994, total deaths were at a 10-year low, but they jumped 30 percent in the non-residential category from 1989 to 1990. Figure 72 also shows that non-residential dollar loss adjusted to 1994 dollars and total fires trended downward, by a large 30 and 38 percent, respectively; injuries were down 10 percent.

Figure 73 shows the relative magnitudes of the fire problem in non-residential structures by its component property categories for 1994. Storage facilities, institutions, and manufacturing plants were the three leading property types for deaths, by themselves accounting for 50 percent of the non-residential fire deaths. This will surprise many, who might expect nursing homes, dormitories, or nightclubs to be high. However, one large incident in a particular category (such as the Beverly Hills Nightclub fire in May 1977 that took 165 lives or the Oklahoma City bombing in 1995 that claimed 168 lives) could make that category the leader in a particular year because the absolute number of deaths tends to be small for most categories in most years. In 1994, deaths in storage fires

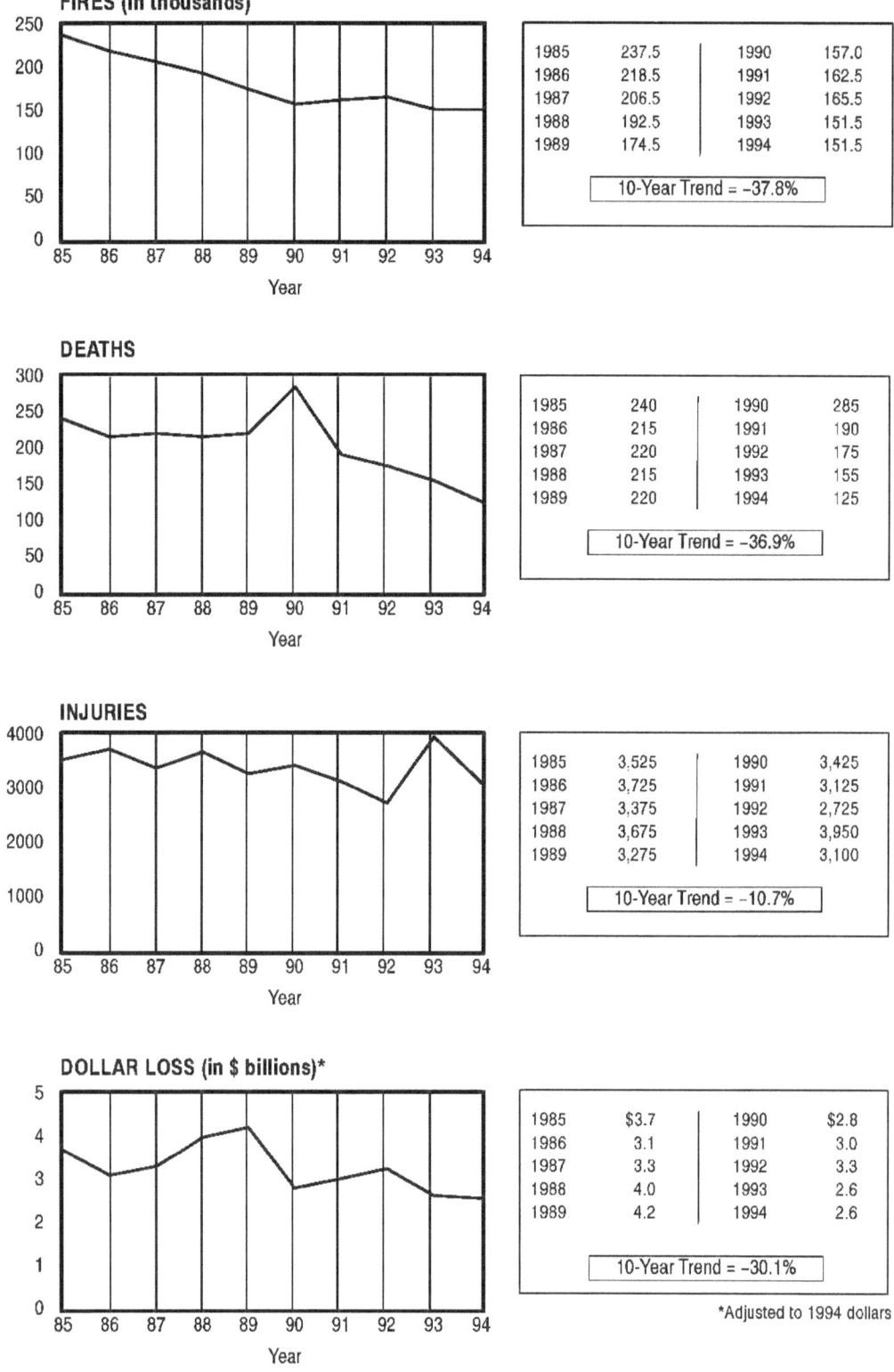

Sources: NFPA Annual Surveys and Consumer Price Index

Figure 72. Trends in Non-Residential Fires and Fire Losses

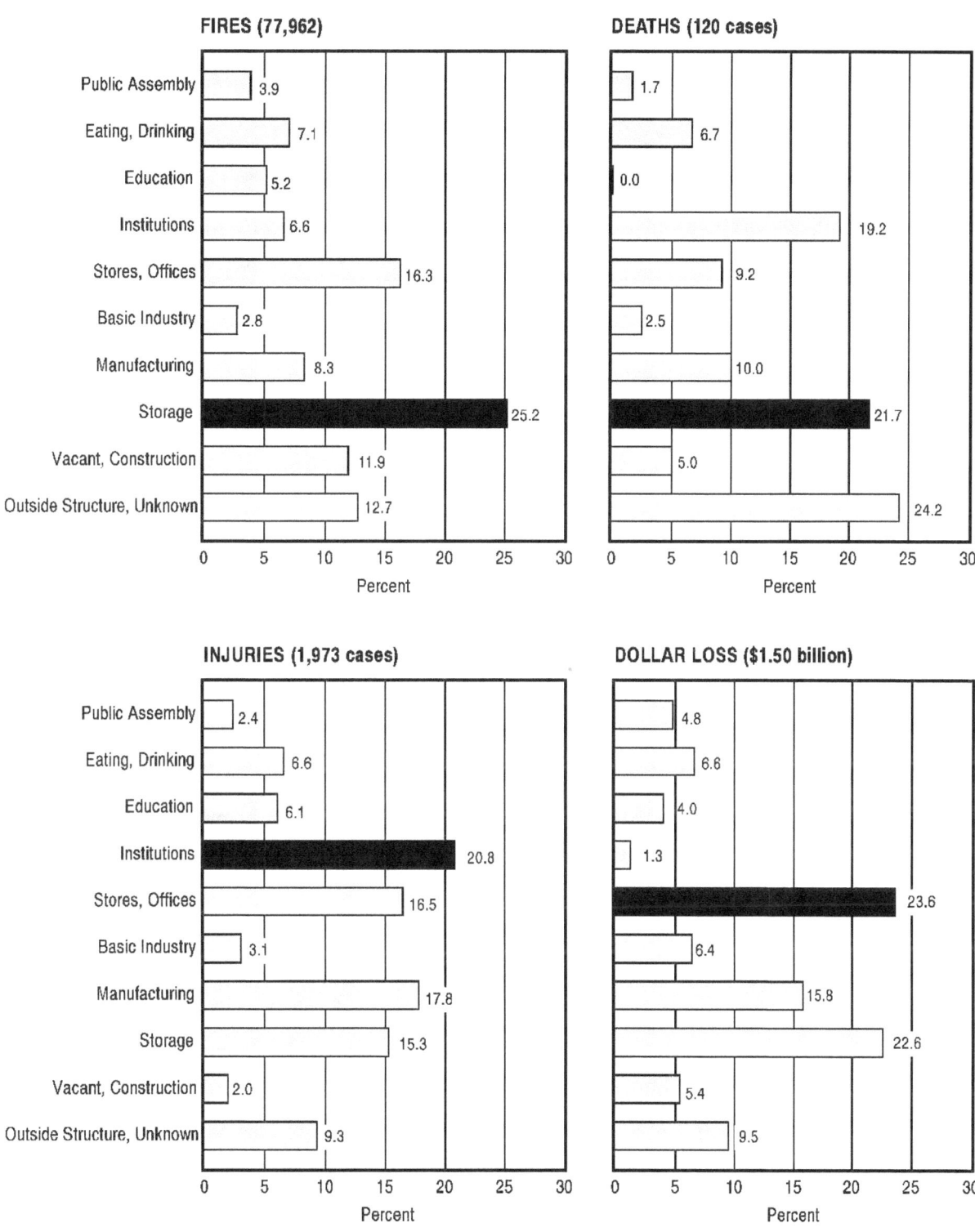

Figure 73. 1994 Non-Residential Fires and Fire Losses by Property Type

123

ranked first, whereas in 1990 it was in fourth place. Manufacturing fire deaths in 1994 dropped significantly from 1990 (23 percent vs. 10 percent).

The rank order of property types for fire injuries is similar to that for deaths, except that storage fires account for a somewhat smaller portion (15 percent injuries vs. 22 percent deaths). Institutional fires account for about 20 percent of both injuries and deaths.

In terms of dollar loss, stores/offices and storage facilities accounted for 46 percent of the total loss, the same percentage as in 1990. Storage facilities and stores/offices also led for fire incidence and represented 42 percent of all non-residential fires.[1]

The low rank ordering of some property categories should not obscure the fact that all of the categories have thousands of fires, multimillions of dollar loss, and hundreds of casualties. All parts of the fire problem need to be addressed. The relative magnitudes might help suggest where the greatest effort is needed.

When Fires Occur

TIME OF DAY. Non-residential fires are a heterogeneous category, and the time of day when each of its different component property types peak may not agree with the overall picture, which is depicted in Figure 74 for fires, deaths, injuries, and dollar loss.

The incidence of all fires has the smoothest shape variation because it is based on the largest sample. Fires peak in the afternoon and evening, from noon to 8 p.m. Perhaps this is when workers are tiring on their job and are more accident prone or careless—but that is speculation.

Fire deaths fluctuate greatly because the sample of deaths in NFIRS is fewer than ten for most 1-hour intervals. The heaviest concentration in 1994 is between 9 and 10 p.m. As in 1990, the second highest peak is between 3 and 4 a.m.

Injuries tend to be at fairly high levels throughout normal work day and evening hours, 7:00 a.m.–10:00 p.m. with the highest concentration during the workday itself, 9:00 a.m.–5:00 p.m. Fire injuries are relatively low in the nighttime and early morning period when the majority of the workforce is at home.

Peak dollar losses occur after hours, especially between 10:00 p.m. and 6:00 a.m. Their leading cause is suspected to be arson.

MONTH OF YEAR. Fires in non-residential properties are relatively uniform throughout the year, with the peak in April and the nadir in September. Nevertheless, the problem exists at a high level throughout the year (Figure 75).

---

[1] The rank ordering from NFPA's annual surveys are generally, but not exactly, similar to the NFIRS results above. See Appendix A for the NFPA-derived trends.

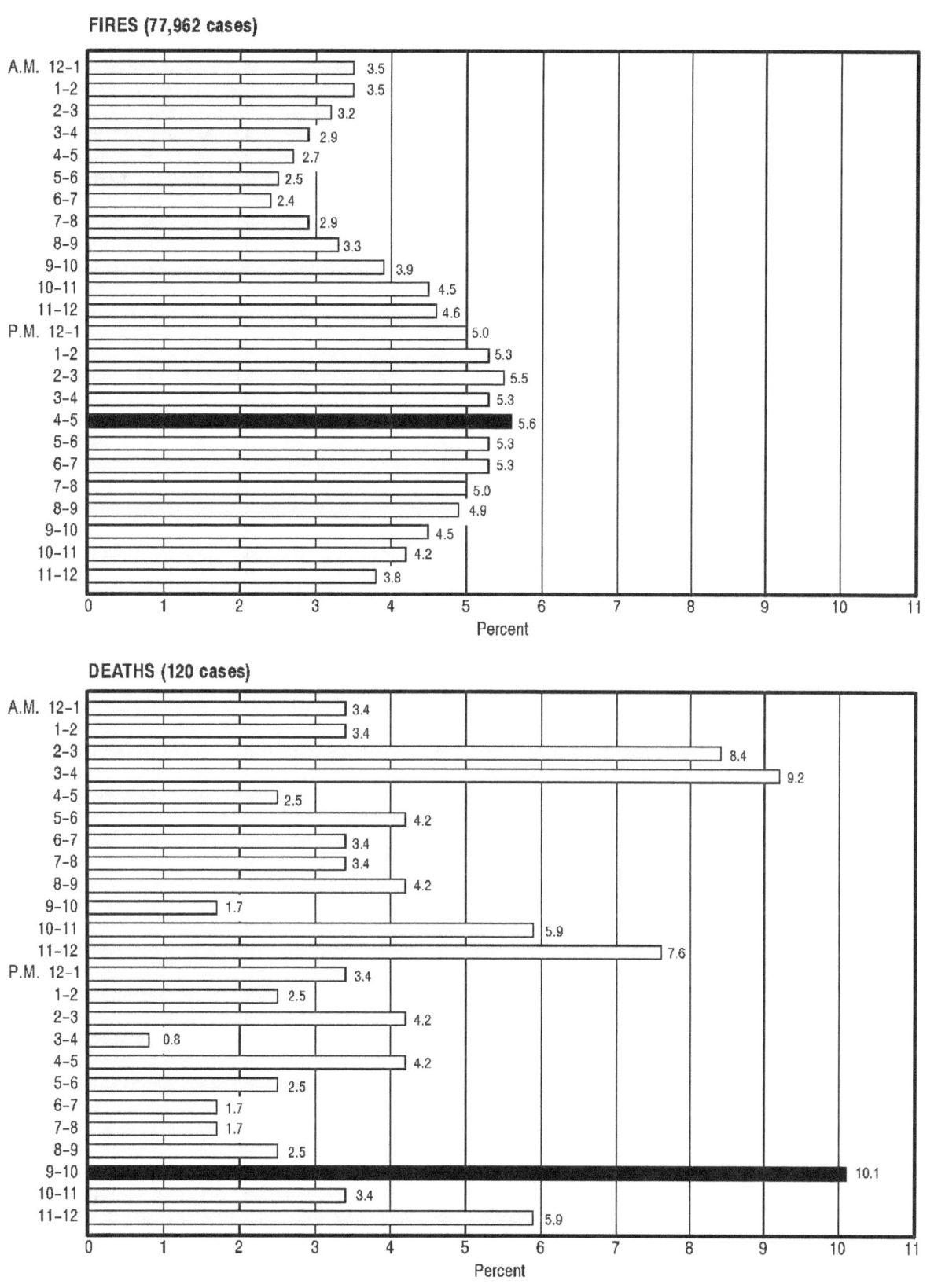

FIRES (77,962 cases)

DEATHS (120 cases)

Continued on next page

Figure 74. Time of Day of 1994 Non-Residential Fires and Fire Losses

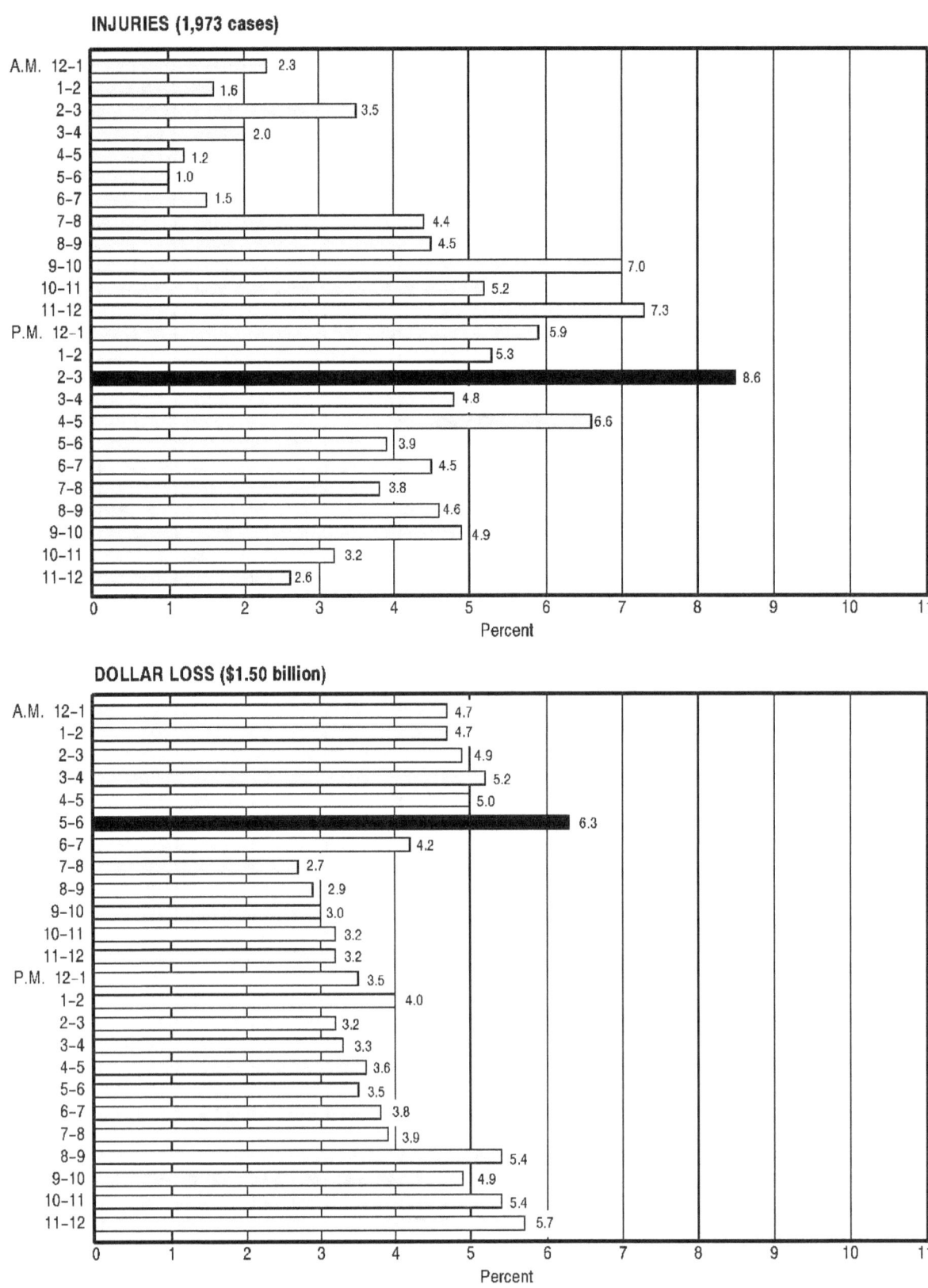

INJURIES (1,973 cases)

DOLLAR LOSS ($1.50 billion)

Source: NFIRS

Figure 74. Time of Day of 1994 Non-Residential Fires and Fire Losses (cont'd)

126

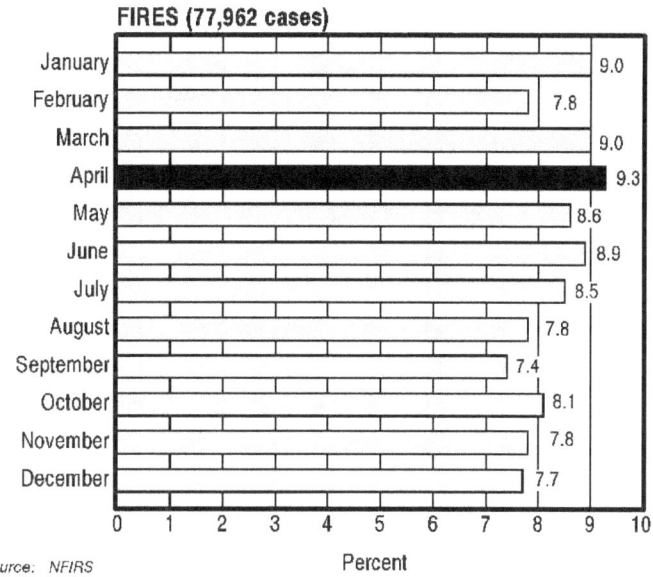

Figure 75. Month of Year of 1994 Non-Residential Fires

DAY OF WEEK. Non-residential fires are almost uniform by day of week, except that there are slightly fewer on Sundays when fewer people are at work (Figure 76). The profile is probably less uniform for subcategories of occupancies such as restaurants. There was no significant change from 1990 percentages.

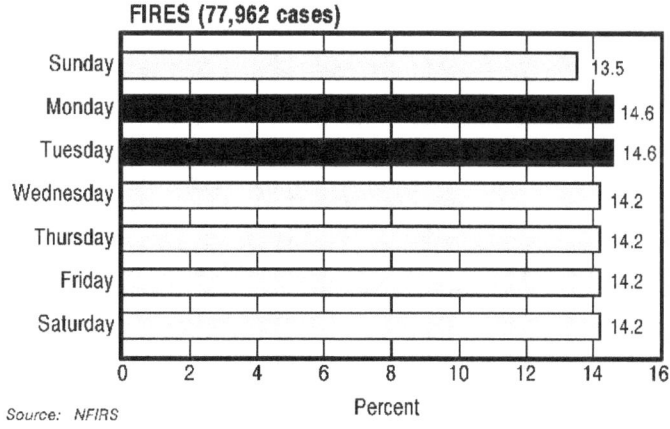

Figure 76. Day of Week of 1994 Non-Residential Fires

Causes

By far the leading cause of non-residential fires in 1994 is arson (Figure 77). Arson also accounts for more than one-third of the dollar loss. Historically, arson has accounted for the highest dollar loss and, like the overall number of fires, it has trended downward. Arson has been the leading cause

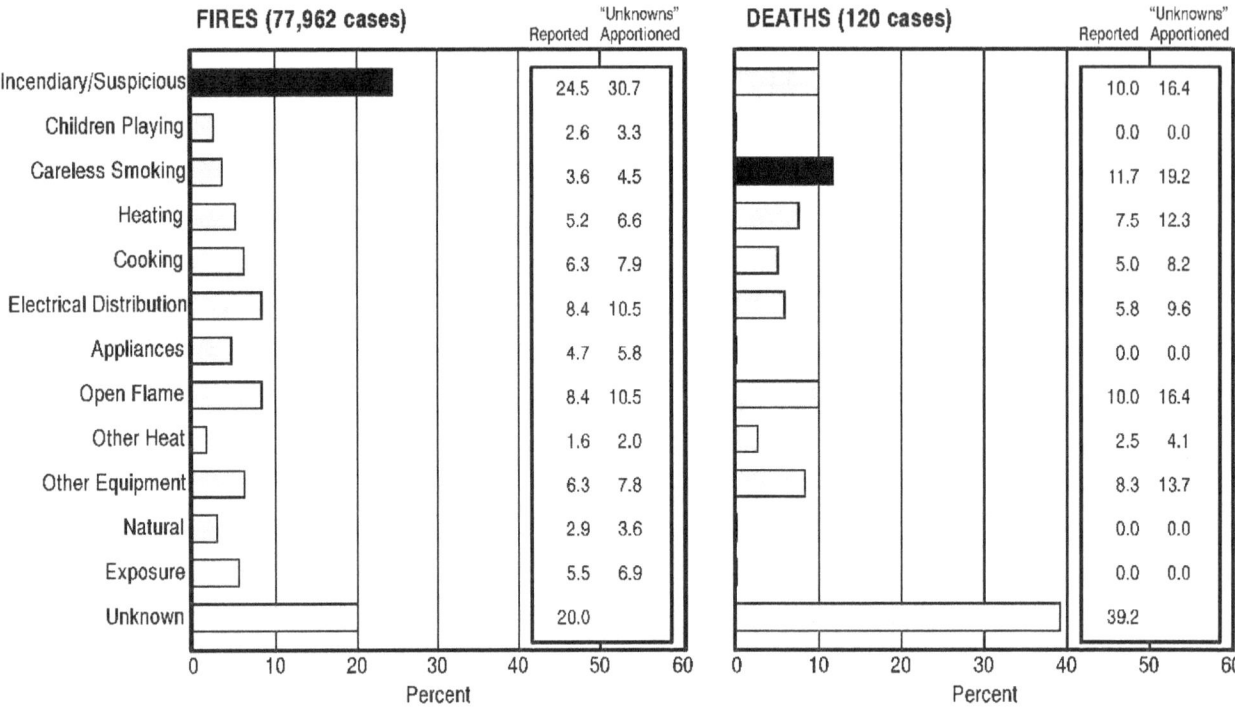

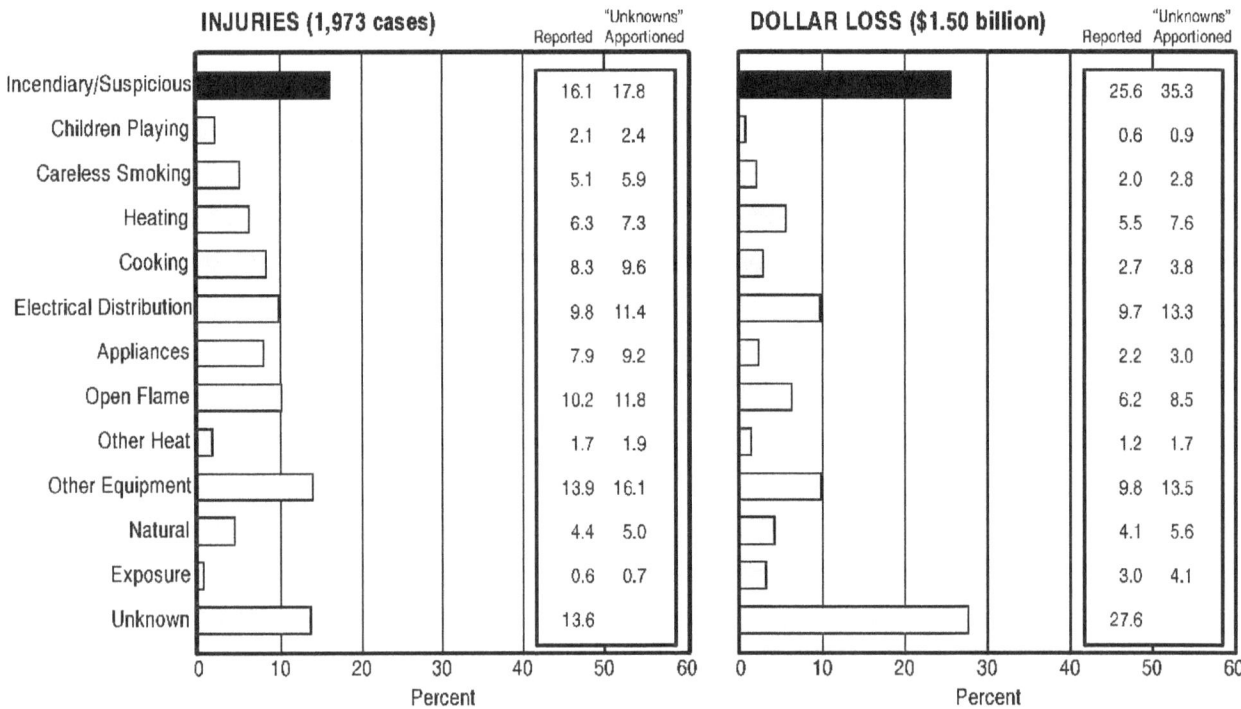

Source: NFIRS

Figure 77. Causes of 1994 Non-Residential Fires and Fire Losses

of non-residential fire deaths for 8 of the past 10 years. In 1994, however, as in 1988, smoking is the leading cause of non-residential deaths, with arson and open flame tied for second. Arson is also the leading cause of injuries, with other equipment slightly behind.

The trends in causes of non-residential fires, deaths, injuries, and dollar loss are shown in Figure 78. In most years, arson was the leading cause in all fire and fire loss categories. The causes for deaths fluctuate considerably because of the smaller numbers of cases involved. No matter how you look at it, arson is the major problem in non-residential occupancies accounting for nearly three times the number of fires in 1994 as any other cause.

Causes by Detailed Property Type

Because the dollar loss and numbers of fires in non-residential occupancies are high, the causes in terms of these two measures are given in Figures 79–88 for each non-residential category. Table 14 summarizes the leading cause of fires and fire dollar loss for each non-residential property type in 1994. Except for the basic industry, eating and drinking, and manufacturing property types, arson was the leading cause in all cases.

Electrical distribution, the leading cause of basic industry fires, is the second leading cause in fires and dollar loss across many property types. These 1994 data are quite similar to 1990 findings. Other high causes are generally related to the type of activity or equipment being used. For example, cooking is the leading cause in eating and drinking establishments. Arson is clearly the leading fire problem for non-residential properties. This has been the case for at least the past 20 years, since NFIRS was started.

Table 14. Leading Causes of 1994 Non-Residential Fires and Dollar Loss

Property Type	Fires	Dollar Loss
Public Assembly	Arson	Arson
Eating, Drinking	Cooking	Arson
Education	Arson	Arson
Institutions	Arson	Arson
Stores, Offices	Arson	Arson
Basic Industry	Electrical Distribution	Electrical Distribution
Manufacturing	Other Equipment	Other Equipment
Storage	Arson	Arson
Vacant, Construction	Arson	Arson
Outside Structures, Unknown	Arson	Arson

Source: NFIRS

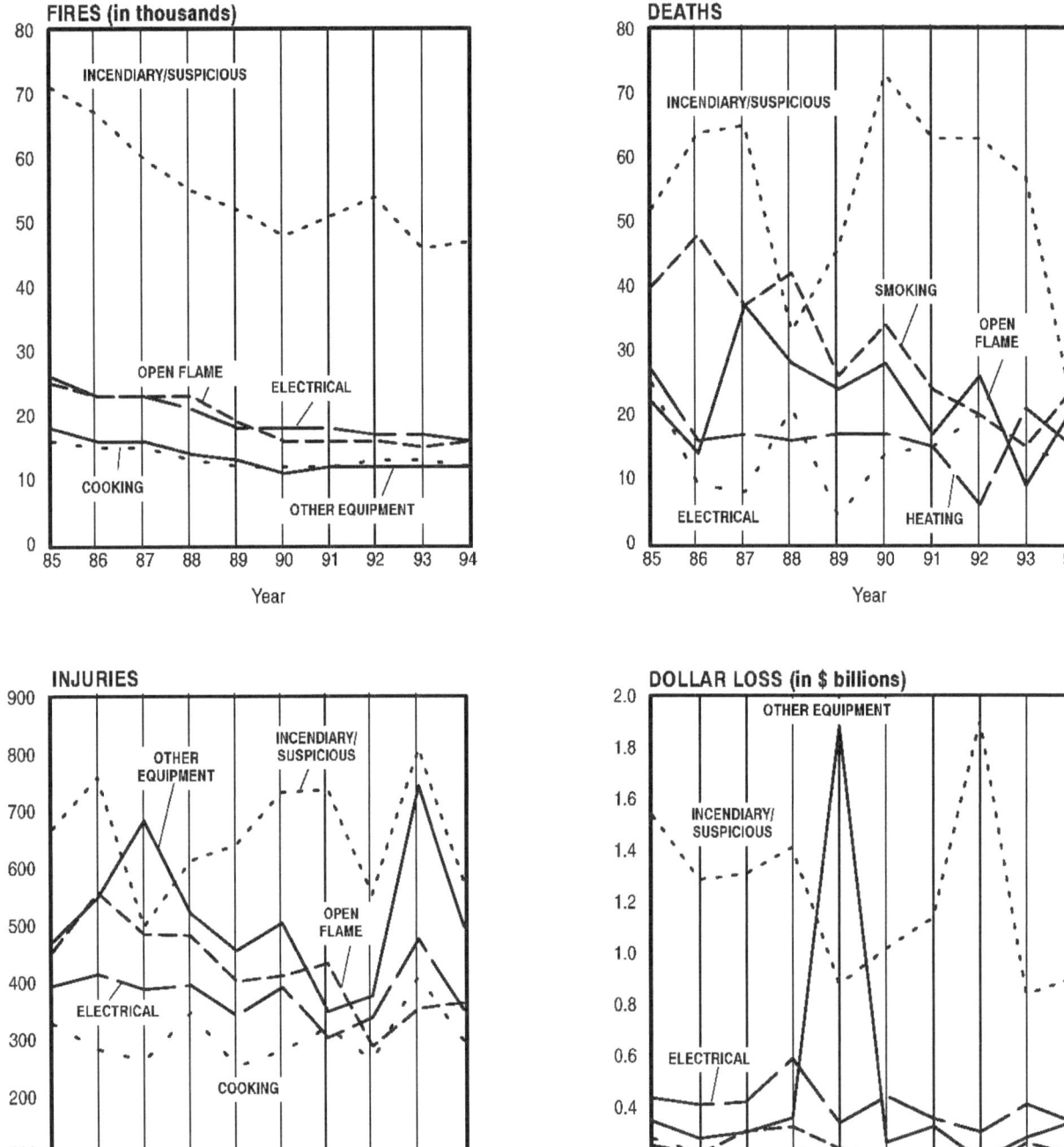

Note: Data for all 12 causes are provided in Appendix B, Table B–5.

Sources: NFIRS and NFPA Annual Surveys

Figure 78. Trends in Leading Causes of Non-Residential Fires and Fire Losses

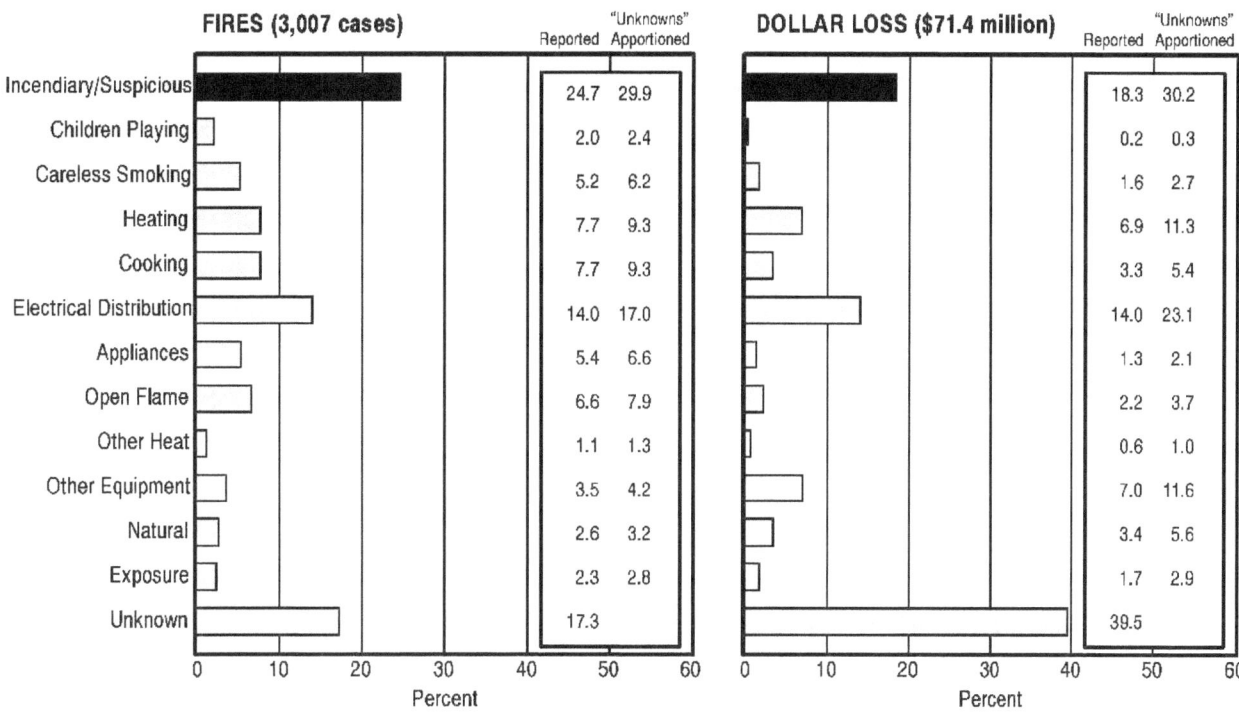

Figure 79. Causes of 1994 Public Assembly Structure Fires and Dollar Loss

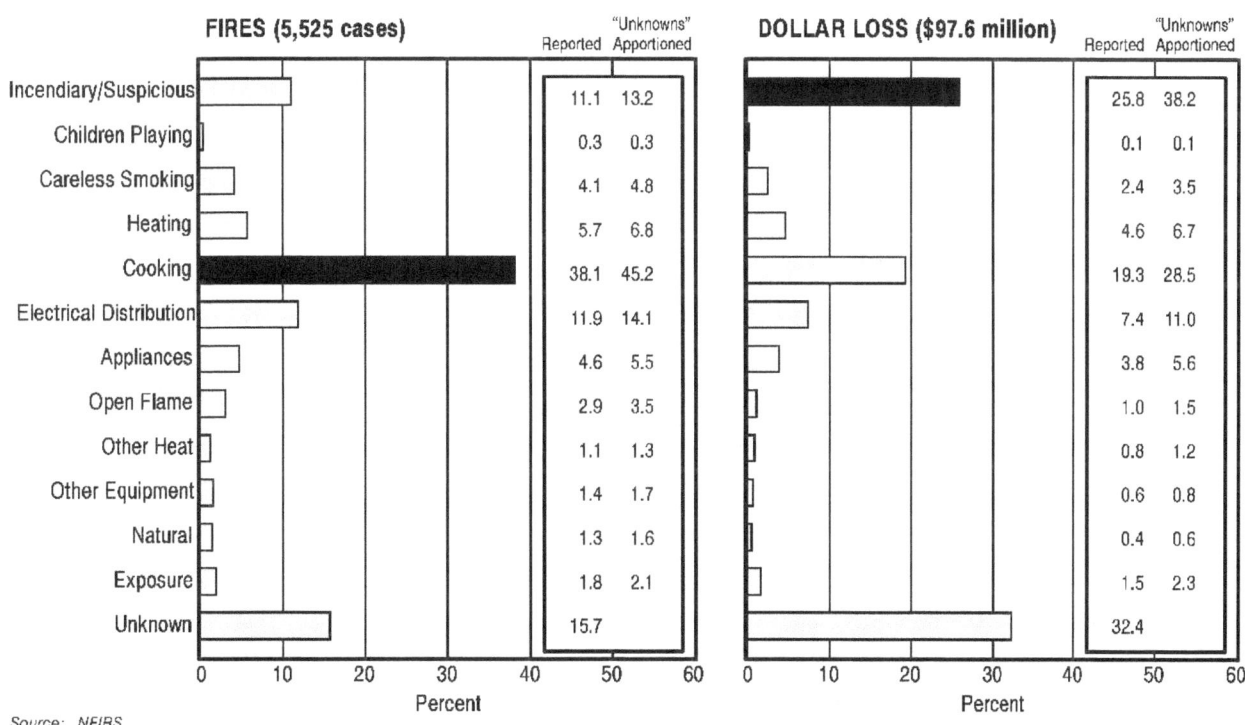

Figure 80. Causes of 1994 Eating and Drinking Establishment Fires and Dollar Loss

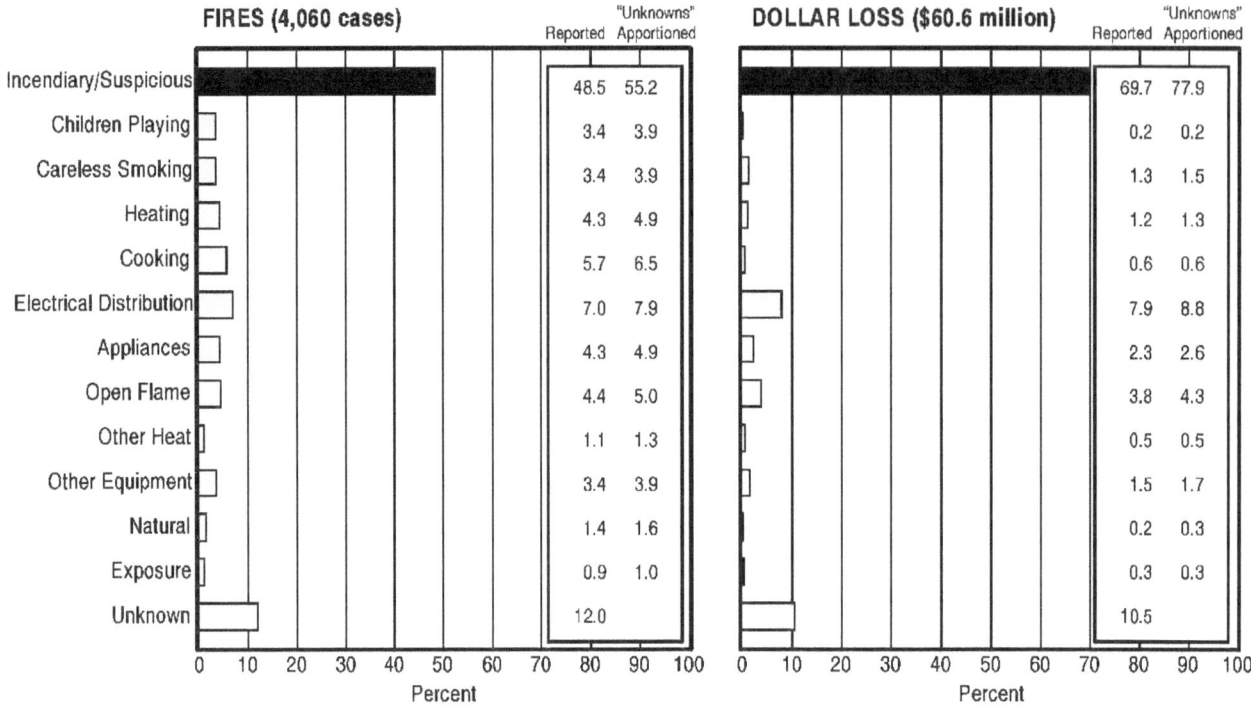

Figure 81. Causes of 1994 Education Structure Fires and Dollar Loss

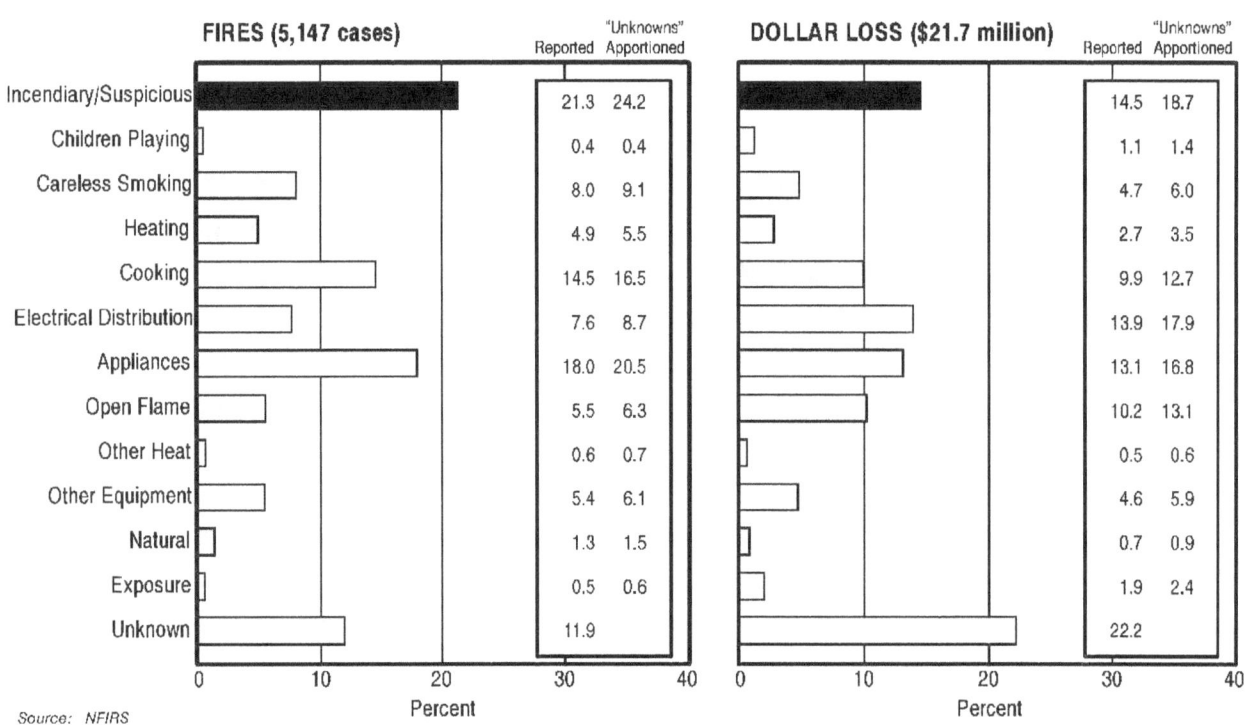

Figure 82. Causes of 1994 Institutional Structure Fires and Dollar Loss

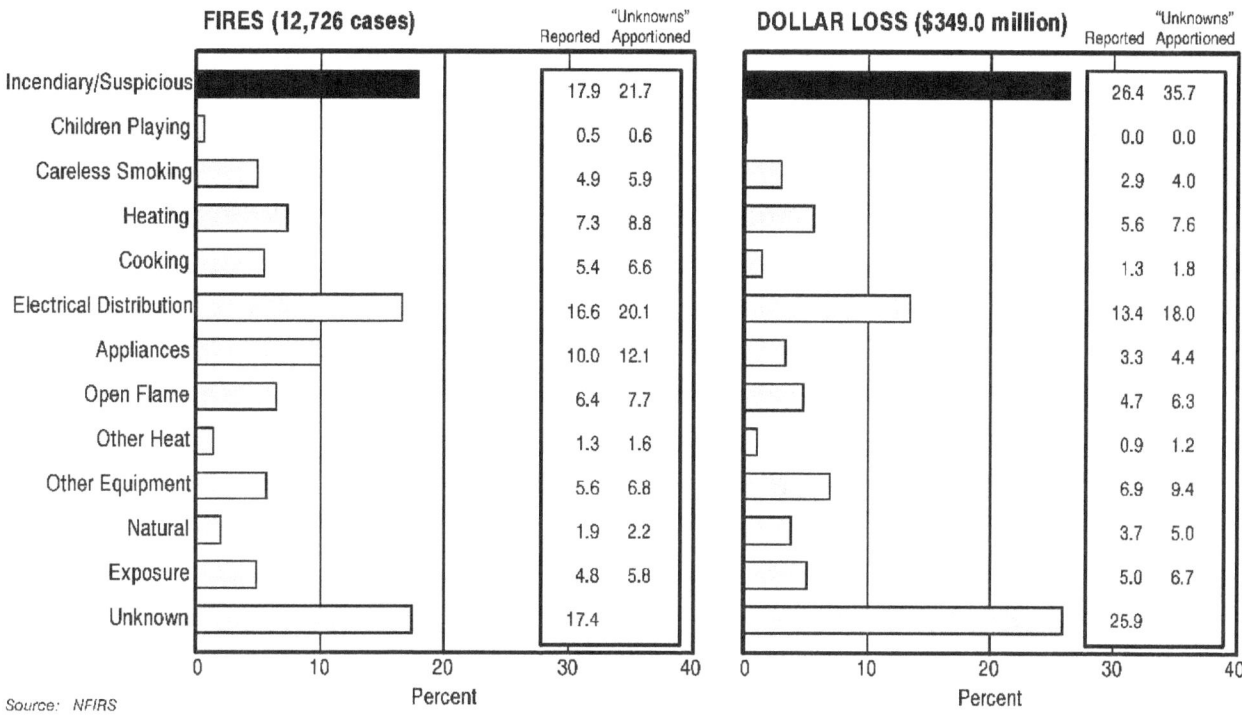

Figure 83. Causes of 1994 Store and Office Fires and Dollar Loss

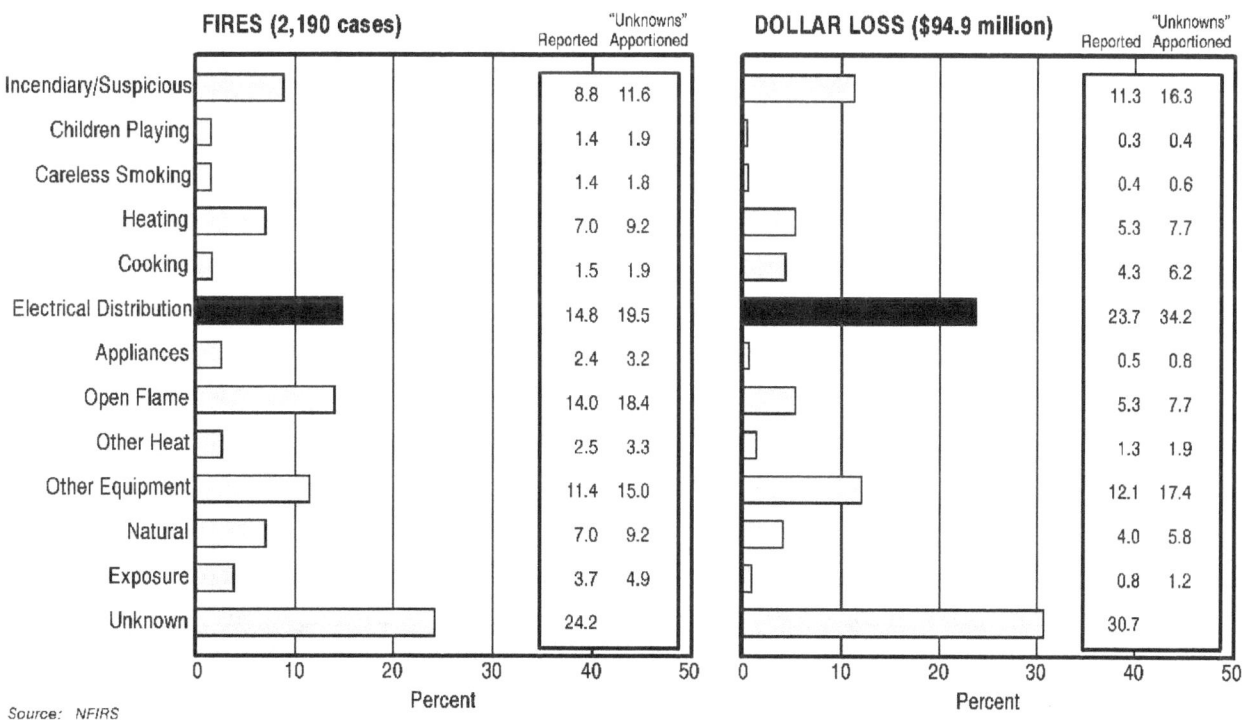

Figure 84. Causes of 1994 Basic Industry Structure Fires and Dollar Loss

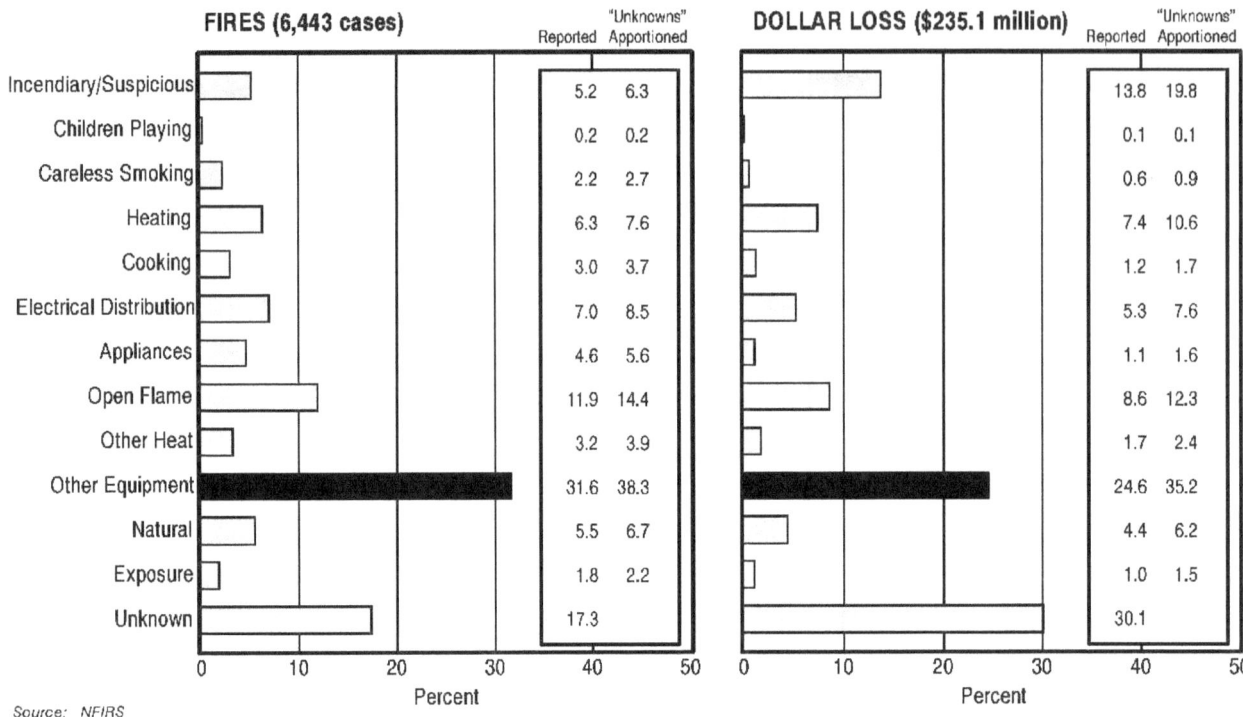

Figure 85. Causes of 1994 Manufacturing Structure Fires and Dollar Loss

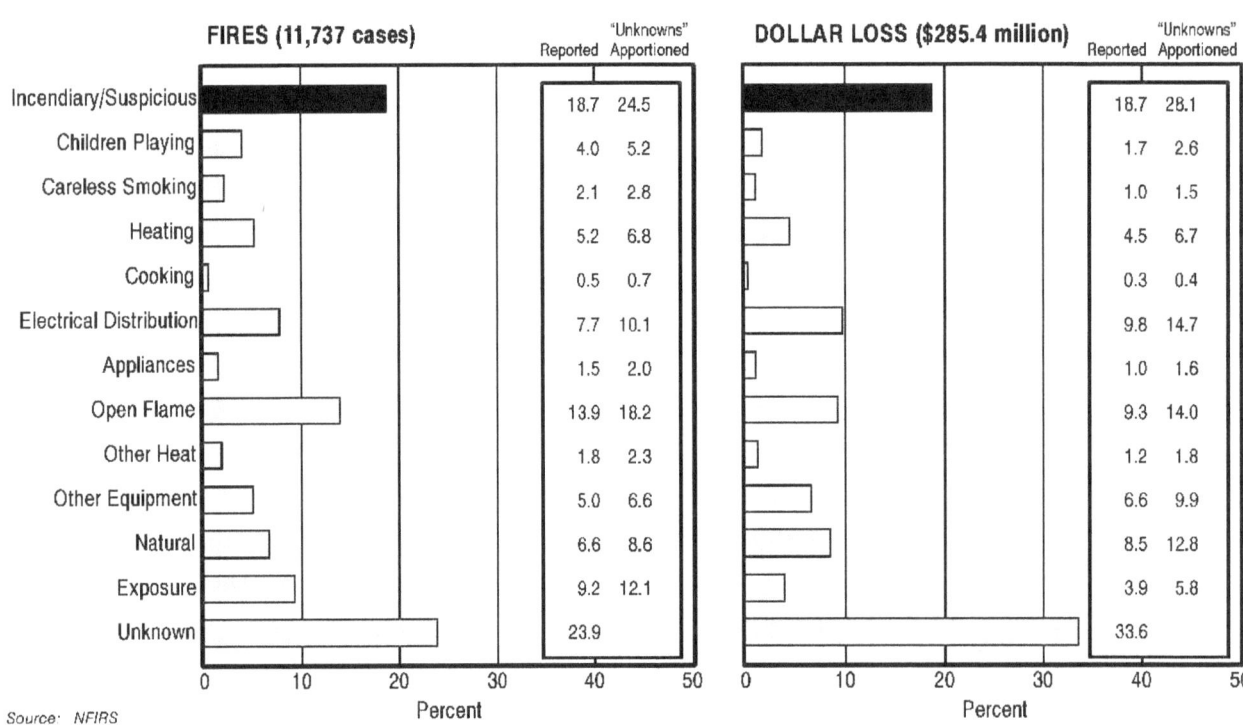

Figure 86. Causes of 1994 Storage Structure Fires and Dollar Loss

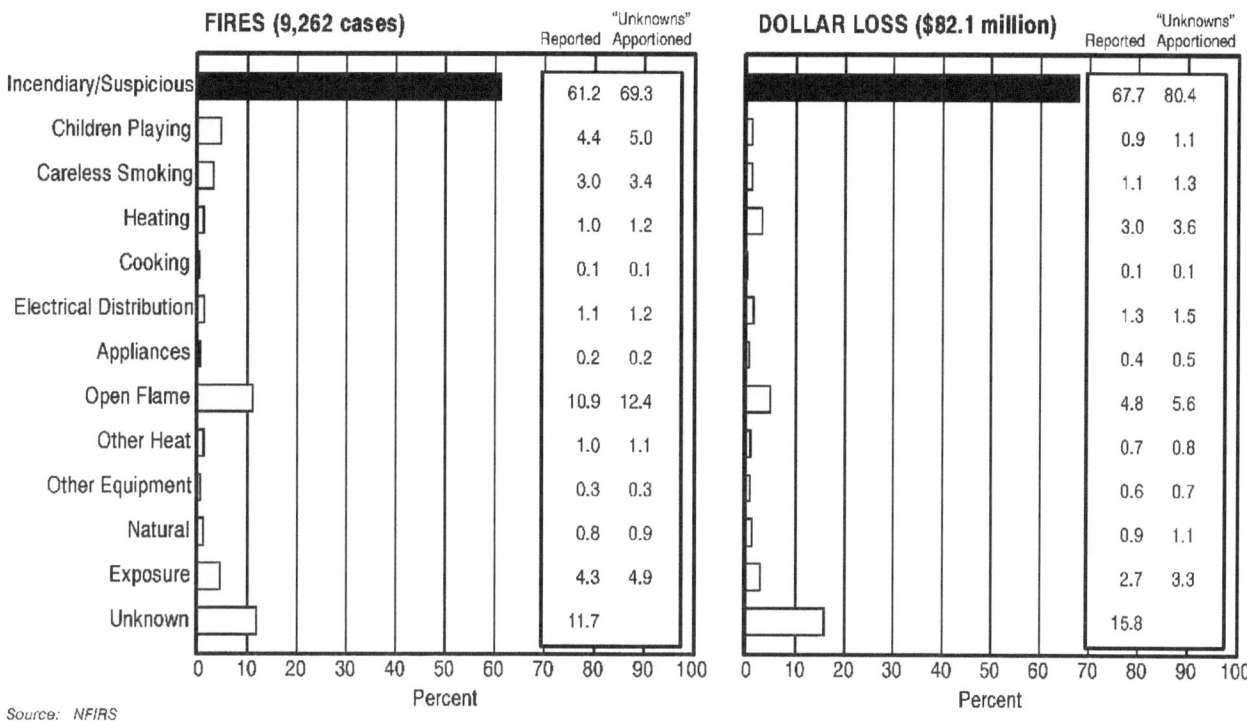

Figure 87. Causes of 1994 Vacant and Construction Structure Fires and Dollar Loss

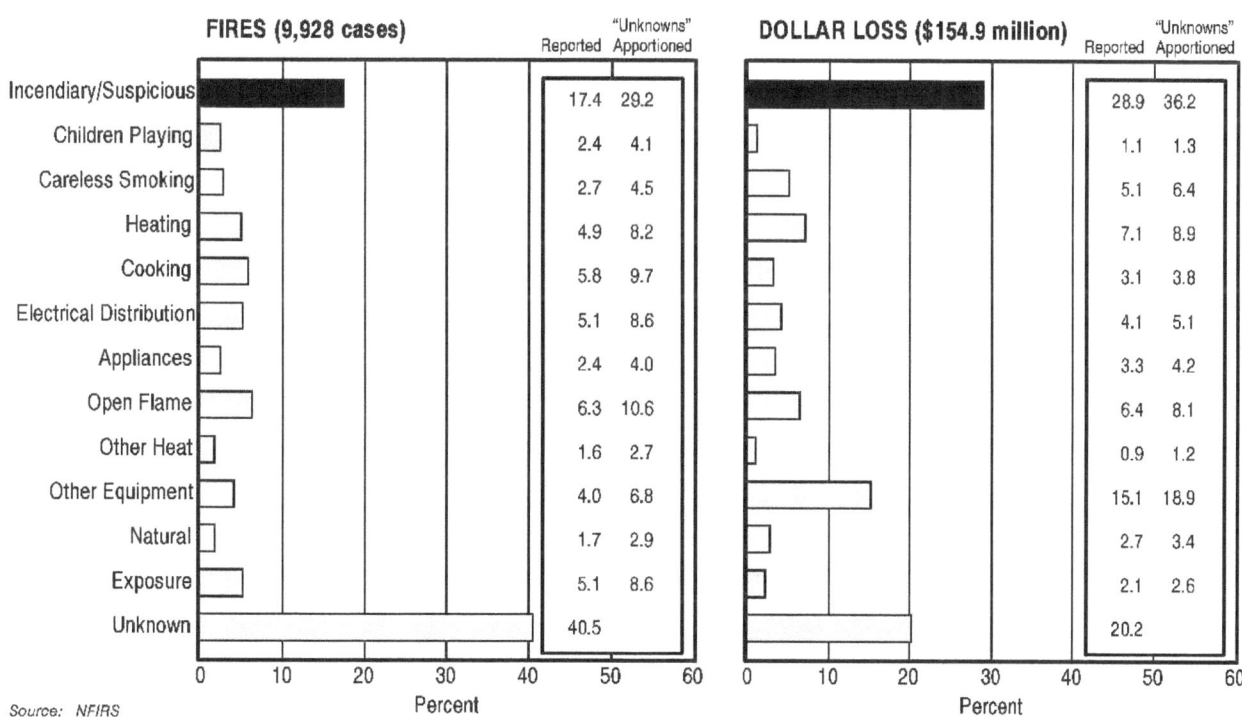

Figure 88. Causes of 1994 Outside Structure and Unknown Fires and Dollar Loss

Sprinkler Performance

Sprinkler systems with partial or complete coverage were reported being present in just 13 percent (unadjusted) of all non-residential structure fires in 1994 (Figure 89). Nevertheless, an encouraging development is that sprinkler installation is 23 percent more prevalent in 1994 than 4 years ago (from 10 to 13 percent overall). This is an increase of about 43 percent from 10 years ago and a 26 percent increase since 1990 (Figure 90). With unknowns allocated, the actual percentage of sprinkler systems present in 1994 fires might be as high as 17 percent.

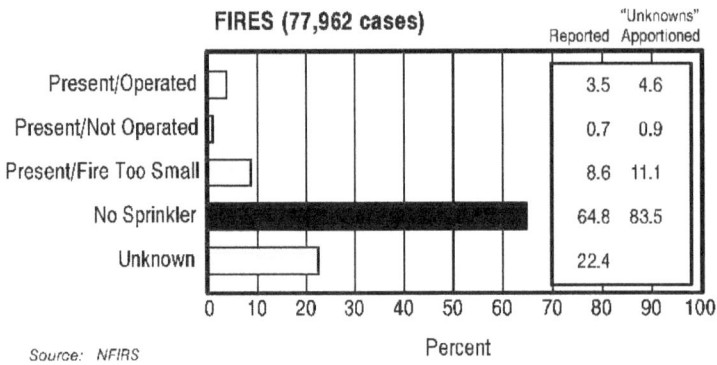

Source: NFIRS

Figure 89. Sprinkler Performance in 1994 Non-Residential Structure Fires

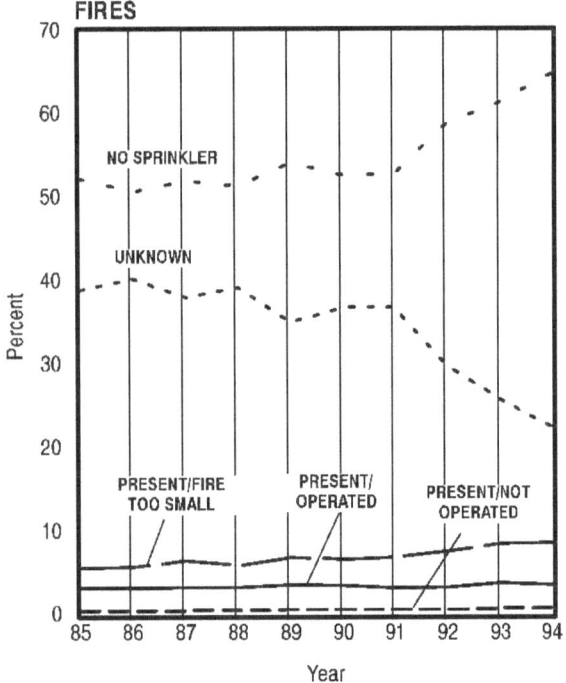

Source: NFIRS

	Present/ Operated	Present/ Not Operated	Present/ Fire Too Small	No Sprinkler	Unknown
1985	3.1%	0.4%	5.5%	52.1%	38.9%
1986	3.1	0.4	5.7	50.6	40.2
1987	3.2	0.5	6.4	51.9	38.0
1988	3.2	0.5	5.8	51.3	39.2
1989	3.5	0.6	6.8	54.1	35.0
1990	3.4	0.6	6.6	52.6	36.8
1991	3.1	0.6	6.9	52.8	36.7
1992	3.2	0.6	7.6	58.9	29.7
1993	3.7	0.7	8.5	61.4	25.7
1994	3.5	0.7	8.6	64.8	22.4

Figure 90. Trends in Sprinkler Performance in Non-Residential Fires

Over the past 10 years, sprinklers were reported to have operated in only 3−4 percent of fires. When sprinklers were present, two out of every three fires were too small to set them off or were in a part of the building away from the sprinklered area.

How effective are sprinklers? It is hard to tell from the NFIRS data alone because the comparisons need to be made for similar properties with similar fire loads, with and without sprinklers. Since NFIRS combines properties of different size and values in the same fixed property class, the data need to be viewed cautiously. Sprinkler systems are more likely to be installed in large and highly valued properties than in small, inexpensive ones. The sprinkler system in a large warehouse may do an excellent job of containing a fire and yet the loss for the fire may be larger than for a fire in an unsprinklered small storage building.

One way around this problem is to compare losses when sprinklers were present and operated versus when they were present and did not operate for a reason other than the fire being too small (that is, the cases where the sprinkler failed or the fire was not near the sprinklered area). The presumption is that the places with sprinklers, whether they went off or not, are more similar to each other than to the places that did not have sprinklers. Figure 91 shows that when sprinklers operated the losses per fire were only about one-half what they were when sprinklers were present and did not operate. This suggests that sprinklers are highly effective.

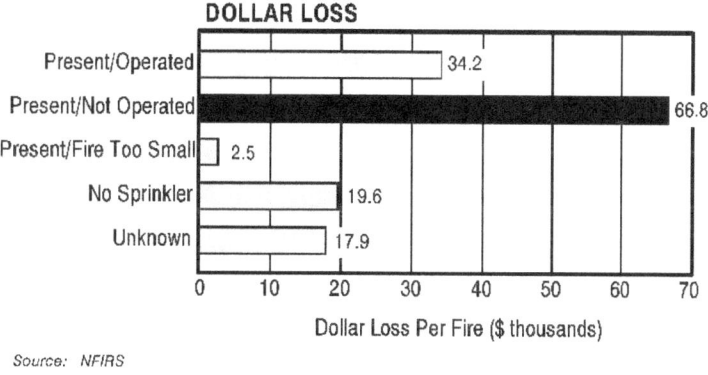

Figure 91. Sprinkler Performance in 1994 Non-Residential Dollar Loss Per Fire

VEHICLES AND OTHER MOBILE PROPERTIES

Vehicles and other types of mobile properties include all means of transportation. They account for a larger portion of the fire problem than most people realize. From 1985 to 1994, vehicles have averaged 16 percent of fire deaths, 11 percent of fire injuries, 12 percent of fire losses, and 24 percent of all reported fires, nearly one in four fires. These percentages are somewhat lower than reported in the Eighth Edition of *Fire in the United States* (1990), which is an encouraging trend.

The vast majority of fires, casualties, and property loss from mobile property involves cars and trucks, with cars clearly dominating this group. Fire departments go to about as many fires involving vehicles as they do involving residences.

Overview of Trends

The trends in fires, fire deaths, injuries, and property loss are shown in Figure 92. Mobile property total fires decreased 12 percent over the 10-year period 1985–1994 according to the NFPA annual surveys. Mobile property fire deaths and injuries trended downward sharply (26 percent and 19 percent, respectively). Mobile property loss decreased by 2 percent.

Types of Vehicles

Figure 93 shows that the vast majority of mobile property deaths, fires, injuries, and dollar loss are highway vehicles. The complexity and ambiguity in counting plane and boat fires associated with accidents are described in a later section titled "Special Data Problems."

Although the 10-year trend in highway vehicle fires, deaths, and injuries show substantial decreases (from 12 to 29 percent), the dollar loss is modest (down 4 percent). The increases in the death and injury trends in the "other" category are due to the small numbers involved.

Figure 94 gives more details on the relative proportions of the reported fire problem by type of vehicle. Automobiles and other passenger vehicles such as vans and buses outnumber trucks by three to one in fire deaths and four to one in both injuries and property loss. And automobiles have eight times as many fires as trucks. On a per-incident basis, trucks have the more serious problem, but there are vastly more car fires than truck fires.

Ignition Factors

For the most part, vehicle fires have one of four origins: the aftermath of a collision, the result of a mechanical failure, the result of an act of carelessness, or the result of arson.

In 1994, most vehicle fire deaths (63 percent) follow collisions, even though collisions are the cause of only 2 percent of vehicle fires (Figure 95). These numbers are largely unchanged from 1990. Preventing such fires is largely the purview of the U.S. Department of Transportation, state and local motor vehicle agencies, and the police, but fire departments are almost always called to the scene when there is a fire or the potential of a fire.

Adjusted for the unknowns, 69 percent of all fires in vehicles and 43 percent of the associated injuries come from mechanical or design problems such as broken fuel lines, faulty catalytic converters, overheating, etc.

Fires of incendiary or suspicious origin account for one in six automobile and mobile property fires. Many vehicle fires are not even investigated for arson, though some insurance companies are at least investigating the most suspicious or obviously incendiary fires before paying insurance claims. However, the arson problem may well be understated from the noninvestigation of these fires.

138

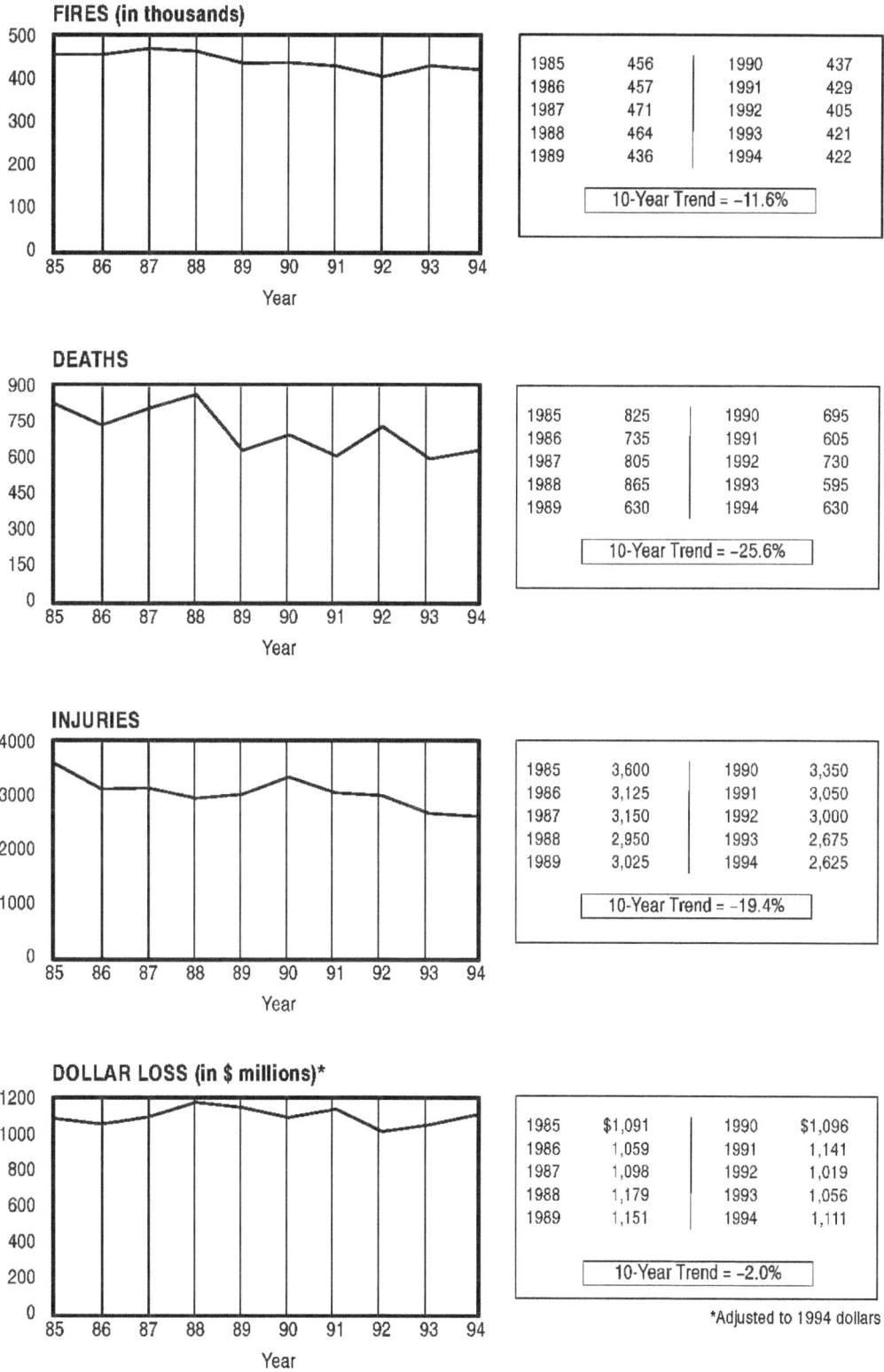

FIRES (in thousands)

1985	456	1990	437
1986	457	1991	429
1987	471	1992	405
1988	464	1993	421
1989	436	1994	422

10-Year Trend = –11.6%

DEATHS

1985	825	1990	695
1986	735	1991	605
1987	805	1992	730
1988	865	1993	595
1989	630	1994	630

10-Year Trend = –25.6%

INJURIES

1985	3,600	1990	3,350
1986	3,125	1991	3,050
1987	3,150	1992	3,000
1988	2,950	1993	2,675
1989	3,025	1994	2,625

10-Year Trend = –19.4%

DOLLAR LOSS (in $ millions)*

1985	$1,091	1990	$1,096
1986	1,059	1991	1,141
1987	1,098	1992	1,019
1988	1,179	1993	1,056
1989	1,151	1994	1,111

10-Year Trend = –2.0%

*Adjusted to 1994 dollars

Sources: NFPA Annual Surveys and Consumer Price Index

Figure 92. Trends in Mobile Property Fires and Fire Losses

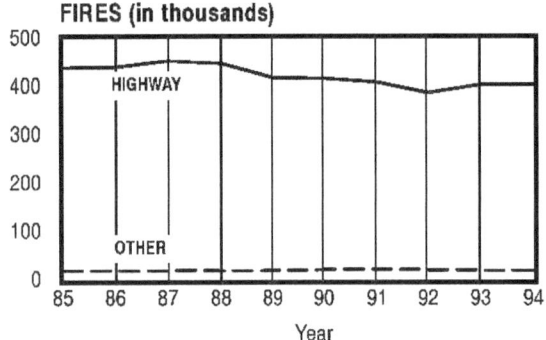

FIRES (in thousands)

	Highway	Other		Highway	Other
1985	437	19	1990	415	22
1986	438	19	1991	407	22
1987	451	20	1992	386	20
1988	446	19	1993	402	19
1989	416	20	1994	402	20

10-Year Highway Trend = –12.4%
10-Year Other Trend = +6.6%

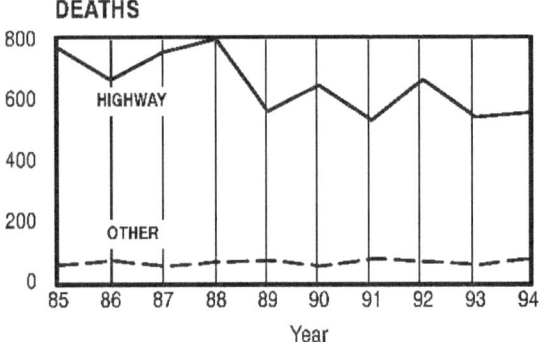

DEATHS

	Highway	Other		Highway	Other
1985	770	55	1990	645	50
1986	665	70	1991	530	75
1987	755	50	1992	665	65
1988	800	65	1993	540	55
1989	560	70	1994	555	75

10-Year Highway Trend = –28.7%
10-Year Other Trend = +14.9%

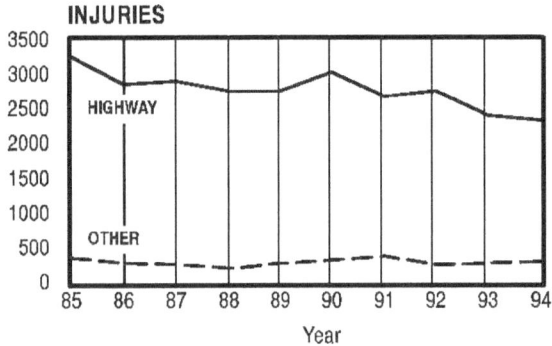

INJURIES

	Highway	Other		Highway	Other
1985	3,250	350	1990	3,025	325
1986	2,850	275	1991	2,765	375
1987	2,900	250	1992	2,750	250
1988	2,750	200	1993	2,400	275
1989	2,750	275	1994	2,325	300

10-Year Highway Trend = –21.4%
10-Year Other Trend = +2.4%

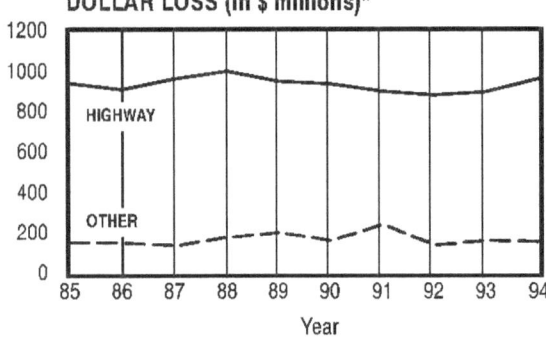

DOLLAR LOSS (in $ millions)*

	Highway	Other		Highway	Other
1985	$ 941	$150	1990	$936	$161
1986	910	149	1991	900	242
1987	963	136	1992	881	138
1988	1,000	179	1993	897	159
1989	950	201	1994	961	150

10-Year Highway Trend = –3.6%
10-Year Other Trend = +7.9%

*Adjusted to 1994 dollars

Sources: NFPA Annual Surveys and Consumer Price Index

Figure 93. Trends in Highway vs. Other Mobile Property Fires and Fire Losses

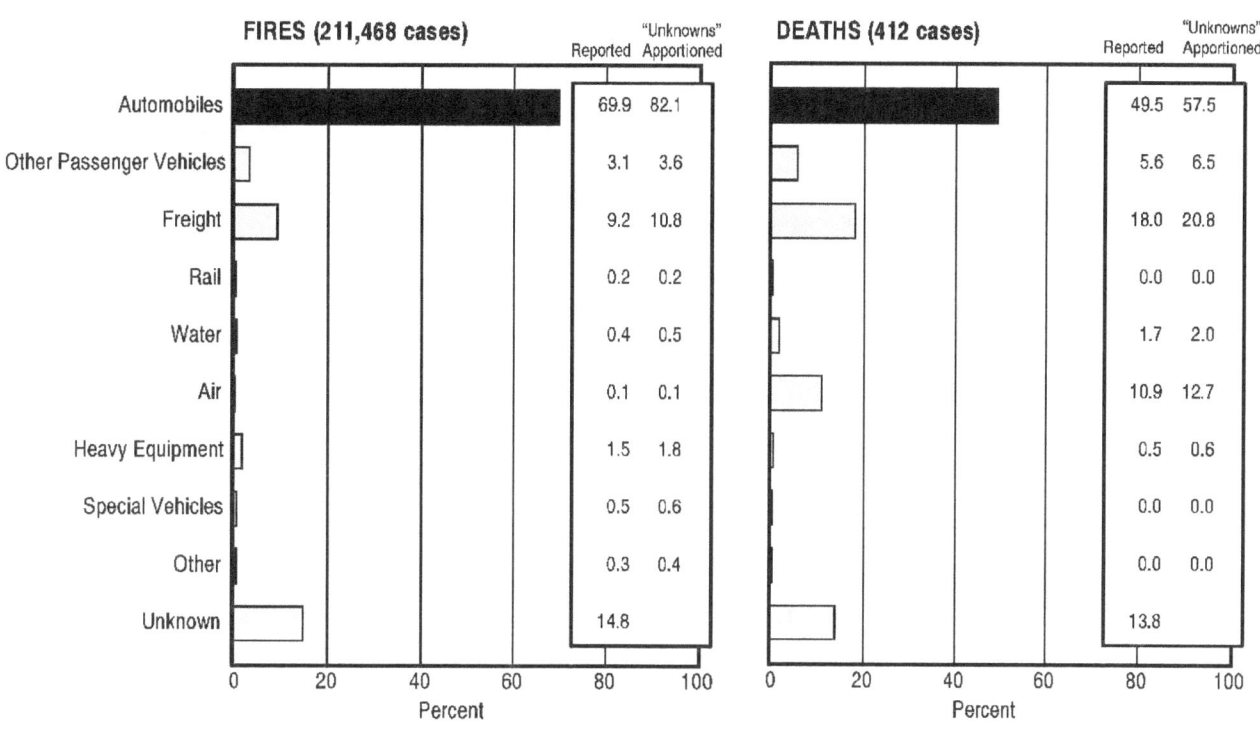

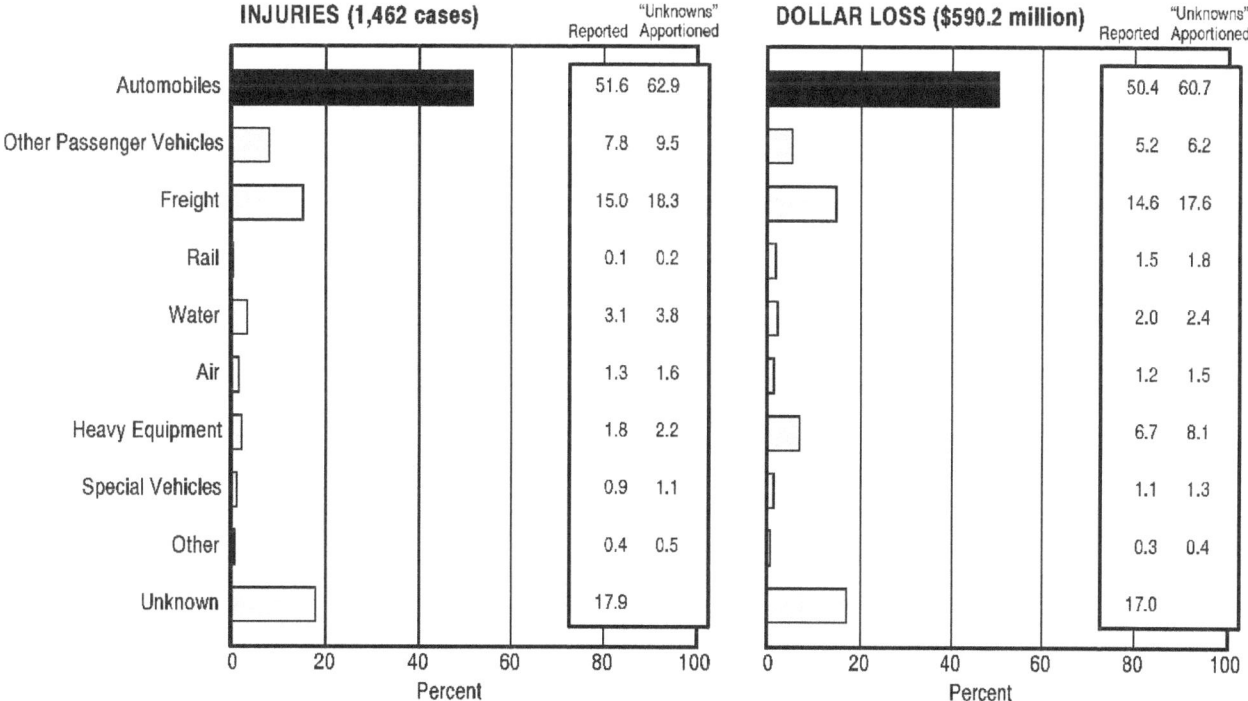

Source: NFIRS

Figure 94. 1994 Mobile Property Fires and Fire Losses by Vehicle Type

141

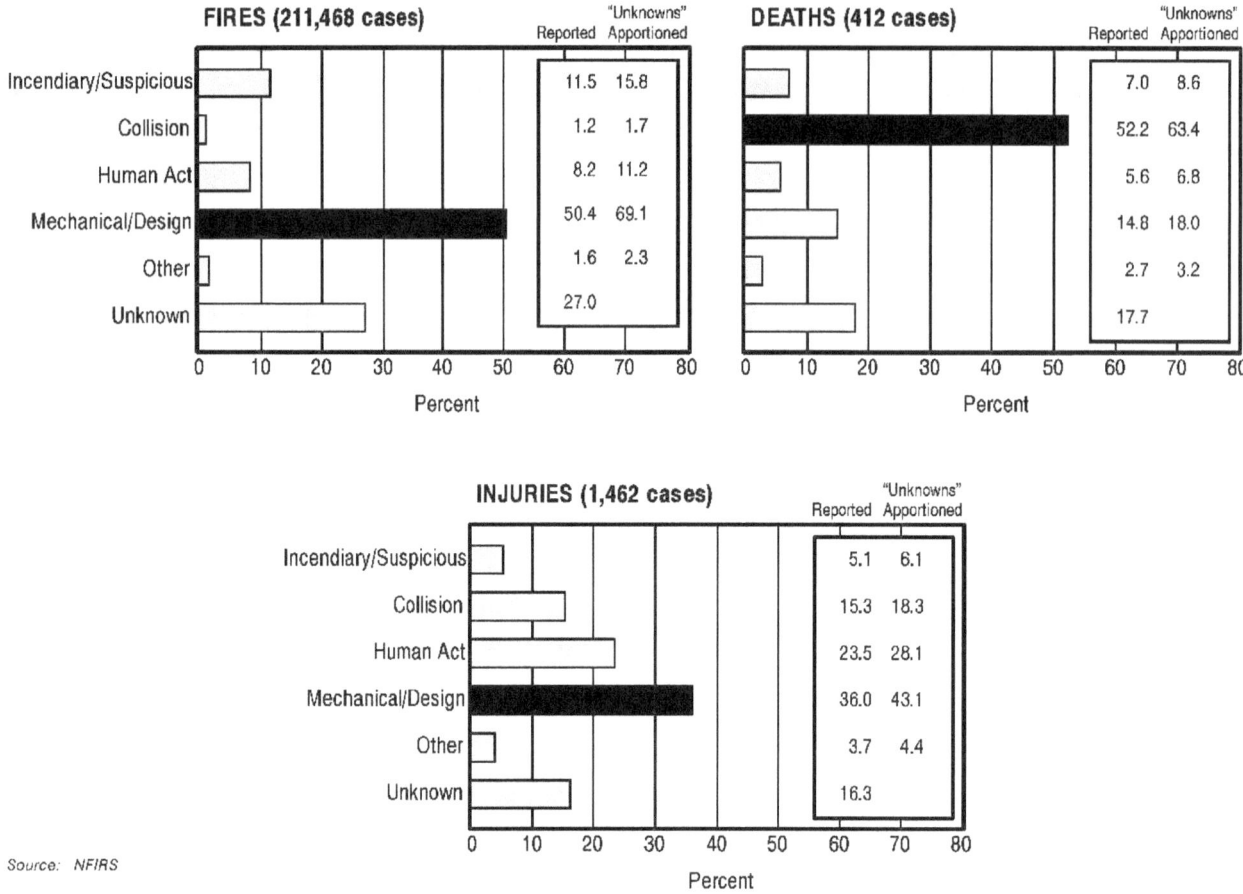

Figure 95. Ignition Factors for 1994 Mobile Fires and Fire Casualties

Carelessness (human act), including causes such as dropped or discarded cigarettes on the upholstery, parking over dry leaves with a hot catalytic converter, and misuse of flammable liquids, especially gasoline, while servicing or maintaining the car, is another major cause of vehicle fires. Carelessness accounts for 28 percent of vehicle fire injuries, though only 11 percent of fires.

In each of the past 10 years, the top ignition factor for fires (mechanical/design), deaths (collision), and injuries (mechanical/design) has remained the same—by a wide margin (Figure 96). The upward trend of deaths from collisions was interrupted in 1989 which saw a sharp drop, but deaths increased again in 1990. Injuries from mechanical or design factors reached their highest level in 1993 but dropped to their lowest level in 1994. The 10-year arson-related trend has remained somewhat stable.

FIRES (in thousands)

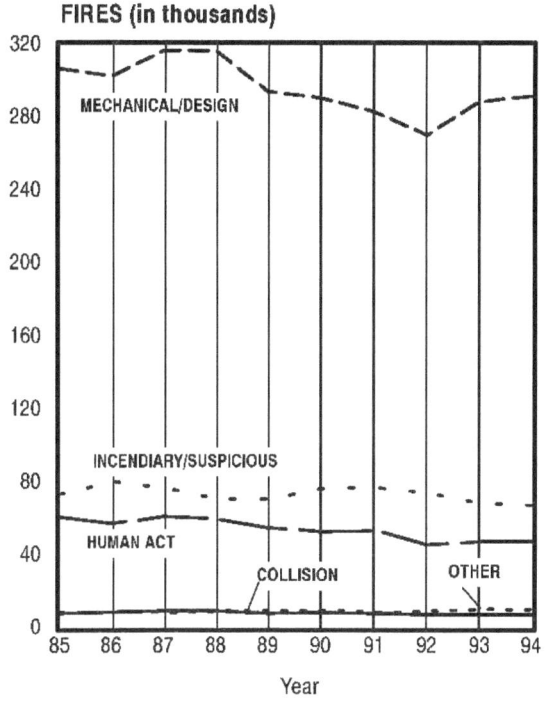

	Incendiary/Suspicious	Collision	Human Act	Mechanical Design	Other
1985	73.0	7.9	60.3	306.3	8.1
1986	80.0	8.6	57.1	302.3	8.6
1987	76.4	9.1	60.9	316.5	8.2
1988	70.4	9.2	59.3	316.4	9.1
1989	70.1	7.9	54.7	293.7	9.2
1990	76.2	8.1	52.5	290.3	9.4
1991	76.9	7.6	53.1	282.6	8.4
1992	73.4	6.9	45.4	270.1	9.2
1993	68.1	6.9	46.9	288.4	10.2
1994	66.5	7.1	47.3	291.6	9.5

DEATHS

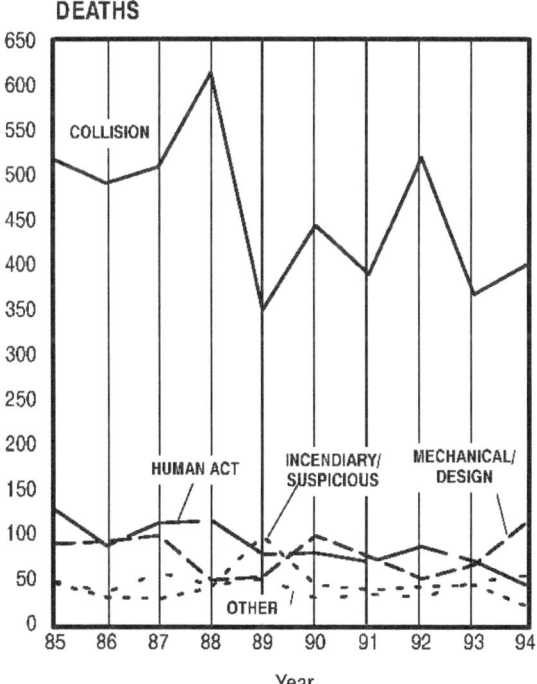

	Incendiary/Suspicious	Collision	Human Act	Mechanical Design	Other
1985	47	517	127	89	45
1986	35	491	87	92	30
1987	57	509	112	99	28
1988	44	615	116	49	41
1989	51	350	78	53	98
1990	29	444	79	98	44
1991	33	390	69	75	38
1992	30	520	86	51	42
1993	48	367	70	66	44
1994	54	400	43	113	20

Continued on next page

Figure 96. Trends in Ignition Factor Causes of Mobile Property Fires and Fire Casualties

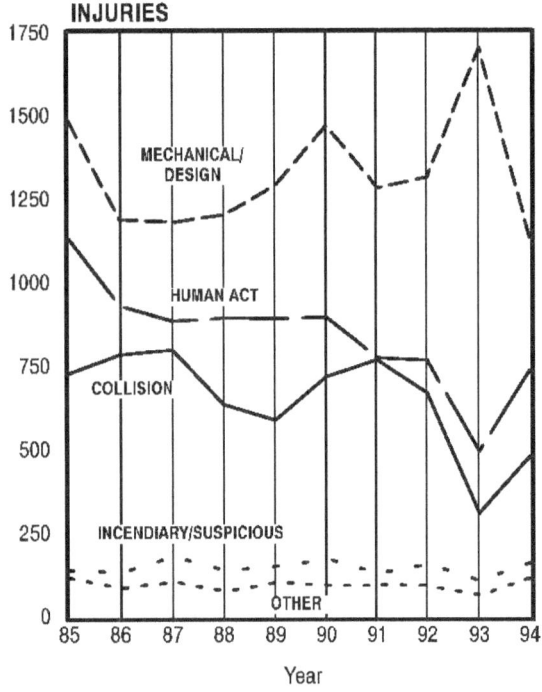

	Incendiary/ Suspicious	Collision	Human Act	Mechanical Design	Other
1985	138	728	1,131	1,486	115
1986	132	784	930	1,189	86
1987	179	799	886	1,181	105
1988	139	635	895	1,203	77
1989	151	588	892	1,292	103
1990	172	717	896	1,471	94
1991	128	769	774	1,284	95
1992	158	669	766	1,316	92
1993	107	308	494	1,703	63
1994	161	480	738	1,130	116

Sources: NFPA Annual Surveys and NFIRS

Figure 96. Trends in Ignition Factor Causes of Mobile Property Fires and Fire Casualties (cont'd)

Because automobile fires are such a large part of the entire mobile property fire problem, the cause profiles for automobile fires in 1994 are extremely similar to those for mobile properties (Figure 97 compared to Figure 95).

Fire departments might well consider adding tips on vehicle fire prevention to the rest of their prevention program in light of the large magnitude and the significant work burden that vehicle fires put on fire departments. The problem has not been given enough attention.

Special Data Problems

When there are fatalities associated with a mobile property accident such as a collision between two cars, it is often difficult to determine whether the fatalities were the result of the mechanical forces or the fire that ensued. Because of the very large number of vehicle fatalities occurring in this country each year and the frequency of fires associated with these accidents, there can be a significant error in estimating the total number of fire deaths if this problem is not carefully addressed. A fire fatality should be counted only if a person was trapped and killed by the fire, rather than killed on impact and subsequently exposed to the fire.

In plane crashes, it is thought that fewer people would die each year if the fire hazard could be reduced. It is not clear how well plane crashes are reported to NFIRS. In 1985, the NFIRS-based plane fire deaths was 57, while the number from the Federal Aviation Administration count was

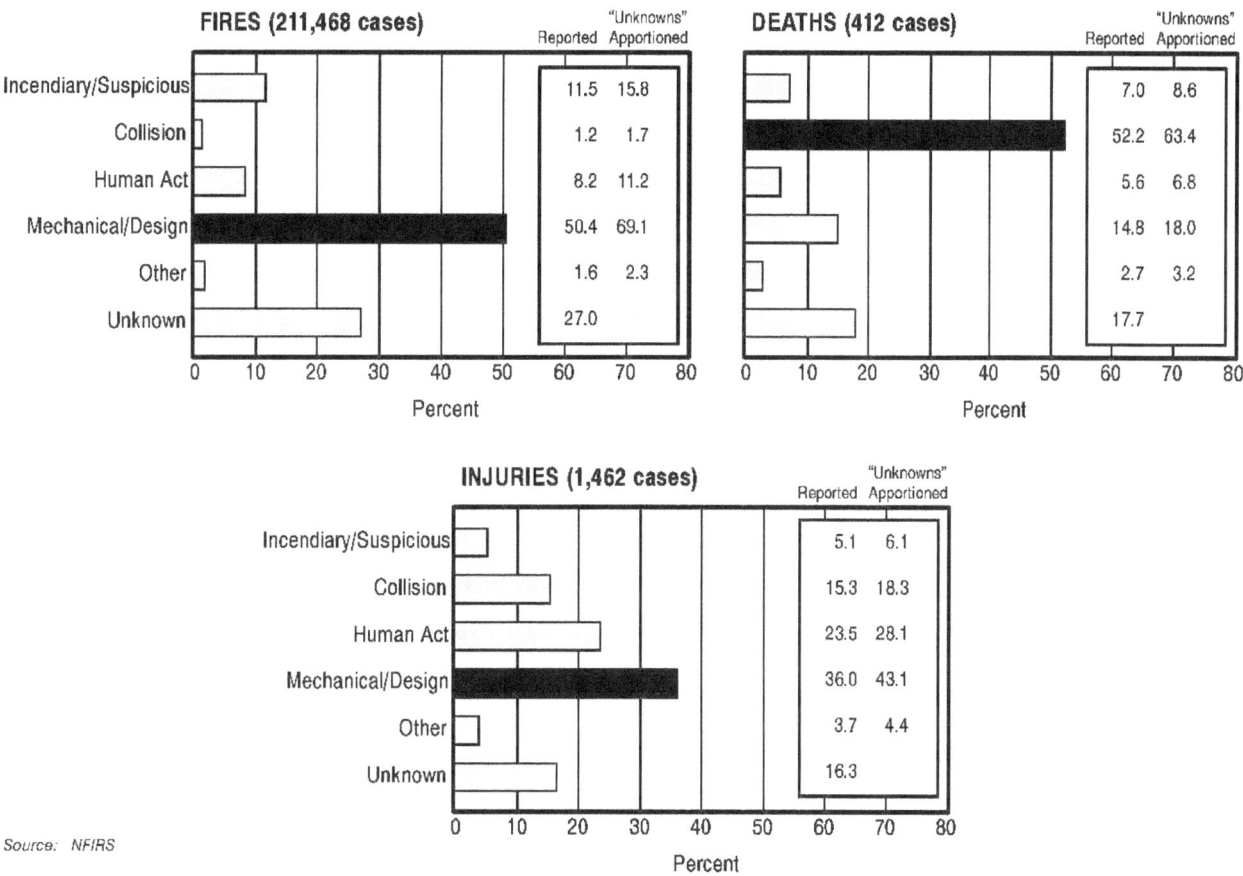

Figure 97. Ignition Factors for 1994 Automobile Fires and Fire Casualties

72—a difference that is not large. However, missing one large crash with fire fatalities could change the mobile property fire death statistics by 20 – 40 percent in a given year, so it is important that these all be reported.

OUTSIDE PROPERTIES

The "outside properties" category includes all fires outside of structures other than vehicle fires. In NFIRS terminology, this includes fires where the type of situation found either is outside of the structure where the burning material has a value or are tree, brush, grass fires, or refuse fires. Grouped in the "other" category are fires whose situation found is not classified, flammable liquid spills out of doors with ensuing fires, and explosions. A subset of outside fires (wildland fires) is discussed in Chapter 6.

Outside fires comprise almost half of all fires (44 percent in 1994) and have been increasing slowly in number and as a proportion of the total number of fires, as was seen in Figure 28, Chapter 2. While large in number, they account for only 3 percent of fire deaths, 4 percent of reported injuries, and, in most years, 3 – 5 percent of reported property loss. These numbers may not, however, reflect the true nature of the problem because of underreporting and the difficulty in setting a price tag on outside fires. Also, many wildland fires are not reported to agencies reporting to NFIRS or to the NFPA annual survey.

Overview of Trends

Figure 98 shows the trends in outside fires. The numbers of fires are enormous—from 800,000 to 1 million. Deaths from outside fires plus miscellaneous other properties not covered elsewhere number 70–130 a year; injuries range from 950 to 1,575. Deaths have a significant 10-year downward trend but this is due to the small numbers; injuries have a slight upward trend. Over the 10 years, adjusted dollar loss has continued to climb for outside properties with value, with an exceptional high agricultural loss in 1988 and high timber loss in 1992.

Estimating dollar loss for outside fires is difficult. To illustrate this problem, consider Figure 99, derived from unscaled (raw) NFIRS data. The estimates are low except for 1986, when one fire in a pineapple plantation in Hawaii was valued at half a billion dollars. (It is not reflected in NFPA data for that year.) That fire represents the largest dollar loss ever entered in NFIRS and may or may not be a fair assessment. Note that the large timber fire reported by NFPA is not reported in NFIRS.

Property Types

Figure 100 shows the relative proportions of the three components of reported outside fires for 1994. Trees, brush, and grass fires account for the most numbers of fires, deaths, injuries, and dollar loss. A majority of deaths, injuries, and dollar loss, however, are unknown. Better reporting to NFIRS is required for outside fires.

When Fires Occur

TIME OF DAY. Over half of all outside fires occur from 1:00 to 9:00 p.m. They are very low in the early hours of the morning, when few people are outside (Figure 101).

MONTH OF YEAR. Outside fires are usually lowest in the fall and winter months and high during late spring and the summer (Figure 102). In 1994, April and July were the months with the highest fire rate, whereas July and June were first and second in 1990. In recent years, local and state governments have placed more rigorous restrictions on burning leaves, which might account for the low autumn numbers. Wetter-than-usual weather, too, may have played a role. The details of the monthly variations are not obviously explained and need more study by the type of outdoor property and by cause.

DAY OF WEEK. As in 1990, outside fires in 1994 are highest on the weekend, when more people are outdoors (Figure 103). Monday was also a peak day.

FIRES (in thousands)

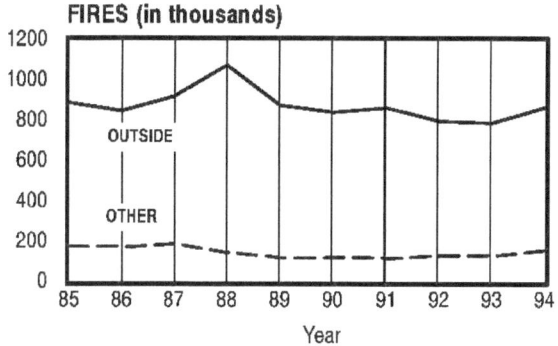

	Outside	Other		Outside	Other
1985	884.0	172.0	1990	838.5	120.0
1986	845.0	170.0	1991	859.5	113.0
1987	916.5	184.5	1992	793.5	128.5
1988	1,072.0	142.0	1993	783.5	127.0
1989	873.5	118.0	1994	861.5	157.0

10-Year Outside Trend = –11.3%
10-Year Other Trend = –26.5%

DEATHS

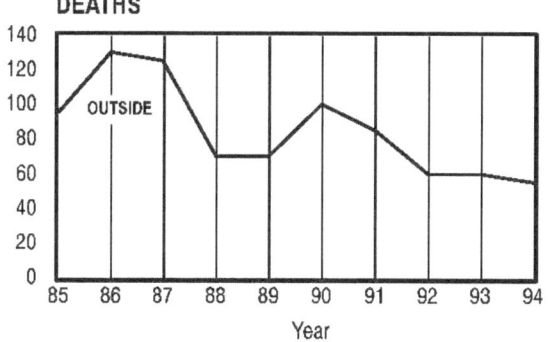

1985	95	1990	100
1986	130	1991	95
1987	125	1992	60
1988	70	1993	60
1989	70	1994	55

10-Year Outside Trend = –47.7%

INJURIES

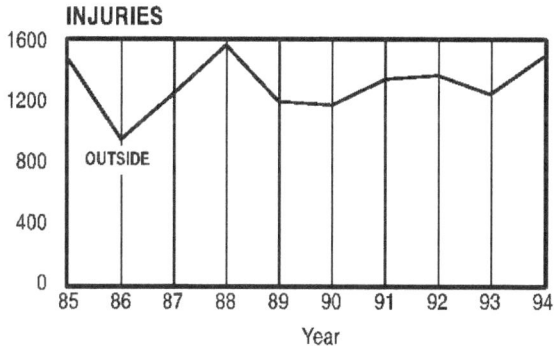

1985	1,475	1990	1,175
1986	950	1991	1,350
1987	1,250	1992	1,375
1988	1,575	1993	1,250
1989	1,200	1994	1,500

10-Year Outside Trend = +9.8%

DOLLAR LOSS (in $ millions)*

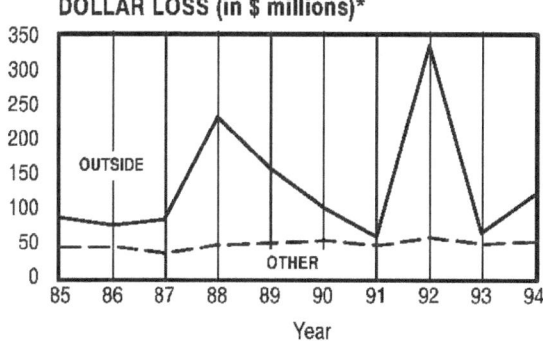

	Outside	Other		Outside	Other
1985	$ 86.8	$44.1	1990	$102.1	$54.4
1986	75.7	44.6	1991	59.8	46.8
1987	83.5	35.2	1992	335.9	58.1
1988	231.8	47.6	1993	64.6	48.2
1989	157.8	50.2	1994	120.0	53.0

10-Year Outside Trend = +46.5%
10-Year Other Trend = +28.7%

*Adjusted to 1994 dollars

Sources: NFPA Annual Surveys and Consumer Price Index

Figure 98. Trends in 1994 Outside and Other Property Type Fires and Fire Losses

DOLLAR LOSS (in $ millions)*

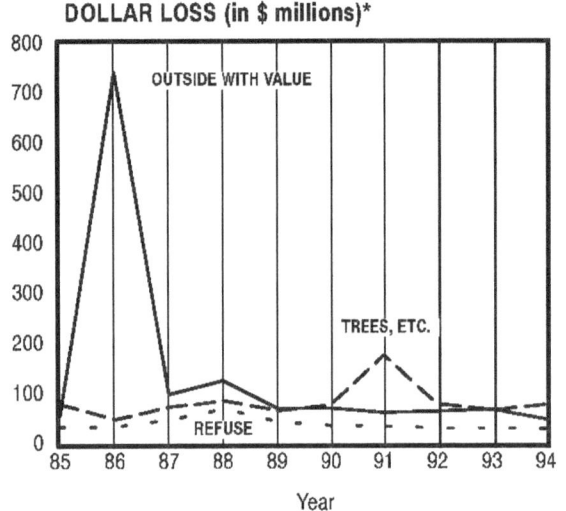

	Outside With Value	Trees, Etc.	Refuse
1985	$ 54.7	$ 77.2	$ 31.9
1986	740.4	47.4	33.7
1987	96.8	72.7	46.5
1988	125.1	84.1	68.1
1989	69.7	65.2	41.6
1990	69.1	76.8	34.5
1991	61.3	175.2	33.7
1992	63.0	77.1	29.4
1993	67.5	65.5	29.4
1994	46.4	77.0	27.5

10-Year Outside With Value Trend = –100.0%
10-Year Trees, Etc. Trend = +34.0%
10-Year Refuse Trend = –32.0%

Sources: NFIRS and Consumer Price Index

*Adjusted to 1994 dollars

Figure 99. Trends in Outside Fire Dollar Loss by Property Type

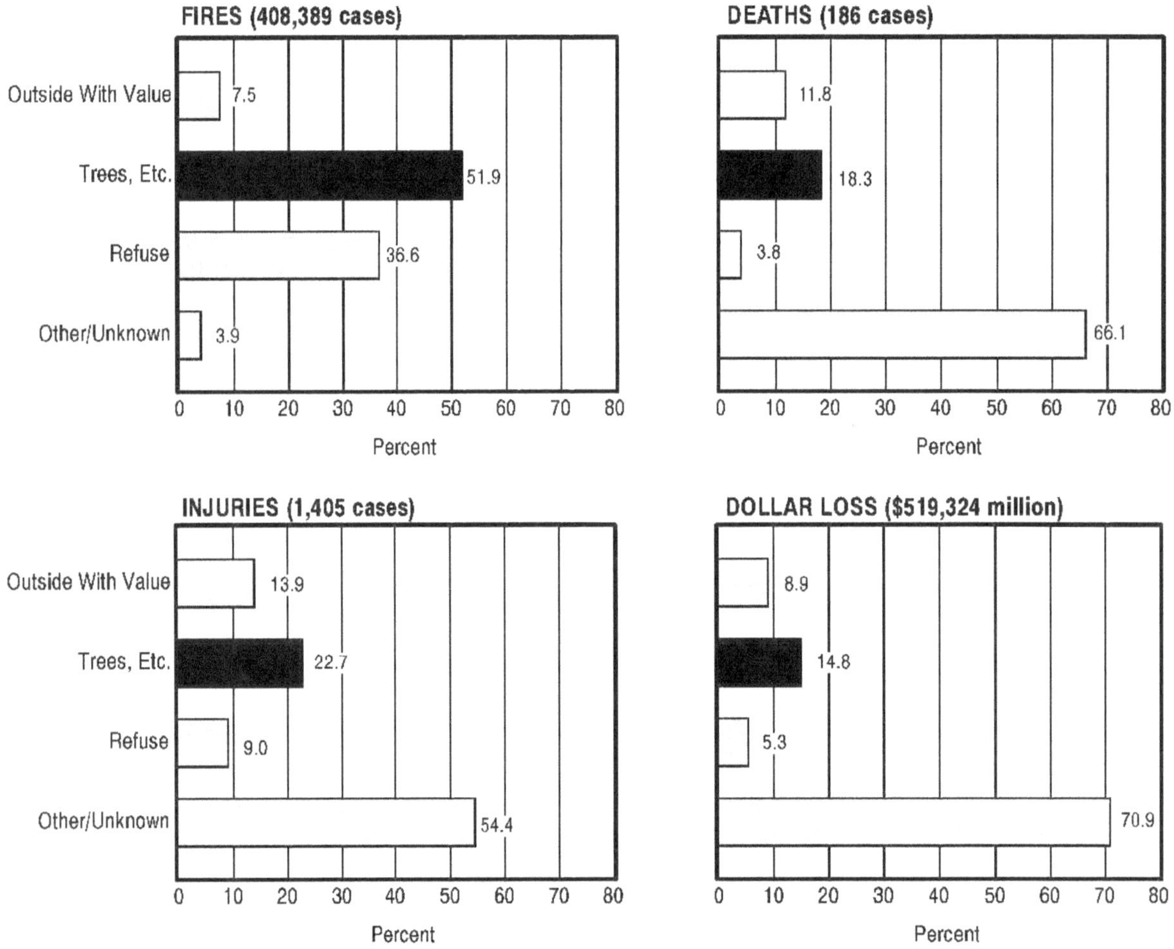

FIRES (408,389 cases)

	Percent
Outside With Value	7.5
Trees, Etc.	51.9
Refuse	36.6
Other/Unknown	3.9

DEATHS (186 cases)

	Percent
Outside With Value	11.8
Trees, Etc.	18.3
Refuse	3.8
Other/Unknown	66.1

INJURIES (1,405 cases)

	Percent
Outside With Value	13.9
Trees, Etc.	22.7
Refuse	9.0
Other/Unknown	54.4

DOLLAR LOSS ($519,324 million)

	Percent
Outside With Value	8.9
Trees, Etc.	14.8
Refuse	5.3
Other/Unknown	70.9

Source: NFIRS

Figure 100. 1994 Outside Fires and Fire Loss by Property Type

FIRES (408,389 cases)

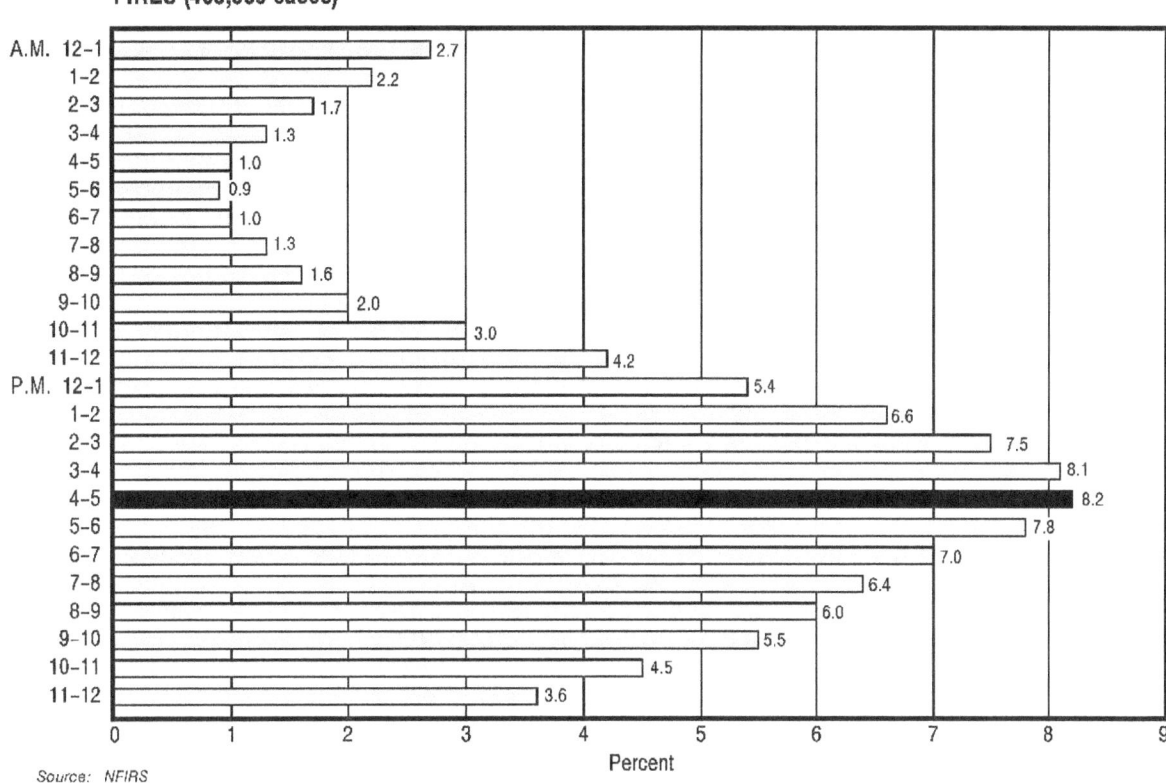

Source: NFIRS

Percent

Figure 101. Time of Day of 1994 Outside and Other Fires

FIRES (408,389 cases)

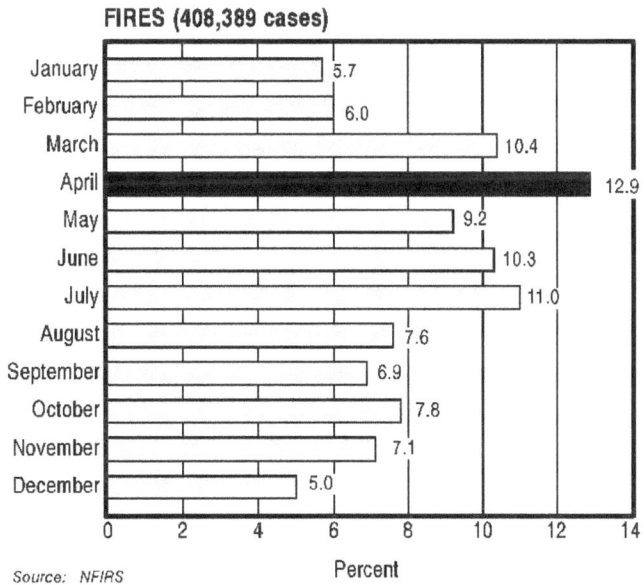

Source: NFIRS

Percent

Figure 102. Month of Year of 1994 Outside and Other Fires

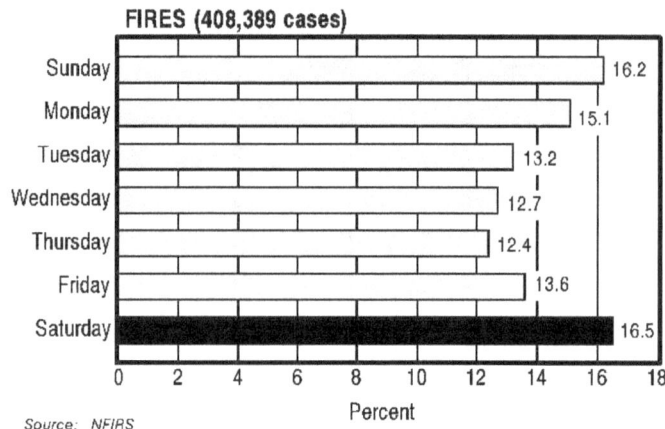

Figure 103. Day of Week of 1994 Outside and Other Fires

Causes

The leading cause of outside fires is arson, with many thought to be set by children. Figure 104 shows the cause profiles for each outside fire category. Again, a large percentage of outside fires have unknown causes.

For outside fires with value, 34 percent of fires are thought to be arson. The rest of the fires are scattered across many categories, with open flame and electrical problems in second and third place, respectively.

More than half of tree, brush, and grass fires had unknown cause. Among the known causes, the two that stand out are arson and open flame, which includes open fires used for cooking. These two causes account for over half of the fires with cause. Following these, but at much lower rates, are children playing and careless smoking. This is the same pattern as was reported in 1990.

For refuse fires, again over half had no reported cause. As in 1990, over half of the reported causes in 1994 were reported as arson, with another 17 percent from open flame (e.g., matches), 11 percent from children playing, and 7 percent from smoking (e.g., a discarded cigarette). Note that refuse fires set inside buildings are structural fires, even if they do no damage, and are reported as part of the property type in which they occur.

Special Data Problems

Setting a value for outside fire damage is a perennial problem. Although it is difficult to assign a dollar value to grass, tree, and rubbish fires, the damage from these fires often requires labor beyond that of the fire department to clean up and restore the area and also causes esthetic problems that are an intangible. Also, some outside fires spread to structural properties and may be reported as structural fires rather than an outside fire with exposure to structures.

Forest fires and other wildfires to which local departments are not called will not be reported to NFIRS if the state or federal agency with principal authority for fighting the fire is not participat-

150

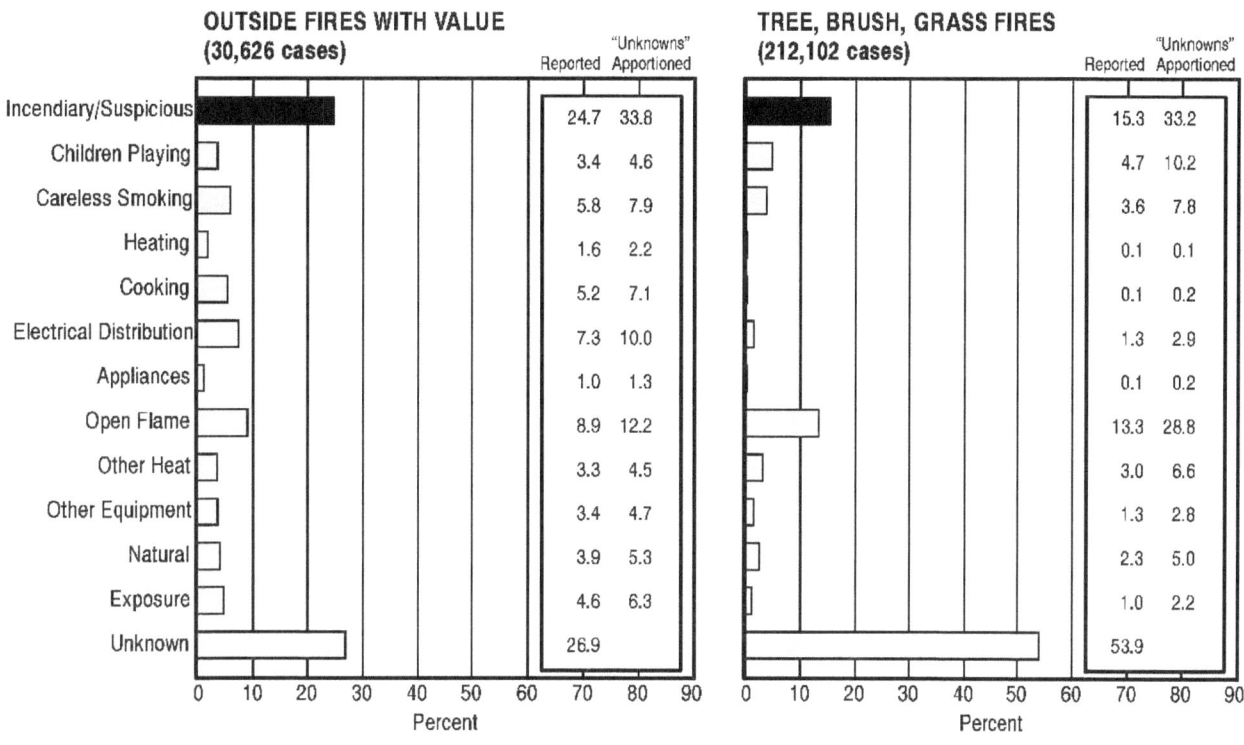

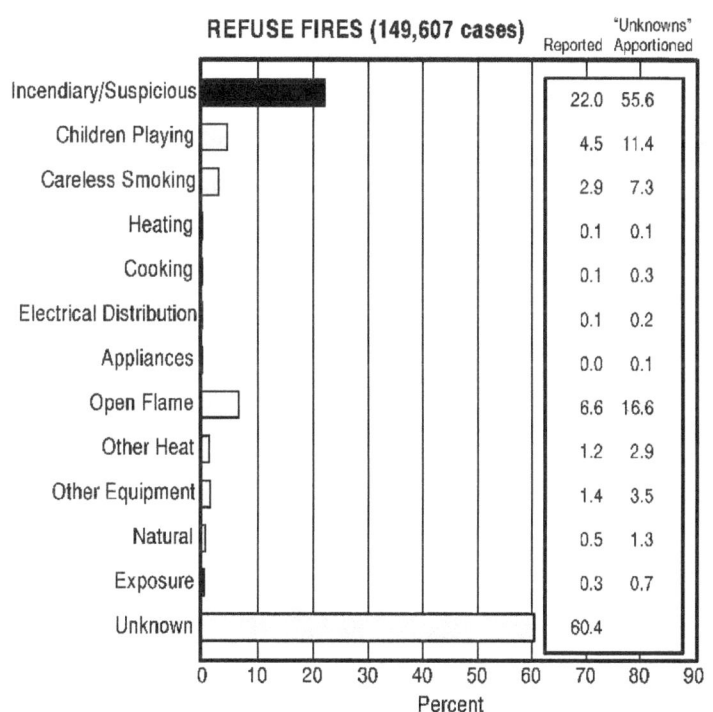

Source: NFIRS

Figure 104. Causes of 1994 Outside Fires by Type

ing in NFIRS. Data from the Departments of Agriculture and Interior are needed to complete the picture.

Another significant problem with data on outside fires is determining their cause. Often the area of origin is obliterated, the people involved have fled, and one is not sure exactly what caused the fire—an unattended campfire, a discarded match or cigarette, lightning strikes, children playing, or even an intentionally set fire. Thus the percent of causes left as unknown is especially high for this category of fires.

USFA RESOURCES ON FIRES IN NON-RESIDENTIAL STRUCTURES

USFA conducts special studies to address specific problems and current issues facing the nation's fire and rescue service. The technical reports produced under the Major Fires Investigations series analyze major or unusual fires with emphasis on sharing lessons learned. They are directed primarily to chief fire officers, training officers, fire marshals, and investigators as a resource for training and prevention.

Major Fire Investigations reports on fires in non-residential properties include the following: *Evacuation of Nanticoke, Pennsylvania Due to Metal Processing Plant Fire, March 1987* (#005); *Fire and Explosions at Rocket Fuel Plant, Henderson, Nevada, May 1988* (#021); *Industrial Plastics Fire Major Triage Operation, Flint, Michigan, November 1988* (#025); *High-Rise Office Building Fire, One Meridian Plaza, Philadelphia, Pennsylvania, February 1991* (#049); *Indianapolis Athletic Club Fire (Two Firefighter Fatalities), February 6, 1991; Major Propane Gas Explosion and Fire, Perryville, Maryland, July 1991* (#053); *Twenty-five Fatality Fire at Chicken Processing Plant, Hamlet, North Carolina, September 1991* (#067); *New York City Bank Building Fire: Compartmentation vs. Sprinklers, January 1993* (#071); *The World Trade Center Bombing: Report and Analysis* (#076); *Four Firefighters Die in Seattle Warehouse Fire, Seattle, Washington* (#077); *California Interstate Bank Building Fire, Los Angeles, May 1988; California—Conservative Approach to Chemical Plant Fire, Ventura County, April 1989; Georgia—Five-Fatality High-Rise Office Building Fire, Atlanta, June 1989; Massachusetts—Swimming Pool Chemical Plant Fire, Springfield, June 1988; Ohio— Sherwin–Williams Paint Warehouse Fire, Dayton, May 1987;* and *Texas—Philips Petroleum Chemical Plant Explosion and Fire, Pasadena, October 23, 1989.*

These publications are available by writing to:

U.S. Fire Administration
Federal Emergency Management Agency
Publications Center, Room N310
16825 S. Seton Avenue
Emmitsburg, MD 21727

Documents may also be ordered via the World Wide Web: *http://www.usfa.fema.gov.* USFA publications are free.

5

FIREFIGHTER CASUALTIES

There has been much progress in reducing on-duty firefighter deaths and injuries over the 10 years 1985–1994 (down 35 percent). Since 1988, firefighter deaths have dropped from 136 to 104. Deaths in 1992 and 1993 reached all-time lows at 75 and 78, respectively. Injuries declined 2 percent and ranged from 95,400 (1994) to 103,300 (1991). These deaths and injuries include casualties from fires, training, and all other on-duty activities. This chapter addresses only casualties associated with fire incidents, with emphasis on information available from NFIRS data. NFIRS does not collect data on injuries from training or non-fire incidents.

Previous editions of *Fire in the United States* focused primarily on firefighter injuries. This edition examines both firefighter deaths and injuries in separate sections.

DEATHS

In 1994, 104 firefighters died while on duty.[1] This is a significant rise after two consecutive years that produced the lowest number of firefighter fatalities since the USFA began keeping records. The increase is attributable primarily to wildland fire fatalities, up from 8 deaths in 1993 to 38 deaths in 1994, including 14 firefighters killed in a single incident on Storm King Mountain in Colorado. Even with this exceptional incident, the total of 104 fatalities is still the third lowest number of fatalities recorded in the 18 years that these data have been collected and continues the long-term trend of reduced fatalities that began in 1979.

Figure 105 shows firefighter deaths increased 33 percent from 1993 to 1994 (26 additional deaths) The number of fatalities involving wildland firefighters increased by 30 deaths from the previous year. The fatalities by type and gender of firefighter are presented in Table 15.

Of the 38 deaths associated with wildland firefighting, 22 occurred on the fireground and 14 involved personnel en route to wildland fires. The other two deaths involved career individuals who

---

[1] A total of 107 firefighter fatalities were reported in 1994. Three of these deaths were attributed to incidents that occurred in prior years. Two firefighters died in 1994 of complications from prior injuries, one from an accident in 1979, and one from a heart attack in 1991. Another firefighter died of AIDS believed to have been contracted while performing emergency medical services in the 1980s. Since the exact date of exposure was not documented, his death has been attributed as a 1991 statistic, the year in which he was first diagnosed with HIV. Those deaths have been added to the statistics for the year in which the incident occurred, consistent with past USFA analyses. The data from two other incidents that occurred in 1994 are included in this analysis, although the firefighters actually died in 1995.

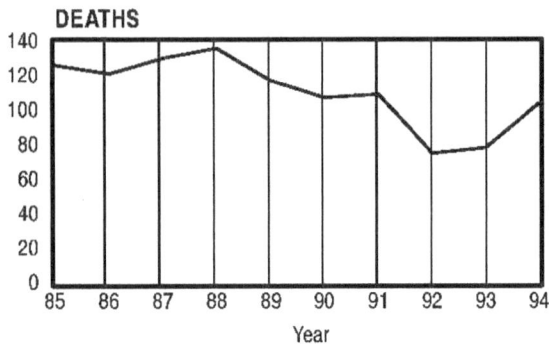

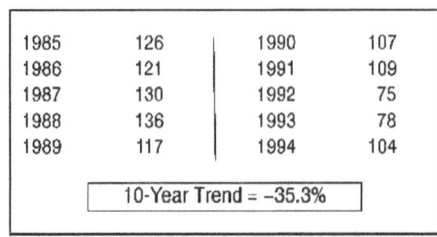

1985	126	1990	107
1986	121	1991	109
1987	130	1992	75
1988	136	1993	78
1989	117	1994	104

10-Year Trend = –35.3%

Sources: NFPA Annual Surveys and the United States Fire Administration

Figure 105. Trends in 1994 Firefighter Deaths

were deployed to positions primarily dedicated to wildland fire protection, although their deaths were from injuries sustained while performing other duties.

Table 15. 1994 Firefighter Deaths

Firefighter Type/Gender	Fatality
Firefighter	
Volunteer	38
Career	34
Wildland Firefighter	
Seasonal	20
Career	4
Contract Aircraft Crew	6
Industrial Emergency Response Team Members	1
Civilians Assigned to Military Reserve Units	1
Men	97
Women	7

Source: United States Fire Administration

The 36 deaths attributed to wildland fireground operations or response to wildland fires is considerably higher than in any previous year that has been analyzed. Fireground deaths at wildland incidents have varied considerably from year to year, but have averaged between 8 and 10 in the most recent years. Although the loss of 14 firefighters in the Storm King Mountain incident represents a significant portion of the total deaths for the year, the overall increase is significant even if this event were excluded from the total.

Activity and Type of Duty

In 1994, 84 percent of firefighter on-duty deaths were associated with an emergency incident (Figure 106). This number includes firefighters who died responding to an emergency, at the emergency scene, and returning from the emergency.

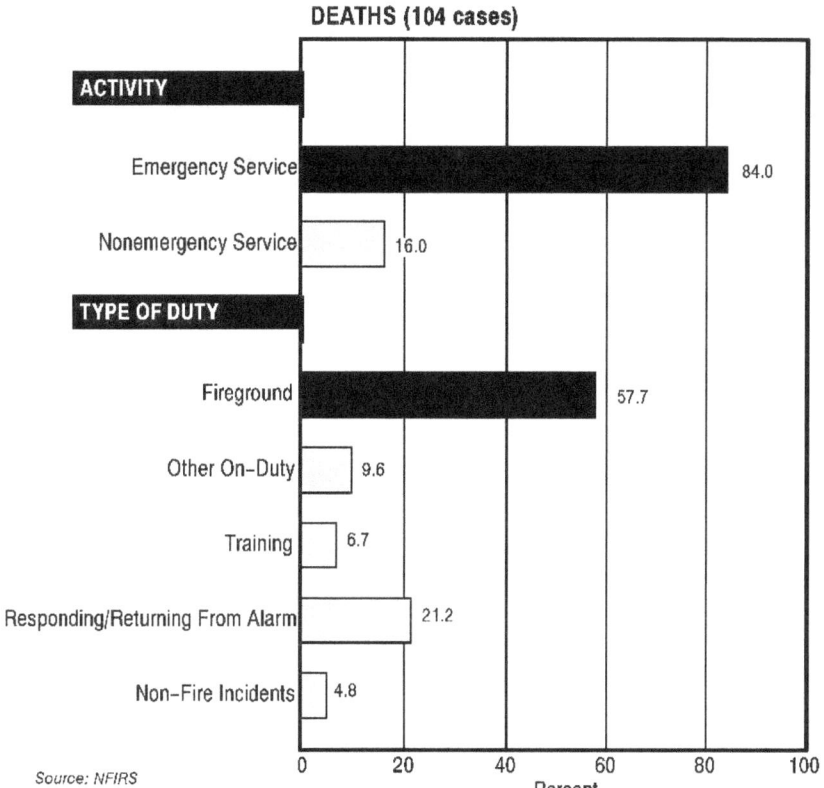

DEATHS (104 cases)

ACTIVITY

Emergency Service — 84.0

Nonemergency Service — 16.0

TYPE OF DUTY

Fireground — 57.7

Other On–Duty — 9.6

Training — 6.7

Responding/Returning From Alarm — 21.2

Non–Fire Incidents — 4.8

Source: NFIRS

Percent

Figure 106. 1994 Firefighter Deaths by Activity and Type of Duty

The remaining 16 percent of firefighter fatalities included training and administrative activities and other functions not related to an emergency. Two on-duty firefighters died in their sleep of heart attacks, and one volunteer firefighter suffered a fatal heart attack doing clerical work at the firehouse.

Firefighter deaths by type of duty is also shown in Figure 106. As in previous years, the largest number of deaths resulted from fireground operations, 58 percent. Of the 60 fireground deaths, 29 were attributed to smoke inhalation,[2] 8 to burn injuries, and 3 to trauma (including two pilots who were killed when their plane crashed while dropping retardant on a wildfire). Nineteen heart attack deaths were attributed to fireground operations, including one that occurred after returning from a fire. One firefighter was electrocuted and one died of an embolism after being injured in a fall at a fire.

As in 1993, the second leading category of firefighter deaths after fireground operations was responding to and from emergency incidents, which accounted for 22 deaths in 1994. Two firefighters and a patient were killed while en route to the hospital when their ambulance was struck head-on by a tractor-trailer truck. Eight fatalities occurred in fire apparatus crashes, with rollovers being

---

[2] The 14 firefighters who died at Storm King Mountain were listed as smoke inhalation deaths on their death certificates.

the leading type of fatal accident. Six deaths resulted from aircraft that crashed during wildland fire-fighting operations. Five firefighters suffered fatal heart attacks while en route or returning from alarms. One volunteer firefighter was killed in a crash of a personal vehicle.

Five deaths were related to activities at the scene of emergency medical (non-fire) incidents. Two were struck by vehicles—one firefighter was killed while on an emergency medical call and the other, a fire-police officer, died while directing traffic at the fire scene. Three firefighters died of heart attacks—one after extricating a patient from an automobile accident, another after returning from an EMS incident in very hot weather, and a third after carrying a patient to an ambulance.

Seven deaths occurred during training, although none involved live fire training. Five of the training deaths were from heart attacks and one of a cerebral aneurysm. A fire management officer was killed when a 106-mm recoilless rifle exploded during avalanche control training. This training was part of his collateral duties.

Ten deaths occurred during other nonemergency duty activities.

Cause and Nature of Fatal Injury or Illness

The term *cause* refers to the action, lack of action, or circumstances that directly result in the fatal injury, while the term *nature* refers to the medical nature of the fatal injury or illness, or what is often referred to as the cause of death. The fatal injury usually is the result of a chain of events, the first of which is recorded as the cause. For example, if a firefighter is struck by a collapsing wall, becomes trapped in the debris, runs out of air before being rescued, and died of asphyxiation, the cause of the fatal injury is recorded as "struck by collapsing wall" and the nature of the fatal injury is "asphyxiation." Likewise, if a wildland firefighter is overrun by a fire and dies of burns, the cause of the death would be listed as "caught/trapped," and the nature if death would be "burns." This follows the convention used in NFIRS casualty reports, which are based on NFPA fire reporting standards.

Figure 107 shows the distribution of deaths by cause of fatal injury or illness. As in most previous years, the leading cause is stress or overexertion—36 percent in 1994. The act of firefighting has been shown to be one of the most physically demanding activities that the human body performs. Of the 37 stress-related fatalities in 1994, 36 firefighters died of heart attacks and 1 firefighter died of an aneurysm.

The second leading cause of firefighter fatalities in 1994 was "caught or trapped," accounting for 36 firefighter fatalities (35 percent). Seventeen firefighters were overrun by moving brush or wildland fires. Nine firefighters were trapped by rapidly changing fire conditions inside burning structures, and seven apparently became disoriented and died in building fires. Two firefighters died as a result of structural collapses. A fire chief who was directing operations outside of a burning garage died from burns after it exploded.

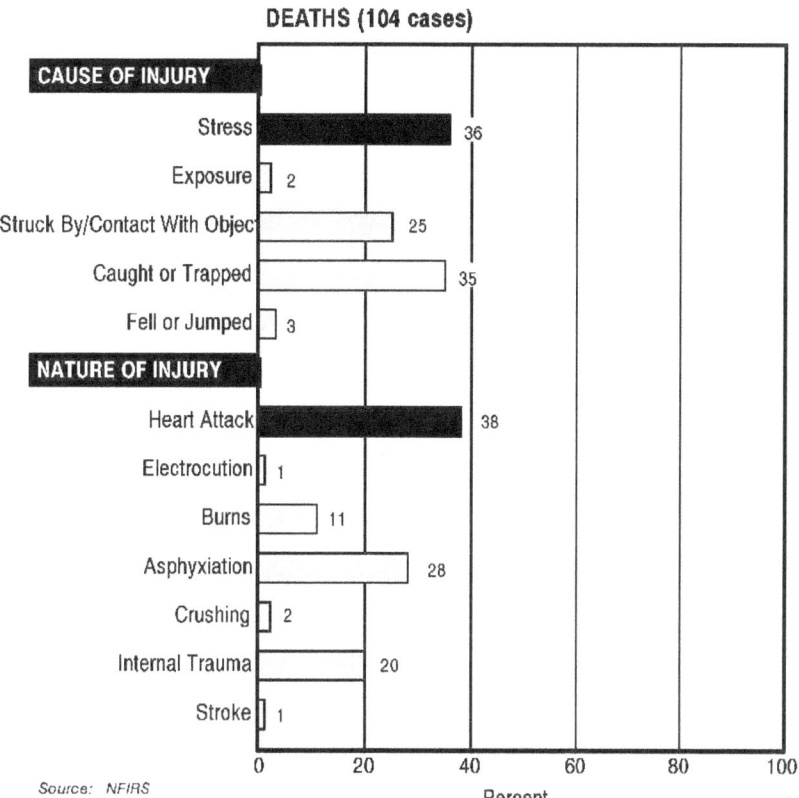

DEATHS (104 cases)

CAUSE OF INJURY

Stress	36
Exposure	2
Struck By/Contact With Object	25
Caught or Trapped	35
Fell or Jumped	3

NATURE OF INJURY

Heart Attack	38
Electrocution	1
Burns	11
Asphyxiation	28
Crushing	2
Internal Trauma	20
Stroke	1

0   20   40   60   80   100

Source: NFIRS

Percent

Figure 107. 1994 Firefighter Deaths by Cause and Nature of Injury

The third leading cause of firefighter fatalities was "struck by" or coming in "contact with an object." Of the 26 firefighters (25 percent) killed, 11 were in vehicle accidents, 8 were in aircraft accidents, 2 were struck by vehicles while at the scene of an emergency, 1 was struck by a falling tree, 1 was struck by a helicopter rotor blade, 1 was hit by shrapnel when a 106-mm recoilless rifle misfired and exploded, 1 was struck by falling debris at a fire, and 1 was electrocuted when he came in contact with an electric power line while carrying a chainsaw down an aerial ladder from the roof of a fire building.

Three firefighters (3 percent) died as a result of falls. One died when he fell through a hole in the floor while checking a room for fire extension; one firefighter, who suffered a broken ankle in a fall down stairs at a fire, died of an embolism; and one died when he fell from the roof of his fire station.

Exposure to smoke or toxic gases was listed as the causal factor of two deaths (2 percent). A chief officer suffered a fatal heart attack after breathing toxic gases while performing overhaul at a house fire. He was not wearing protective gear or SCBA. Another firefighter suffered a fatal heart attack after being overcome by heat and toxic gases while standing by at a controlled burn.

Figure 107 also shows the distribution of the 104 deaths by the medical nature of the fatal injury or illness. Thirty-nine firefighters died of heart attacks in 1994—36 by stress, 1 by an embolism, and 2 by inhalation of toxic gases on the fireground. At least 13 of the firefighters had known high-risk factors for heart attacks, including prior heart conditions, high blood pressure, or obesity, including two who continued to perform firefighting activities after bypass surgery. Autopsy results indicated some coronary artery disease present in most of these cases where medical records were available.[3] Obesity and poor physical fitness were noted as factors in the deaths of several of the heart attack victims, including a 26-year-old, 5½-foot, 275-pound firefighter who died of a heart attack during a training exercise. At least one firefighter had a genetic heart defect that would not have been discovered during a routine physical.

Asphyxiation was the second leading nature of firefighter deaths. A total of 29 firefighter deaths (28 percent) resulted from carbon monoxide poisoning or smoke inhalation: 15 from wildland fires[4] and 14 at structure fires. This includes seven firefighters who were using SCBA, but whose air supply had been depleted.

Internal trauma was responsible for 21 deaths (20 percent). This includes 19 who were involved in vehicle or aircraft accidents.

Burn injuries claimed the lives of 11 firefighters (11 percent). Five died after being caught in flashovers or backdrafts, three died after aircraft accidents, two died after being overrun by wildfires, and one died after being caught in an explosion with no protective clothing.

Two firefighter deaths (2 percent) were attributed to crushing injuries—one from falling debris from a roof collapse, and the second by a falling tree. One firefighter was electrocuted (1 percent). Another firefighter died of a stroke (1 percent).

Ages of Firefighters

Table 16 shows the distribution of firefighter deaths by age and nature and cause of death. Younger firefighters were more likely to have died after becoming caught or trapped during firefighting operations. Stress played an increased role in firefighter deaths as ages increased—all six firefighters who were over 60 years old died of heart attacks. Asphyxiation was the primary medical nature of death among younger firefighters, and heart attacks were much more prevalent among older firefighters.

Fireground Deaths

Fireground deaths in 1994 accounted for 58 percent of total deaths, an increase of 77 percent over 1993. Figure 108 shows the fireground deaths by fixed property use, including the 14 deaths that occurred on Storm King Mountain.[5]

---

[3] Autopsy results and medical records were not available for all heart attack victims.

[4] The autopsy results of the 14 Storm King Mountain fatalities established that their deaths were caused primarily by asphyxiation, secondarily by burns.

[5] Even if the 14 Storm King Mountain fatalities were removed from the statistics, the increase would still be a significant 35 percent.

Table 16. Age of Firefighter at Time of Death by Nature and Cause in 1994

	Age										Total
	Under 21	21 to 25	26 to 30	31 to 35	36 to 40	41 to 45	46 to 50	51 to 55	56 to 60	Over 60	
Cause											
Stress	0	0	1	3	5	4	6	6	6	6	37
Exposure	0	0	0	0	0	0	1	1	0	0	2
Struck/Contact With Object	3	2	5	3	2	3	1	4	3	0	26
Caught/Trapped	0	8	11	5	4	5	3	0	0	0	36
Fell or Jumped	0	0	0	1	0	1	0	1	0	0	3
Nature											
Heart Attack	0	0	1	3	5	3	7	8	6	6	39
Electrocution	0	0	0	1	0	0	0	0	0	0	1
Burns	1	3	0	3	2	0	2	0	0	0	11
Asphyxiation	0	5	11	3	3	6	1	0	0	0	29
Crushing	0	0	1	0	0	0	0	0	1	0	2
Internal Trauma	2	2	4	2	1	3	1	4	2	0	21
Stroke	0	0	0	0	0	1	0	0	0	0	1

Source: NFIRS

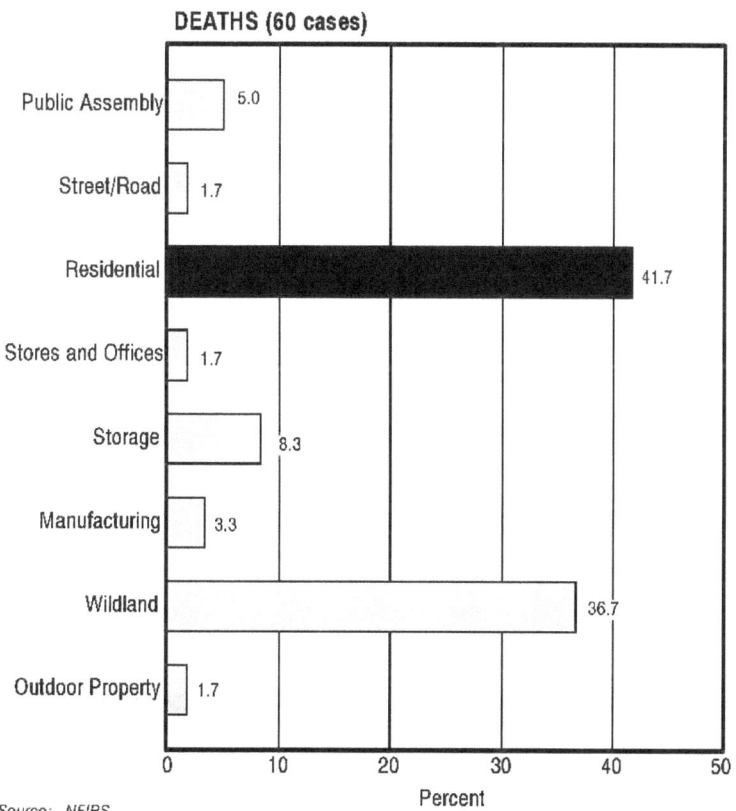

DEATHS (60 cases)

Source: NFIRS

Figure 108. 1994 Firefighter Deaths on Fireground by Fixed Property Use

Residential occupancies accounted for the highest number of fireground fatalities in 1994 with 25 fatalities (42 percent). This is consistent with all prior years. It is not that residential fires are more dangerous, but rather that they are more common. (Residential fires accounted for 72 percent of all structure fires in 1994, but only 42 percent of firefighter fatalities.)

Twenty-two firefighters died while engaged in brush or wildland firefighting in 1994, up from 8 in 1993—an increase of 175 percent. Five firefighters died in storage occupancies, which include warehouses and other types of storage facilities. Three firefighters died in public assembly occupancies—all three in church fires. One died in a commercial use building fire, one died of a heart attack while standing by at an open field fire, and one died of a heart attack after returning to the station from an auto fire.

Figure 109 illustrates the activity the firefighters were engaged in at the time they sustained the injury or illness that caused their death. The activity of cutting fire lines to contain grass, brush, and forest fires accounted for 19 deaths, which is 32 percent of fireground fatalities and reflects the high number of wildland fatalities in 1994.

Search-and-rescue operations in burning structures were in process when 14 deaths (23 percent) occurred. This activity, the second highest category overall, is the highest category for structure fires. Analysis of these deaths reveals that nine died of asphyxiation, four died of burns, and

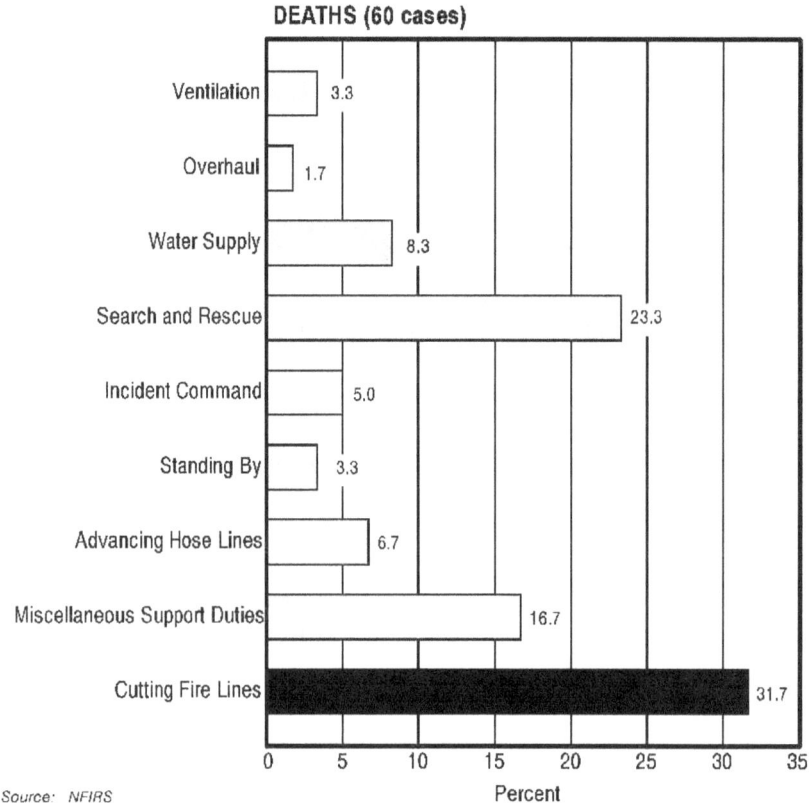

Figure 109. 1994 Firefighter Deaths on Fireground by Type of Activity

one died of a heart attack. At least 12 of these deaths may have involved firefighters who were conducting search operations on or above the fire floor while they or other companies were also ventilating the structure, resulting in rapidly changing fire conditions. This indicates the need for close communication and coordination between ventilation, search, and attack operations, and demonstrates the dangers of conducting search-and-rescue operations without a protective hose line, when the fire location is unknown, or when the fire has not yet been confined.

Support and other duties on the fireground accounted for 10 deaths (17 percent) of the 1994 fireground fatalities. Support duties include utility control, raising ladders, setting up equipment, or engaging in unspecified activities. These deaths include two pilots who were dropping slurry on a wildfire when their plane crashed and an industrial firefighter who died when he ran out of air at a fire in a building at a mine. Several firefighters included in this category suffered fatal heart attacks after arriving on the fireground but had not been assigned to specific activity.

Five heart attack deaths (8 percent) occurred during water supply activities. Three of these individuals were operating fire pumps and two were pulling supply lines.

The remaining 12 fireground deaths (20 percent) involved firefighters performing various other firefighting activities. Four firefighters died while advancing hose lines (7 percent); two officers suffered heart attacks and one officer was burned while they were commanding incidents (5 percent); and two firefighters were performing ventilation (3 percent)—one fell through the roof and the other was electrocuted descending a ladder. Two firefighters (3 percent) died of heart attacks while standing by at controlled burns, and one chief officer (2 percent) died during overhaul when he was exposed to toxic gases and suffered a heart attack.

Of the fireground deaths where firefighters were caught or trapped in burning buildings, at least four were found wearing personal alert safety system (PASS) devices that were in the off position. One firefighter was wearing a PASS device that activated when he became trapped, but rescuers were unable to reach him before he ran out of air. Another firefighter's PASS device was in the armed mode when he was found, but it was not reported if it was sounding. The status of PASS devices was not reported from any of the other fireground fatalities.

Time of Alarm

The distribution of 1994 fireground deaths according to the time of day when the incidents were reported is shown in Figure 110. The highest number of fireground deaths occurred for alarms that were received between 3 p.m. and 5 p.m.[6] The times of alarms where firefighters died were evenly distributed, with higher rates at 1−3 p.m. and 7−9 p.m.

---

[6] The high total is driven by the 14 firefighters killed on Storm King Mountain; the blowup occurred just before 5 p.m., but the fire had been burning for several days.

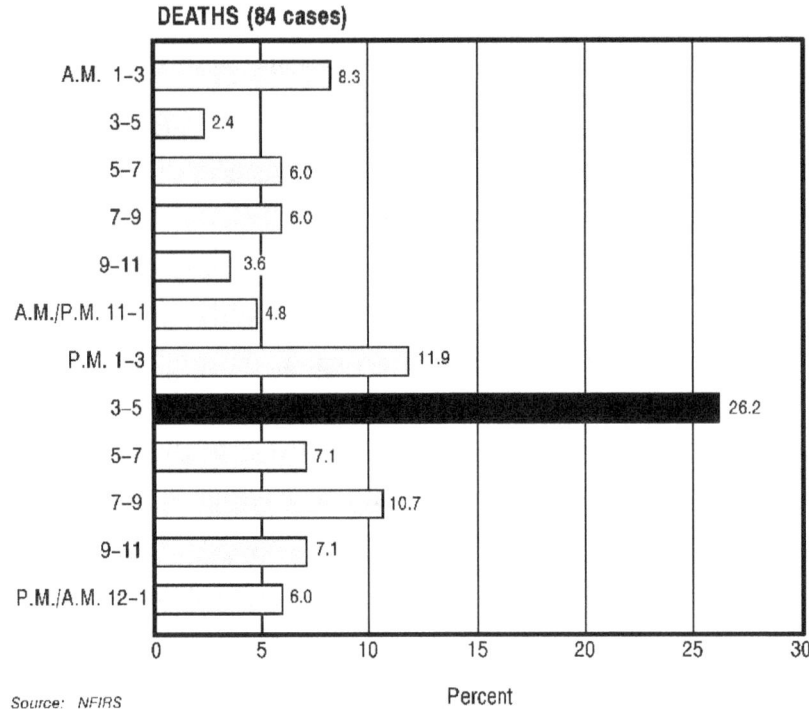

Figure 110. 1994 Firefighter Deaths by Time of Alarm

Month of the Year

Figure 111 illustrates firefighter fatalities by month of the year. Firefighter fatalities peaked in July, the month of the Storm King Mountain fire. Other high months were recorded in January and August. May had the lowest number of deaths.

INJURIES

Fire-related injuries constitute about 55 percent of all firefighter injuries, and numbered about 53,000 in 1994.[7] And in 1994 more than twice as many firefighters were injured as civilians. Figure 112 shows the 13 percent downward trend in the fireground portion of these injuries.

Injuries by Property Type

The majority of firefighter injuries (58 percent) reported to NFIRS are associated with residential fires, largely because that is the largest single subcategory of structural fires (Figure 113). Residential fires have double the number of firefighter injuries as do non-residential structures (25 percent). The proportions have been quite consistent over the 10-year period 1985–1994. Outside, vehicle, and other fires combined are 17 percent of firefighter injuries.

---

[7] NFPA reports fireground injuries as 52,900. To this should be added a portion of the injuries categorized as responding to or from an incident (which includes but is not limited to fires).

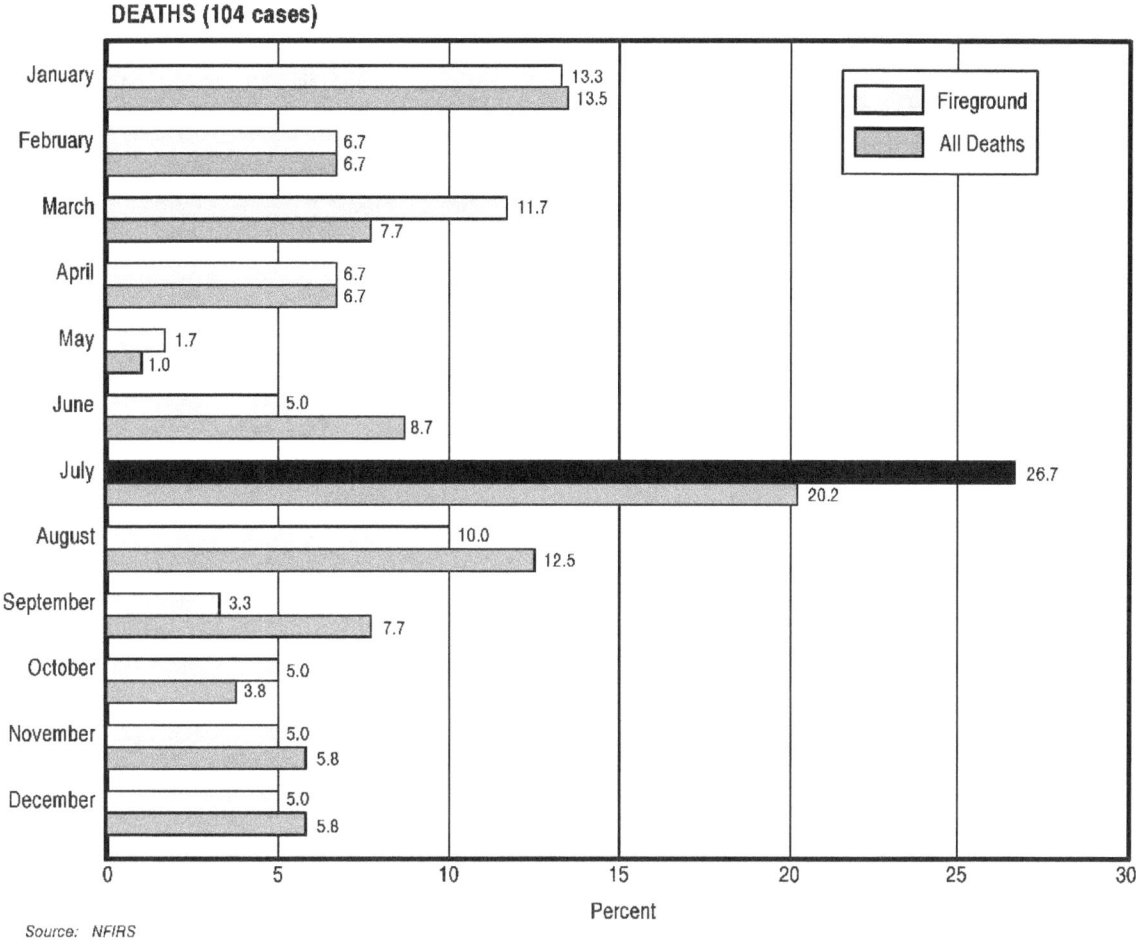

Figure 111. 1994 Firefighter Deaths by Month of Year

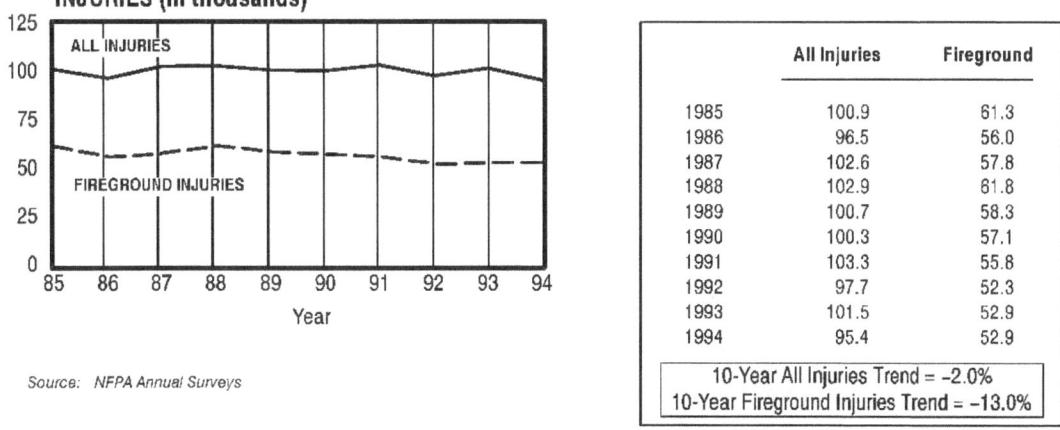

Figure 112. Trends in 1994 Firefighter Injuries

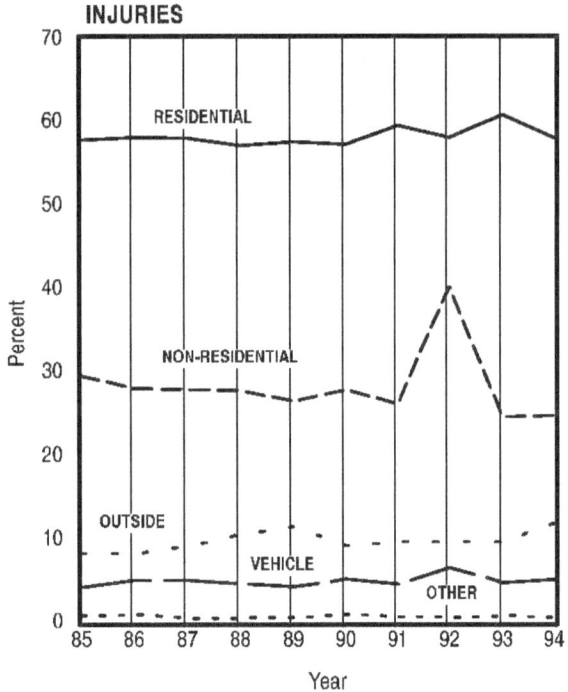

INJURIES

	Residential	Non-Residential	Vehicle	Outside	Other
1985	57.7%	29.4%	4.1%	8.2%	0.7%
1986	58.0	27.9	4.9	8.2	0.9
1987	57.9	27.7	4.9	9.0	0.4
1988	57.0	27.6	4.5	10.4	0.5
1989	57.5	26.4	4.2	11.3	0.5
1990	57.2	27.7	5.1	9.1	0.9
1991	59.4	26.0	4.5	9.5	0.6
1992	58.0	40.1	6.4	9.5	0.5
1993	60.7	24.5	4.6	9.5	0.7
1994	57.9	24.7	5.1	11.8	0.4

Source: NFIRS

Figure 113. Trends in Firefighter Injuries by General Property Type

Figure 114 gives a more detailed look at the relative proportion of firefighter injuries by property type. Half of all firefighter injuries occur in structures at one- and two-family dwelling fires. Apartments account for another 19 percent.

The proportions of injuries by property type were consistent over 1985–1994. Figures 115 and 116 show these proportions for residential and non-residential properties.

Injuries per Fire

Firefighter injuries per fire have been gradually trending downward.[8] The injury rates for structural fires are over ten times those for outside, vehicle, and other fires (Figure 117).

Figure 118 shows that the firefighter injuries per fire for apartments and one- and two-family homes fires have been remarkably constant over the 10-year period. The injury rate for hotel/motel fires and other residential is down sharply, but both of these categories fluctuate considerably from year to year because of small sample sizes.

Figure 119 shows the firefighter injury rates per 1,000 fires for structural occupancies. Several types of non-residential properties, especially storage, manufacturing, vacant/under construction,

---

[8] The 1983–1994 NFPA *Fire Command* and NFPA *Journal* articles on firefighter injuries also show a downward trend in injuries per fire, from 9.4 injuries per 1,000 fires in 1983 to 7.3 in 1994. These rates are half the NFIRS injury rate for all fires. It is not clear why there is such a large difference; it may be a definitional problem.

164

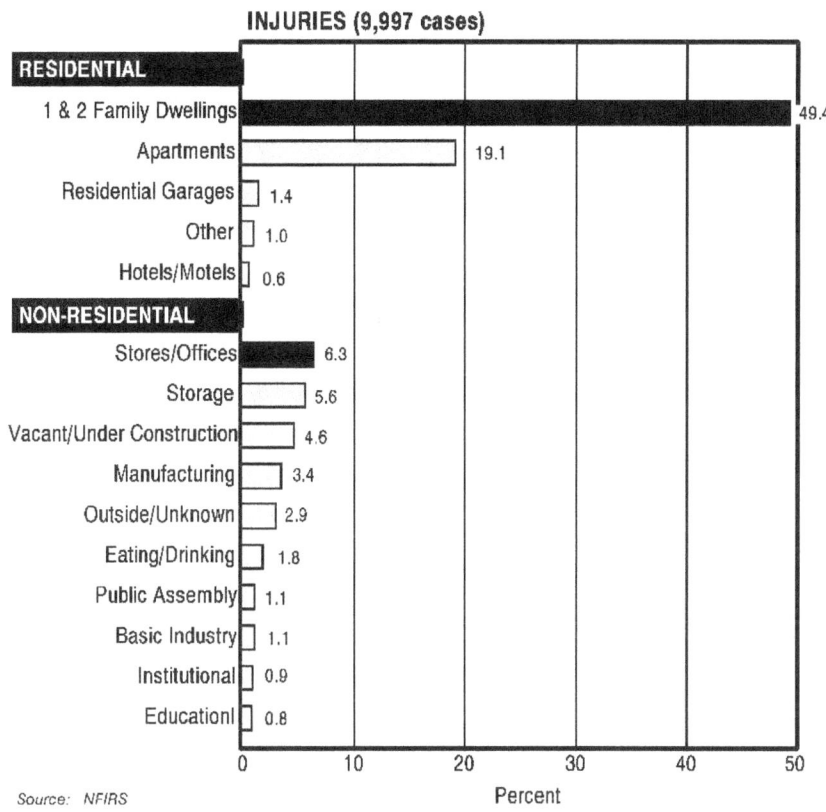

Figure 114. 1994 Firefighter Injuries by Property Type (Structure Fires Only)

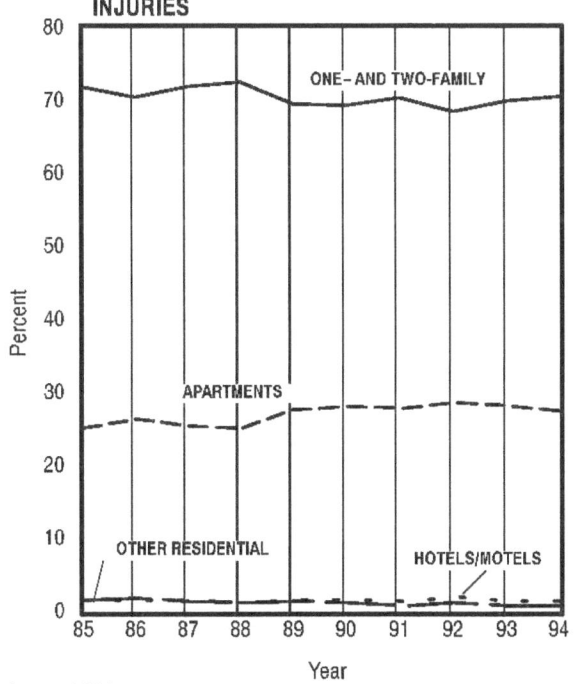

	One- and Two-Family	Apartments	Hotels/Motels	Other Residential
1985	71.8%	25.1%	1.5%	1.5%
1986	70.4	26.3	1.8	1.5
1987	71.9	25.3	1.3	1.5
1988	72.5	25.0	1.2	1.2
1989	69.5	27.6	1.3	1.6
1990	69.3	28.0	1.2	1.5
1991	70.3	27.7	0.7	1.3
1992	68.4	28.5	1.2	1.9
1993	69.9	28.1	0.7	1.3
1994	70.5	27.3	0.8	1.4

Source: NFIRS

Figure 115. Trends in Firefighter Injuries in Residential Structure Fires

165

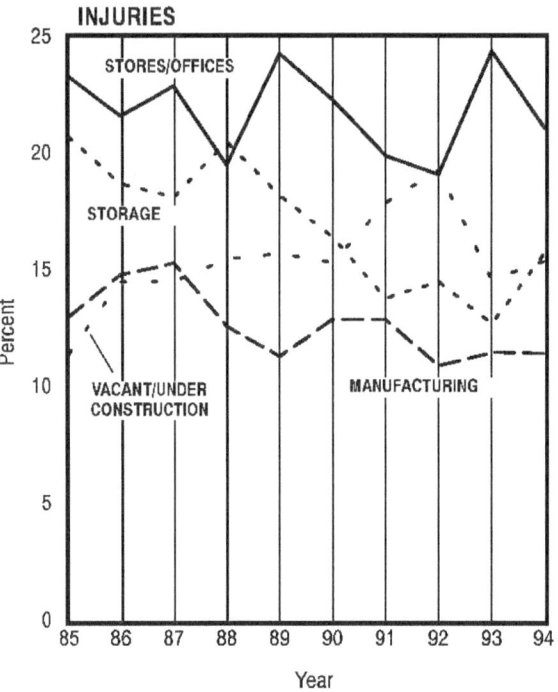

Note: Data for all 11 non-residential structure types are provided in
Appendix B, Table B-6.

Sources: NFPA Annual Surveys and NFIRS

Figure 116. Leading Trends in Firefighter Injuries by Type of Non-Residential Structure Fires

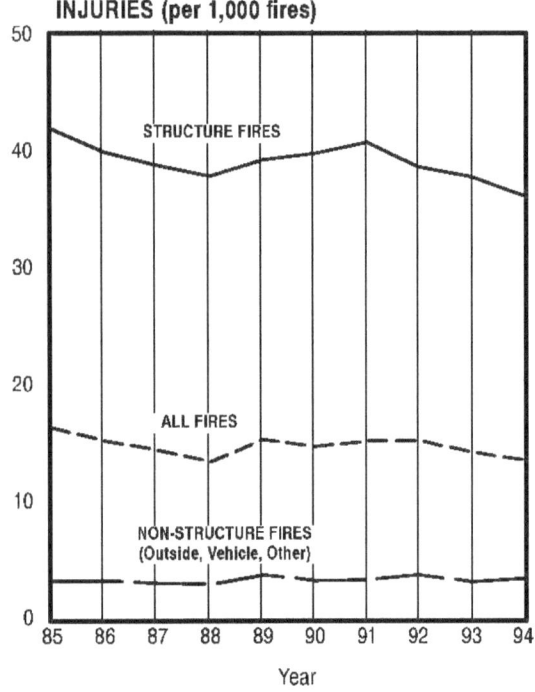

	Structure Fires	All Fires	Non-Structure Fires
1985	41.9	16.3	3.2
1986	39.9	15.2	3.2
1987	38.8	14.4	3.0
1988	37.9	13.4	2.9
1989	39.3	15.3	3.7
1990	39.8	14.7	3.2
1991	40.8	15.2	3.3
1992	38.7	15.2	3.7
1993	37.8	14.2	3.1
1994	36.2	13.5	3.4

Source: NFIRS

Figure 117. Trends in Severity of Firefighter Injuries by Type of Fire

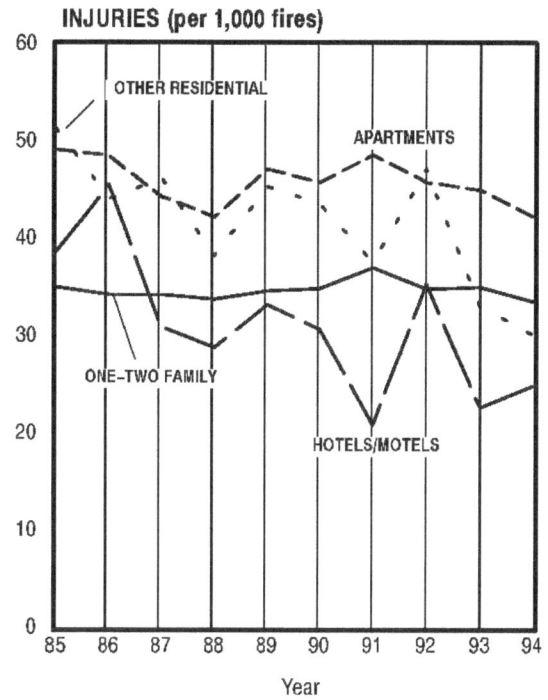

INJURIES (per 1,000 fires)

	One–Two Family	Apartments	Hotels/Motels	Other Residential
1985	35.0	49.1	38.6	51.3
1986	34.2	48.5	45.6	44.1
1987	34.2	44.3	30.9	46.2
1988	33.8	42.2	28.8	38.2
1989	34.6	47.1	33.2	45.3
1990	34.9	45.8	30.7	43.6
1991	37.0	48.6	20.8	37.6
1992	34.8	45.8	35.3	47.1
1993	35.0	45.0	22.6	33.0
1994	33.5	42.2	24.9	30.1

Source: NFIRS

Figure 118. Trends in Severity of Firefighter Injuries in Residential Structure Fires

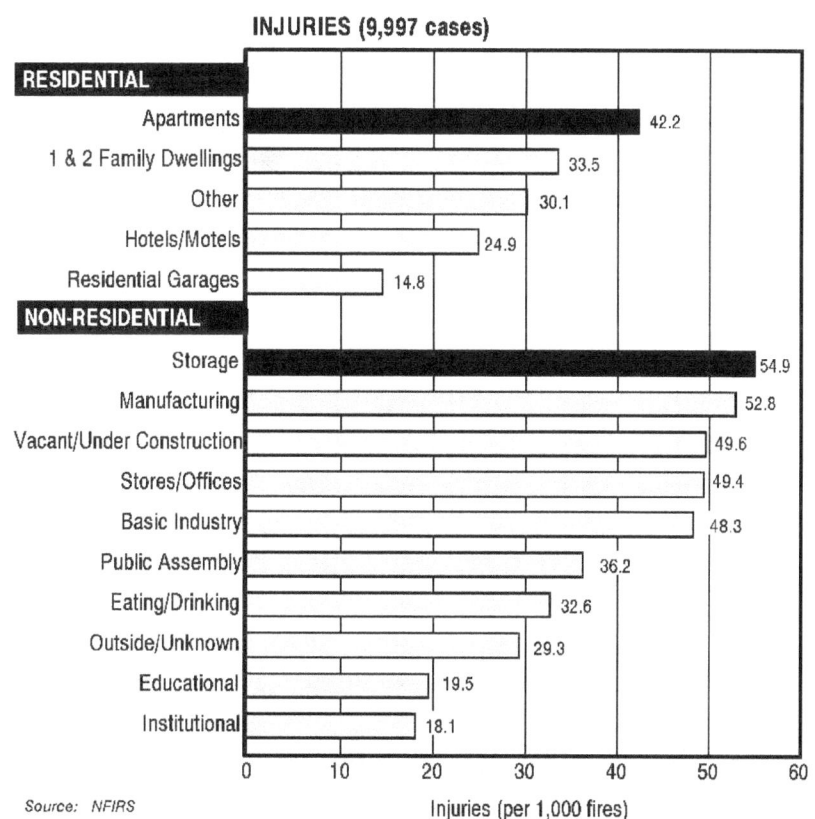

INJURIES (9,997 cases)

Source: NFIRS

Injuries (per 1,000 fires)

Figure 119. Severity of 1994 Firefighter Injuries by Property Type (Structure Fires Only)

167

stores/offices, and basic industry pose the greatest risk. Residential properties pose less risk per fire than these non-residential properties.

Vacant properties have long been a firefighting concern (Figure 120). In the mid 1970s, the most dangerous fires were those in vacant properties and properties under construction. The layout of these structures is often unfamiliar and continually changing from week to week. Fire defenses built into such structures are often not working or only working partially. Also, there are many pitfalls where a misstep can cause serious injury. Many of these fires are started when no one is around and the fire gets considerable headway before the fire department is called. This combination makes them the third-most hazardous non-residential fire in 1994, behind storage (which leaped to first place) and manufacturing. When fighting fires in vacant properties, there is less of an inclination to risk firefighters' lives.

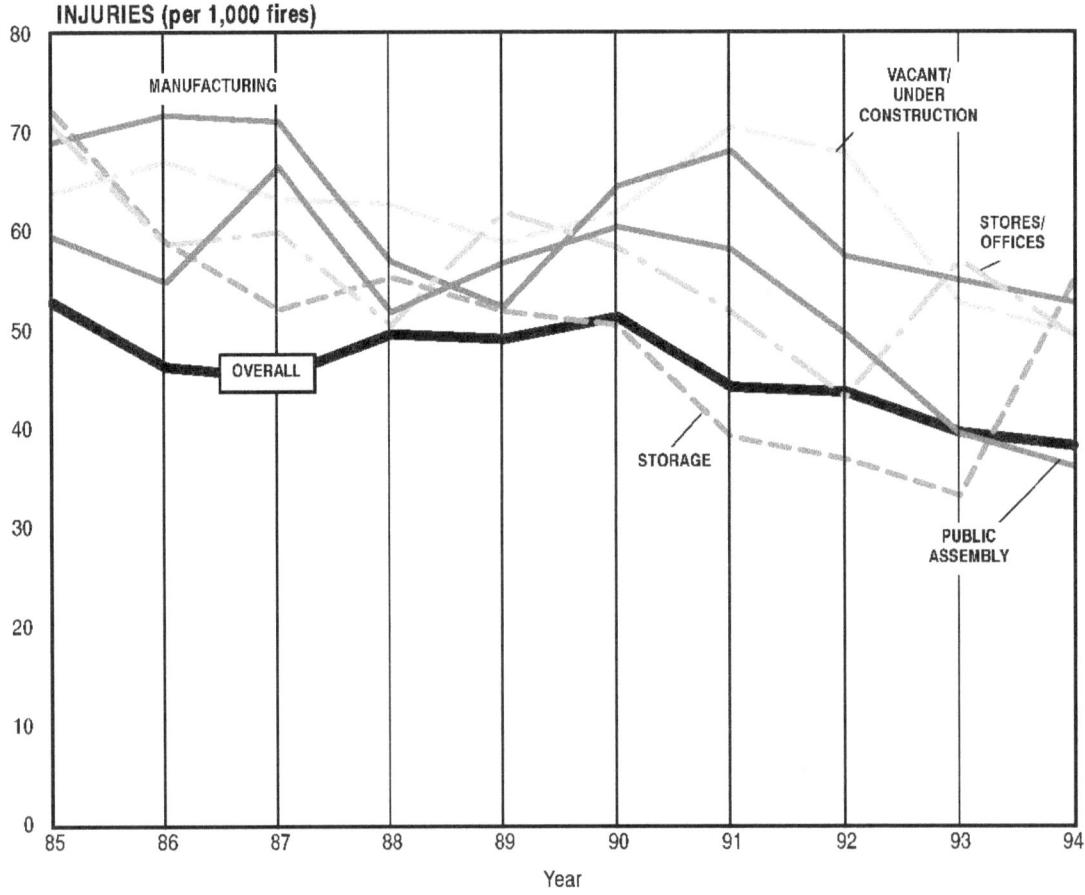

Note: Data on all non-residential fires provided in Appendix B, Table B–6.

Source: NFPA Annual Surveys and NFIRS

Figure 120. Trends in Severity of Firefighter Injuries in Non-Residential Structure Fires

For non-residential properties in general, the injury rate per fire fluctuates widely from year to year, but the four highest risk properties (storage, manufacturing, vacant/construction, and stores) are all trending downward—a promising pattern for firefighter fireground safety.

Characteristics of Injuries

AGE. Figure 121 shows the profile of firefighter injuries by age for all property types. One-third of all injuries occur to firefighters aged 30−39. The types of injuries incurred by firefighters vary with age. Typically, the leading cause of injury among younger firefighters relates to smoke inhalation and for older firefighter it is strains and sprains. These results relate to physical fitness variations with age, to the effect of age on assignments, and perhaps to the bravado of younger firefighters.

TIME OF DAY. More firefighter injuries occur after noon than before. However, there is no sharp peak. The times that are most hazardous to civilians (evening meal times for injuries) are not the same as the times firefighters get injured (Figure 122).

MONTH OF YEAR. Firefighter injuries are somewhat higher in the winter (December−March) when residential fires peak and again in June−August when outside fires are a factor (Figure 123).

PART OF BODY INJURED. The most common firefighter injuries in 1994 were to the torso, followed closely by arm/hand and leg/foot. All areas of the body are vulnerable, including internal injuries from smoke inhalation. The firefighter must protect his or her entire body with a complete protective outfit and be in good physical condition (Figure 124).

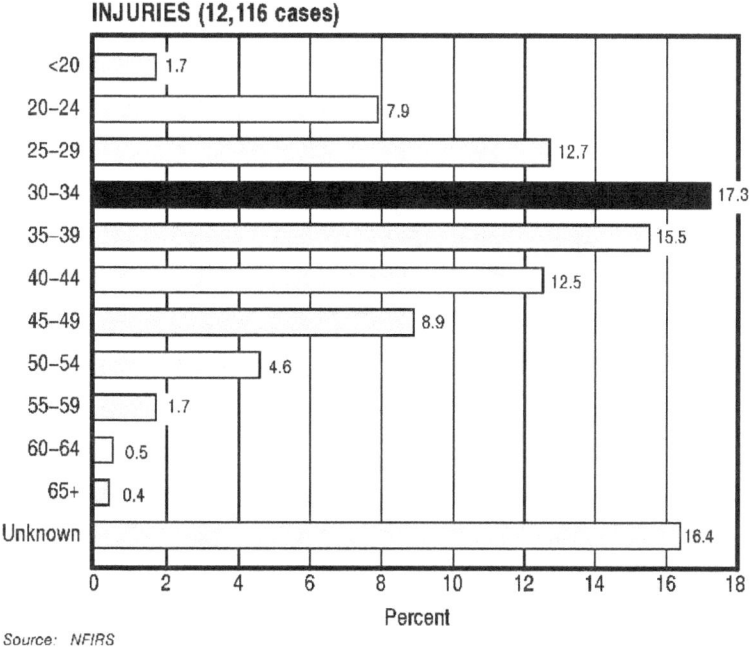

Figure 121. 1994 Firefighter Injuries by Age

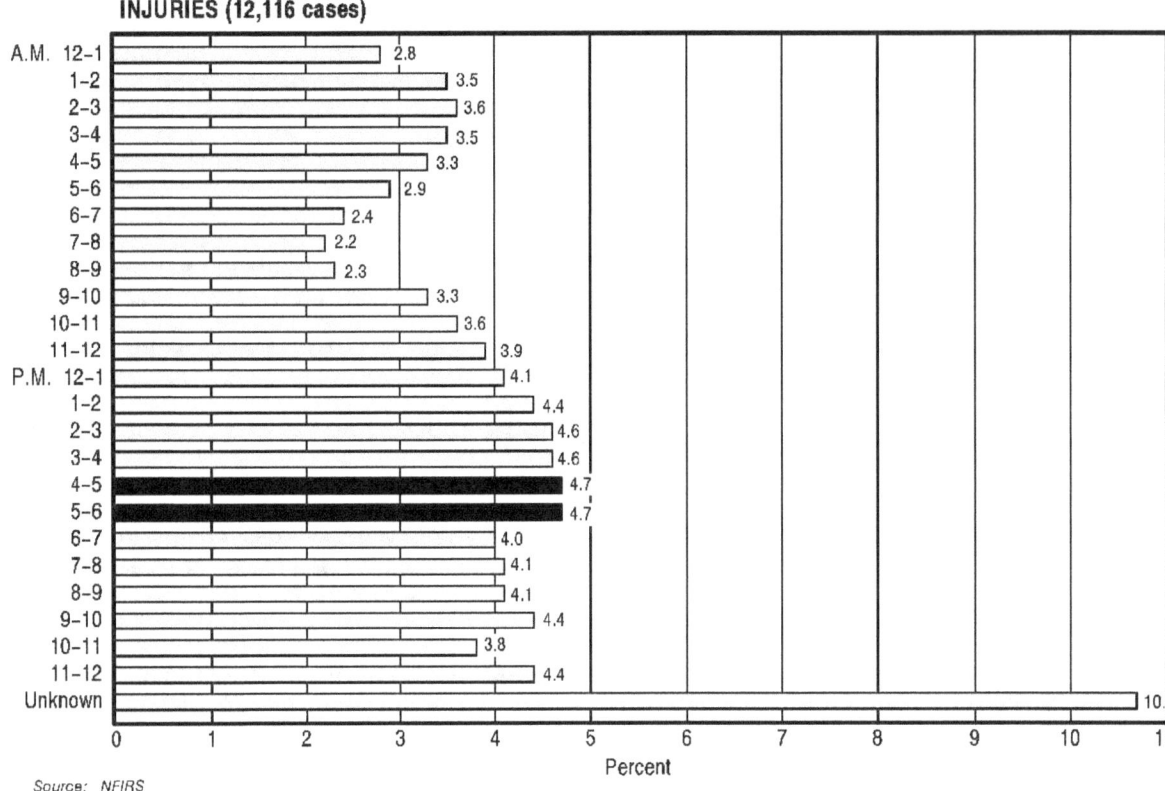

Figure 122. 1994 Firefighter Injuries by Time of Day

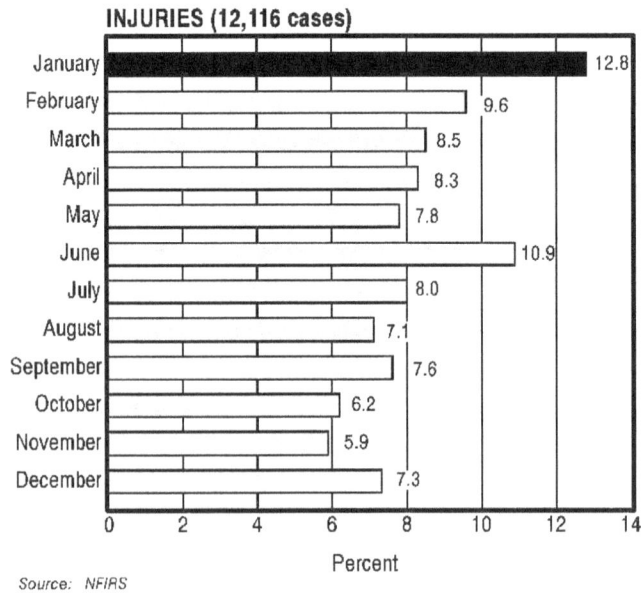

Figure 123. 1994 Firefighter Injuries by Month of Year

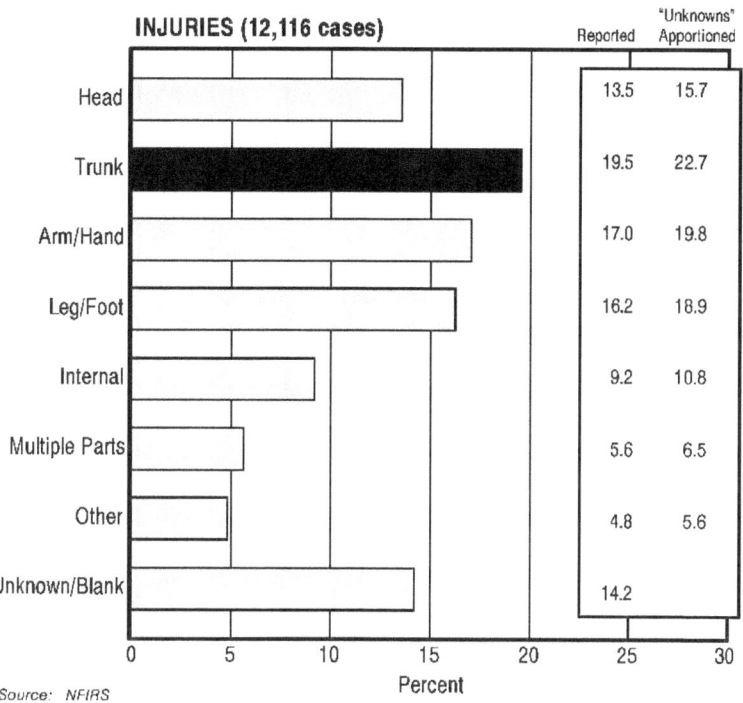

Figure 124. 1994 Firefighter Injuries by Part of Body Injured

CAUSE OF INJURY. By far the largest category of firefighter injuries associated with fires was reported to be contact with or exposure to the flames or to smoke (33 percent of injuries, adjusted) (Figure 125). The second highest category was overexertion and strains (23 percent), followed closely by fell or slipped (22 percent). No cause was reported for 21 percent of the injuries.

WHERE INJURIES OCCUR. According to NFIRS, 91 percent of the firefighter injuries associated with reported fires occur on the scene above ground (Figure 126). This percentage is nearly equally divided between injuries occurring inside and outside the structure. Significantly smaller percentages are reported as occurring en route or below ground level. (As a reminder, there also are many firefighter injuries that are not associated with fires and that are not included here.) One-quarter of injuries did not have a reported location.

The striking point here is that many firefighter injuries (47 percent) occur in areas outside the fire building, where the firefighter might feel relatively safe. There often are more firefighters operating outside the fire building and exposed to injury than there are inside. Outside fires include vehicle fires, which contribute to this high incidence of injuries.

TYPE OF ACTIVITY WHEN INJURED. More than half of firefighter injuries occurred while extinguishing the fire; suppression support accounted for 23 percent (Figure 127).

TYPE OF MEDICAL CARE. Over half of the reported fire injuries associated with fires were treated at hospitals. Another 43 percent were treated but not transported. A small percentage (less than 5 percent) of firefighters were treated elsewhere (Figure 128).

171

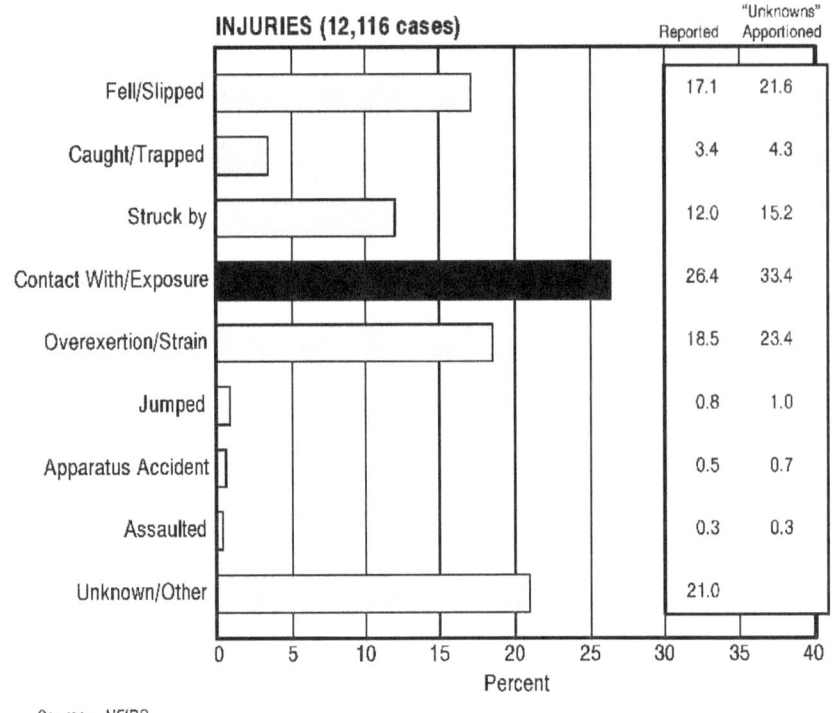

Source: NFIRS

Figure 125. 1994 Firefighter Injuries by Cause

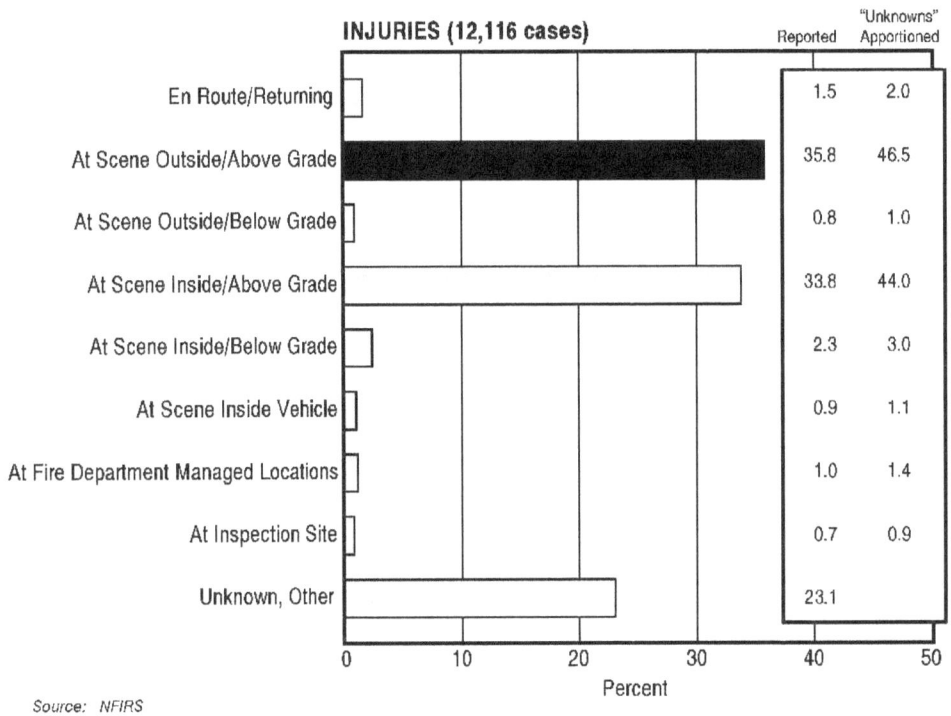

Source: NFIRS

Figure 126. 1994 Firefighter Injuries (All Fires) by Where Injury Occurs

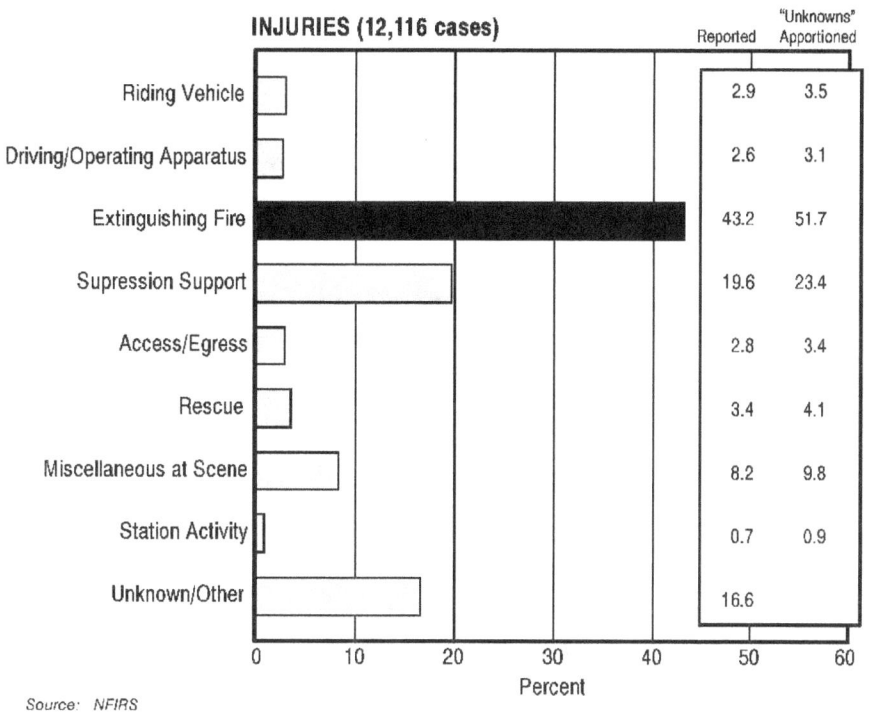

Figure 127. 1994 Firefighter Injuries by Type of Activity

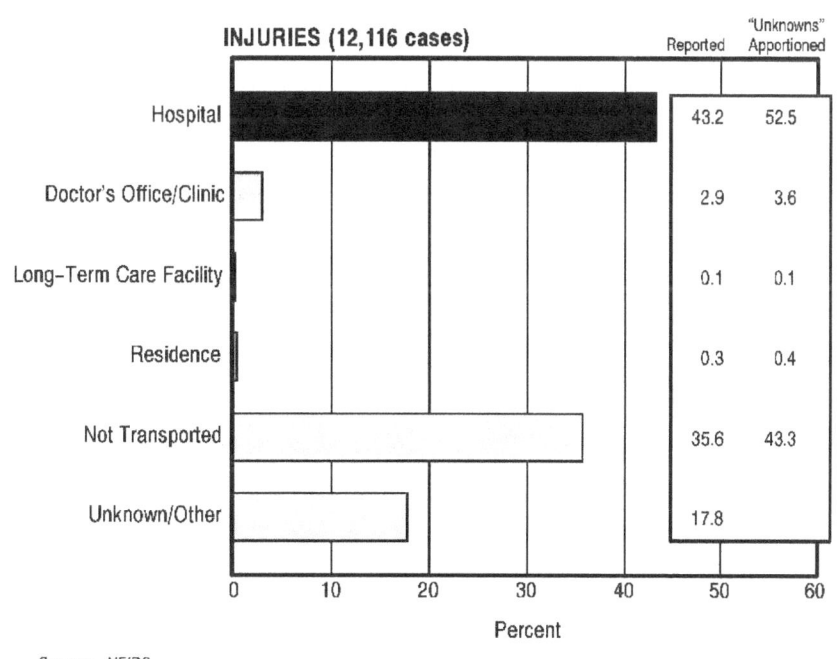

Figure 128. 1994 Firefighter Injuries (Fire Incidents) by Where Treated

USFA RESOURCES ON FIREFIGHTER CASUALTIES

Publications

The U.S. Fire Administration recently revised its NFIRS Firefighter Casualty Report to improve the quality of available data in its annual review of firefighter line-of-duty deaths. This and other USFA-supported research and development are intended to increase the safety and well-being of emergency response personnel. USFA encourages sharing of research findings and incorporation of innovations in equipment available to firefighters and other responders through programs that focus on health and safety studies; research, training and awareness; emergency medical services; search and rescue; and equipment and technology development.

Because accidents involving emergency vehicles are one of the leading causes of firefighter death and injury, USFA has several resources on the subject for fire departments and emergency medical services departments. *Emergency Vehicle Driver Training* (#FA–110) is a 220-page training package that includes both an instructor manual and a student workbook designed to assist fire and emergency medical service (EMS) departments with training in emergency vehicle operations. *Alive on Arrival—Tips of Safe Emergency Vehicle Operations* (#L–195) is a pamphlet detailing actions that emergency vehicle operators, passengers, and officers-in-charge can do to improve safe operation of emergency vehicles.

Also available is a 48-page special report titled *Fire Apparatus/Train Collision* (#FA–104), which presents the investigation of the collision near Catlett, Virginia, on September 28, 1989.

Publications addressing incident response issues have been developed for fire and EMS departments. Among these are *Emergency Incident Rehabilitation* (#FA–114), a short booklet that includes a sample standard operating procedure and guidelines for establishing a rehab area to reduce heat- or cold-related injuries to emergency response personnel operating in labor-intensive or extreme climate conditions.

Guides are also available on recommended safe practices, including response to crashes involving cars equipped with air bags and on a comprehensive safety program designed for fire department safety officers.

USFA also emphasizes research and development of protective clothing for chemical, emergency medical, and search-and-rescue emergencies as well as structural firefighting protective clothing and self-contained breathing apparatus (SCBA). For example, USFA has been involved in the development of a new test method for evaluating the performance of complete firefighter protective clothing ensembles. A suite integrity field test was conducted during hazardous materials training for USFA's study, *Qualitatively Evaluating the Comfort, Fit, Function, and Integrity of Chemical Protective Suite Ensembles* (#FA–107). Three protective clothing ensembles were evaluated in *Physiological Field Evaluation of Hazardous Materials Protective Ensembles* (#FA–109). Another study, the *Non-Destructive Testing and Field Evaluation of Chemical Protective Clothing* (#FA–1060), details a procedure, field tested by the Cambridge, Massachusetts, Fire Department, developed for assessing the presence of contamination before or after decontamination of chemical protective clothing.

USFA has supported research into health hazards faced by firefighters, including the *Northwest Firefighters Mortality Study: 1945–1989* (#FA–105). USFA also supports symposia on the occupational health and hazards of the fire service focusing on emerging firefighter safety and health issues.

A manual has been prepared for emergency response managers on infection control programs based on federal laws, regulations, and standards. The *Guide to Developing and Managing an Emergency Service Infection Control Program* (#FA–112) addresses modes of disease transmission, measures for prevention, incident response and recovery, station issues, and training/role modeling. The 200-page manual provides a step-by-step approach to designing, implementing, managing, and evaluating a fire or emergency medical services department infection control program. The guide also is a key resource in a National Fire Academy course on infection control.

USFA has developed a series of comprehensive manuals for fire service and EMS managers interested in instituting programs for firefighter health promotion and injury prevention. The 80-page *Fire and Emergency Service Hearing Conservation Program Manual* outlines measures to reduce the risk of occupationally induced hearing loss. USFA also is conducting research to identify causes and to develop solutions to reduce the stress level in EMS providers. A 175-page *Stress Management Model Program* (#FA–100) is available.

USFA has also studied major urban search-and-rescue incidents for lessons learned regarding safety of firefighting, EMS, and other rescue personnel. Six reports published in November 1992 describe *Urban Search and Rescue: in Will County, Illinois, Following the 1990 Tornado* (#FA–122); *in Crested Butte, Colorado State Bank, Following an Explosion Collapse* (#FA–120); *in New York City, Following a Commercial Building Collapse* (#FA–121); *in Brownsville, Texas, Following a Commercial Building Collapse* (#FA–123); *in the Santa Cruz Area, Following the Loma Prieta Earthquake* (#FA–124); and *in San Bernadino, California, Following a Major Train Derailment in a Residential Neighborhood* (#FA–125).

Reports produced under USFA's Major Fires Investigations series are directed primarily to chief fire officers, training officers, fire marshals, and investigators as a resource for training and prevention. Recent reports on incidents with firefighter deaths and injuries include: *Four Firefighters Killed, Trapped by Floor Collapse, Brackenridge, Pennsylvania, December 20, 1991* (#061); *Indianapolis Athletic Club Fire (Two Firefighter Fatalities), Indianapolis, Indiana, February 6, 1991* (#063); *Six Firefighter Fatalities in Construction Site Explosion, Kansas City, Missouri, November 1988* (#024); *Three Firefighter Fatalities in Training Exercise, Milford, Michigan, October 1987* (#015); *High-Rise Office Building Fire, One Meridian Plaza, Philadelphia, Pennsylvania, February 1991* (#049); and *Wood Truss Roof Collapse Claims Two Firefighters, Memphis, Tennessee, January 1993* (#069). The report on *Michigan—Industrial Plastics Fire Sends 97 to Hospital, Flint, Michigan, November 1988* (#025) credits the successful outcome of this fire to the incident command system used, including a strict requirement for SCBA use and rotation of personnel.

These publications are available by writing to:

U.S. Fire Administration
Federal Emergency Management Agency
Publications Center, Room N310
16825 S. Seton Avenue
Emmitsburg, MD 21727

Documents may also be ordered via the World Wide Web: *http://www.usfa.fema.gov*. USFA publications are free.

Video Training

FEMA's Emergency Education NETwork (EENET) provides video training and education via satellite for the fire service and emergency management community. EENET programs are satellite-distributed videoconferences broadcast over the "C" band and allow for audience interaction when originally broadcast. Each program is designed as a standalone training activity of 4½ hours, and student materials are provided for each workshop.

Several previous EENET programs dealing with firefighter health and safety include the following:

- *U.S. Fire Administration's Forum on Communicable Diseases*, May 17, 1989

- *Heat Stress Induced by Chemical Protective Clothing*, July 19, 1989

- *Protective Actions for Hazardous Materials*, October 4, 1989

- *Infection Control: Today's Requirements for Fire and EMS Departments*, December 4, 1991

- *Chemical Protective Clothing Standards: An Overview of NFPA 1991, 1992 & 1993*, June 24, 1992

- *Protective Clothing for Emergency Medical Operations: An Overview of NFPA 1999*, August 19, 1992

Tapes of these and other broadcasts from 1989 to the present are available for a modest cost through:

The National Audiovisual Center
Customer Service Section
8700 Edgeworth Drive
Capitol Heights, MD 20743–3701
(301) 763–1896; (800) 788–6282 for credit card orders

In addition, all broadcasts prior to 1989 and all current broadcasts may be borrowed from state emergency management offices or from FEMA regional offices.

For further information on EENET or if you would like to get on the EENET mailing list, contact:

Emergency Education NETwork
National Emergency Training Center
16825 S. Seton Avenue
Emmitsburg, MD 21727
(301) 447–1068; Fax: (301) 447–1363

National Fire Academy Courses

USFA's National Fire Academy (NFA) works to enhance the ability of the fire service and allied professions to deal more effectively with fire and related emergencies. Courses are delivered on campus at the resident facility in Emmitsburg, Maryland, and off campus throughout the nation in cooperation with state and local fire training officials and local colleges and universities. A new initiative begun in 1992 offers NFA resident courses on a regional basis. Expanded opportunities are available for fire service personnel to participate in academy courses that are handed off through NFA's train-the-trainer program. Academy handoff courses are also available through the National Audiovisual Center.

While firefighter health and safety issues are addressed in numerous NFA courses, several offerings include these issues as a major thrust. NFA's course on *Command and Control of Fire Department Major Operations* (Resident Course #R304) is a 2-week on-campus course for fireground managers that links the subjects of fireground operations and safety. Emphasis is placed on increasing the command officer's awareness regarding the causes of firefighter fatalities and types and kinds of injuries.

Protective clothing and breathing apparatus are among the topics covered in NFA's course on *Hazardous Materials Operating Site Practices* (Resident Course #R229), which focuses on the strategies and safe procedures for alleviating the danger of a hazardous materials accident.

Firefighter safety is also an emphasis in NFA's *Volunteer Incentive Program*, a series of on-campus courses designed specifically for volunteer fire officers.

An off-campus course addressing firefighter health and safety issues is *Firefighter Safety and Survival: The Company Officer's Responsibility (FSCO)*. This course examines significant areas of firefighter fatalities and injuries associated with emergency and non-emergency situations and provides recommended solutions and implementation methods. Another off-campus offering is *Firefighter Health and Safety: Program Implementation and Management (FHSP)*, which focuses on the design and implementation of a departmental safety program. *Infection Control for Emergency Response Personnel: The Supervisor's Role and Responsibilities (ICERP)* is a 2-day course covering a broad range of infection control issues.

For information about course offerings, eligibility, and application procedures, write to:

The National Fire Academy
U.S. Fire Administration
16825 S. Seton Avenue
Emmitsburg, MD 21727

National Fire Academy off-campus course materials are available for purchase for locally sponsored delivery from the National Audiovisual Center. Current academy off-campus courses, consisting of an instructor guide, student manual, and supporting audiovisual aids, are also available. Courses available for purchase include *Firefighter Safety and Survival: The Company Officers Responsibility (1988), Firefighter Health and Safety: Program Implementation and Management (1988)*, and *Infection Control for Emergency Response Personnel: The Supervisors Role and Responsibilities (1993)*. For information on how to order courses, contact the National Audiovisual Center at the address and phone number listed earlier.

6

SPECIAL TOPICS

This chapter addresses topics relating to the fire problem that, because of their severity, require special attention. First, the fire problem of the United States is compared to that of 13 industrial nations. Second, because arson is a very serious problem in the United States, a detailed examination is presented. Third, wildland fires have traditionally been of concern to rural residents, but with the rapid expansion of urban population to rural areas, wildland fires have become a more important and visible problem. Finally, the total cost of fires in the United States is enormous. We have attempted to aggregate the "hidden" cost of fires with the known costs in order to better understand the magnitude of the fire problem.

INTERNATIONAL FIRE COMPARISON

The United States historically has had one of the highest fire loss rates of the industrialized world in terms of both fire deaths and dollars loss. This unenviable status has perplexed many experts in the fire world. The United States is health and safety conscious in many areas—automobiles, consumer products, food, medical drugs—and has a vast arsenal of technological resources to combat fire. Why then is such a safety conscious and technologically advanced society a world leader in fire losses?

This comparison deals with the fire death rates for 13 industrialized nations and the United States. Although comparisons of total fires and total fire losses would be preferable, reliable data are not available due to diverse record keeping and fire classification practices in different countries. Loss estimates can then vary within a country, depending on the source of the information. This is especially true for data regarding monetary loss. In the United States, for example, the monetary loss reported by a fire department can vary significantly from that reported by an insurance company. And both of these estimates may differ from the monetary loss as perceived by the owner or occupant. Fire deaths, however, are less controversial as they are more readily identified and consistently counted, although they too have reporting problems.

Death Rates

Figure 129 depicts the average per capita fire death rates for 14 industrialized nations from 1979 to 1992. As this figure demonstrates, the United States ranked only behind Hungary as having the highest per capita fire death rate. At a rate of 26.5 deaths per million population, the United

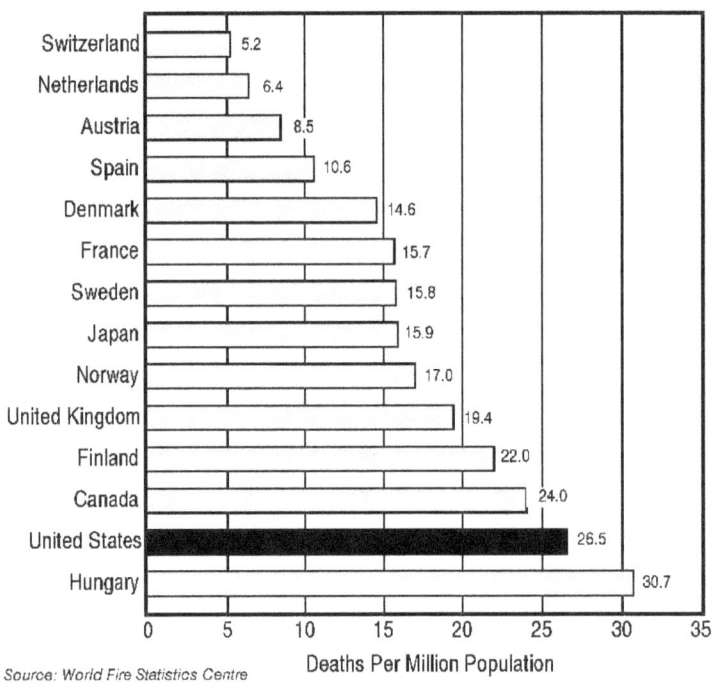

Source: World Fire Statistics Centre

Deaths Per Million Population

Figure 129. Average Fire Death Rate by Country (1979–92)

States' fire death rate was more than five times that of Switzerland, which had the lowest rate of all the countries considered—5.2 deaths per million population.

Figure 130 indicates that, at least in absolute terms, the situation in the United States improved greatly between 1979 and 1992. The U.S. fire death rate fell 46.3 percent, from 36.3 fire deaths per million population in 1979 to 19.5 fire deaths per million population in 1992. This decline, however, was not limited to the United States; rather, it was an international trend. Of the countries considered, only Hungary and Denmark recorded increases in their rates of fire deaths over that period. The reduction of fire deaths for the United States (46 percent, or 16.8 fire deaths per million population) was the largest absolute and relative drop of any of the countries shown, almost twice the size of the next biggest drop—the United Kingdom, with a reduction of 38 percent, or 9 fire deaths per million population.[1]

Despite its impressive gains, the United States still has one of the highest per capita fire death rates among the countries considered, as shown in Figure 131. The most current comparative data (1992) reveal that the United States, while having substantially reduced its fire death rate, is still 30 to 50 percent higher than its peer nations (countries that analysts consider most like the United States). And in the case of Switzerland and the Netherlands, and United States' fire death rate is nearly triple. Many people feel that there is little reason for the United States, which possesses a wealth of advanced fire suppression technologies and fire service delivery mechanisms, to lag so

---

[1] Canada is not considered in this comparison.

180

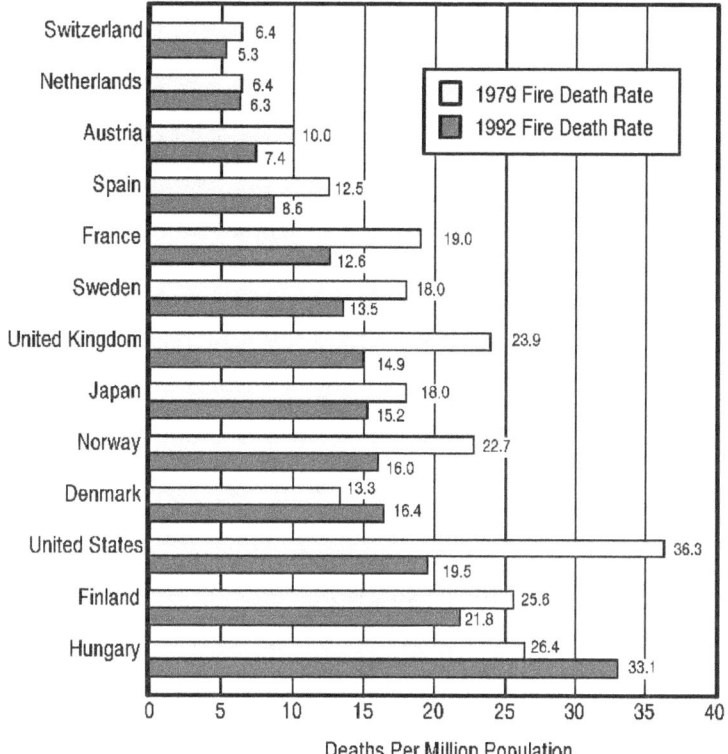

Note: Canada is not included as 1979 data were not available

Source: World Fire Statistics Centre

Figure 130. 1979 and 1992 Fire Death Rates Ranked by Percent of Decrease

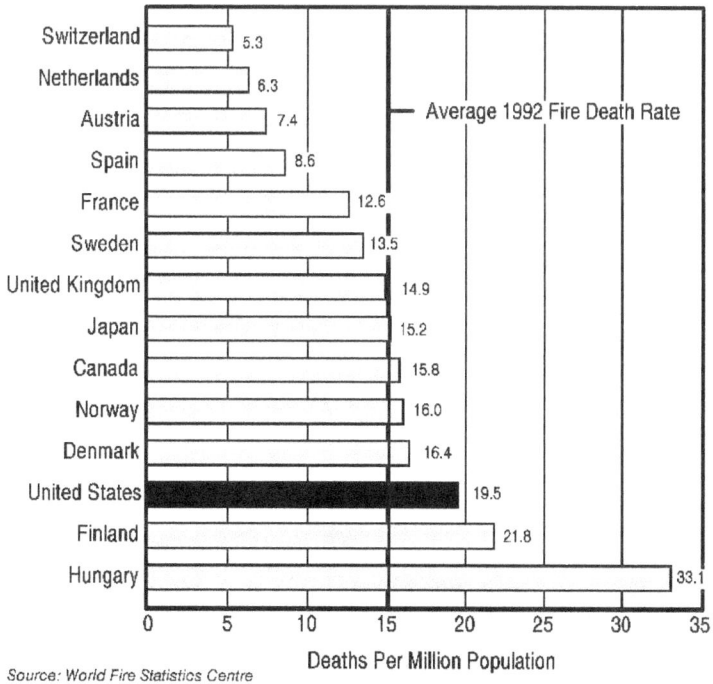

Source: World Fire Statistics Centre

Figure 131. 1992 Fire Death Rate by Country

far behind other nations in terms of fire safety. However, most of the advanced fire technology used in the United States is installed in public places and most fire deaths occur in the home.

Death Rate Trends

Figure 132 compares fire death rate trends for the 14 countries divided into five regional groups: North America (Canada and the United States), Western Europe (Austria, Denmark, France, the Netherlands, Spain, Switzerland, and the United Kingdom), Scandinavia (Finland, Norway, and Sweden), Hungary, and Japan over a 13-year timespan.[2] In this figure, the trend line for North America has a much steeper downward slope than that of the other regions, indicating that North America has reduced its fire death rate significantly more than other regions. In fact, North America experienced a 12.8 percent annual reduction in its fire death rate, compared to a 4.4 percent increase for Hungary, a 2.3 percent reduction for both Scandinavia and Japan, and a 3.8 percent reduction for Western Europe. The trends for Hungary, Scandinavia, and Japan, however, must be viewed with caution, as the data series are for smaller populations and the fit between the trend lines and the data is not as good as those for North America and Western Europe. The poorer fit may well be an indication that the fire loss data gathered in those three areas are less reliable than the data from North America and Western Europe.

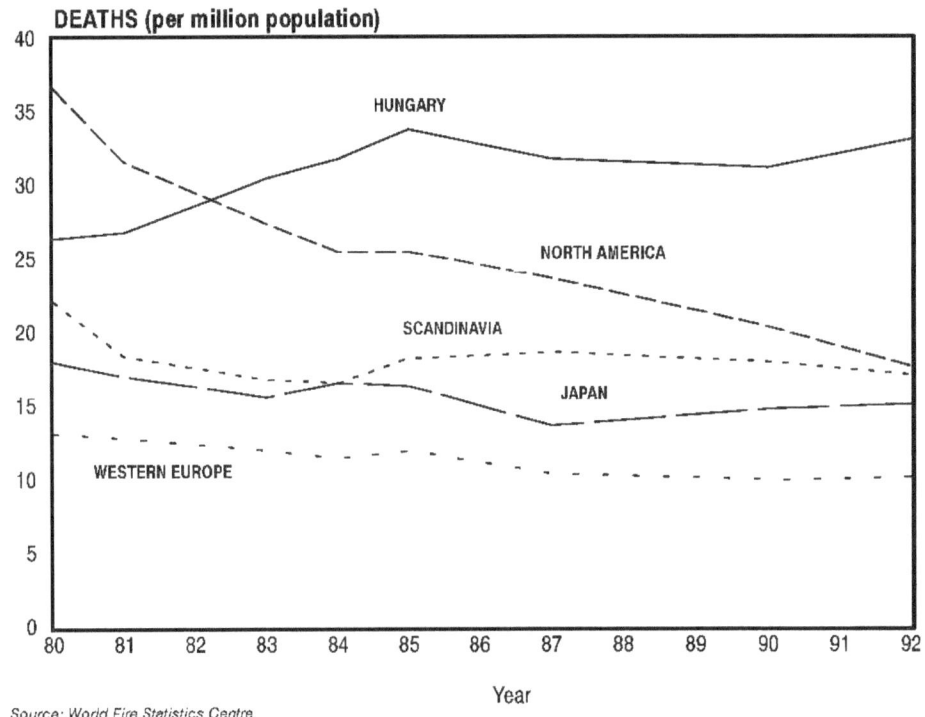

Figure 132. Trends in Fire Deaths by Region

Source: World Fire Statistics Centre

---

[2] Hungary and Japan are reported as "regions" to provide bases for comparison of Eastern Europe and Asia, respectively. The use of a single country to denote the region is necessitated by a paucity of data from those regions.

Conclusions

What can explain the persistently high fire death rates in the United States compared to other countries? The fires of greatest concern are residential fires, since this is where the majority of fire deaths in all countries occur. Although statistical data are not available, the United States is widely believed to have many more residential fires on a per capita basis than the other countries studied. This higher fire rate, as well as the United States' higher fire death rates, are likely a product of several factors:

- *The United States commits fewer resources, both in terms of dollars and staff time, to fire prevention activities than other industrialized countries.* The vast majority of fire department resources are focused on fire suppression rather than on fire prevention. It is estimated that less than 3 percent of all U.S. fire department budgets are allocated to prevention activities, whereas other industrialized countries are spending between 4 and 10 percent of their budgets on prevention.[3]

- *There is greater tolerance in the United States for "accidental" fires.* In other countries, particularly countries in the Pacific Rim, fire safety is explicitly taught in the schools, by other public institutions, and in the home. Fire safety education concentrates on preventing fires that in the United States might be considered accidental such as cooking and heating fires. In many countries, families are stigmatized if a serious fire breaks out in their home and potentially threatens their neighbors' homes.

- *Whether through ignorance or a false sense of confidence, Americans practice riskier and more careless behavior than people in other countries.* Examples of these behaviors include careless smoking, leaving cooking unattended, and improper use of space heaters. These types of fires occur in other countries, but high relative rates of fire deaths suggest that either fires occur much more frequently in the United States or, at a minimum, they are much more severe.

In sum, industrialized countries in Europe and Asia can provide the United States with valuable lessons on reducing the incidence of residential fires and residential fire deaths through fire prevention. U.S. fire departments have many strengths, particularly in the development and use of highly advanced fire suppression technologies and fire service delivery mechanisms. But U.S. fire departments' overall effectiveness in minimizing total dollar and human losses can be significantly enhanced by copying fire prevention practices from other countries.

---

[3] Schaenman, Philip, "Reinventing Fire Protection," *Firefighter's News*, February – March, 1994, p. 44.

ARSON

Arson is the leading cause of all fires, and it annually kills hundreds of Americans, injures thousands, and equates to billions of dollars in damage to property. More than 500,000 arson fires have been recorded in each of the last 10 years. During 1994, the total number of arson fires was estimated at over 548,000. More than 107,000 of these arson fires occurred in structures, both residential and non-residential. Arson fires also accounted for an estimated 560 fire deaths, 3,440 fire injuries, and $3.6 billion in property damage (Figure 133). The 560 fire deaths in 1994 resulting from arson fires, however, was at a 10-year low, a 37 percent decrease over 1993.[4]

Magnitude of the Arson Problem

In 1994, arson fires accounted for 28 percent of all fires occurring in the United States, making arson the leading cause of fire (see Figure 31, Chapter 2). In comparison, the second and third leading causes (open flame fires and cooking fires) combined accounted for less than arson fires.

Each year approximately 65–70 percent of arson fires are outdoor fires, 20–25 percent are structure fires, and 10–15 percent are vehicle fires. The 1994 distribution of arson fires and fire losses is shown in Figure 134. Trends in arson fires and deaths are shown in Figure 135. Outdoor arson fires do not kill or injure as many people as structure fires, but they are a cause for serious concern for at least two reasons. First, the proximity of occupied or unoccupied structures to vacant lots or other areas where outdoor arson fires are set increases the risk of exposure fires. Second, outdoor fires are often "gateway" fires, particularly for juvenile firesetters. They begin setting fires in trash cans, fields, or empty lots but then move on to targets that bear increasing risk to persons and property.

Each year arson accounts for a high proportion of all losses due to fire in the United States in terms of lives and property. As shown in Figure 31, Chapter 2, arson fires were responsible for 16 percent of all fire deaths in 1994, making arson the second leading cause of fire deaths and ranking behind only careless smoking as the most deadly cause of fire. Arson was also the second leading cause of fire injuries, accounting for 14 percent of all those injured in fires. Only cooking fires injured more people than arson fires. For direct property losses, arson is the leading cause of property damage due to fire. In 1994 the dollar losses attributable to arson accounted for 28 percent of all fire property losses, twice that of the second leading cause, electrical distribution (14 percent).

ARSON FIRES IN STRUCTURES—ALL TYPES. Structure fires, while comprising about one-quarter to one-fifth of all arson fires each year, account for a majority of all arson losses (deaths, injuries, and damage to property) (Figure 134). In 1994, 20 percent of all arson fires occurred in structures. These fires alone accounted for 90 percent of all arson fire deaths, 89 percent of all arson fire injuries, and 89 percent of all arson dollar losses.

---

[4]While "arson" is technically a legal term, it is used here to refer to all fires of incendiary or suspicious origins.

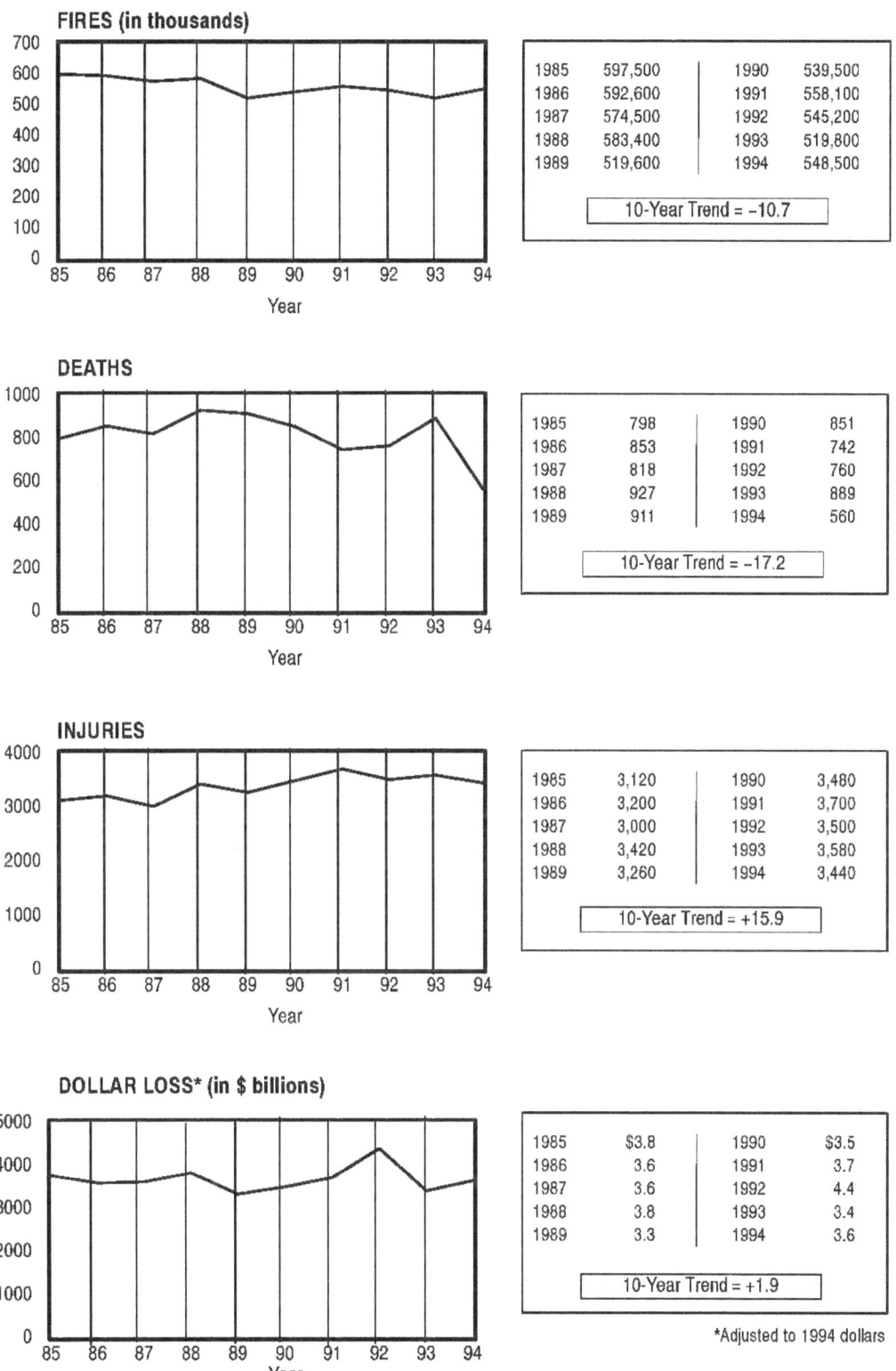

FIRES (in thousands)

1985	597,500	1990	539,500
1986	592,600	1991	558,100
1987	574,500	1992	545,200
1988	583,400	1993	519,800
1989	519,600	1994	548,500

10-Year Trend = –10.7

DEATHS

1985	798	1990	851
1986	853	1991	742
1987	818	1992	760
1988	927	1993	889
1989	911	1994	560

10-Year Trend = –17.2

INJURIES

1985	3,120	1990	3,480
1986	3,200	1991	3,700
1987	3,000	1992	3,500
1988	3,420	1993	3,580
1989	3,260	1994	3,440

10-Year Trend = +15.9

DOLLAR LOSS* (in $ billions)

1985	$3.8	1990	$3.5
1986	3.6	1991	3.7
1987	3.6	1992	4.4
1988	3.8	1993	3.4
1989	3.3	1994	3.6

10-Year Trend = +1.9

*Adjusted to 1994 dollars

Sources: NFIRS, NFPA Annual Surveys, and Consumer Price Index

Figure 133. Trends in Arson Fires and Fire Losses

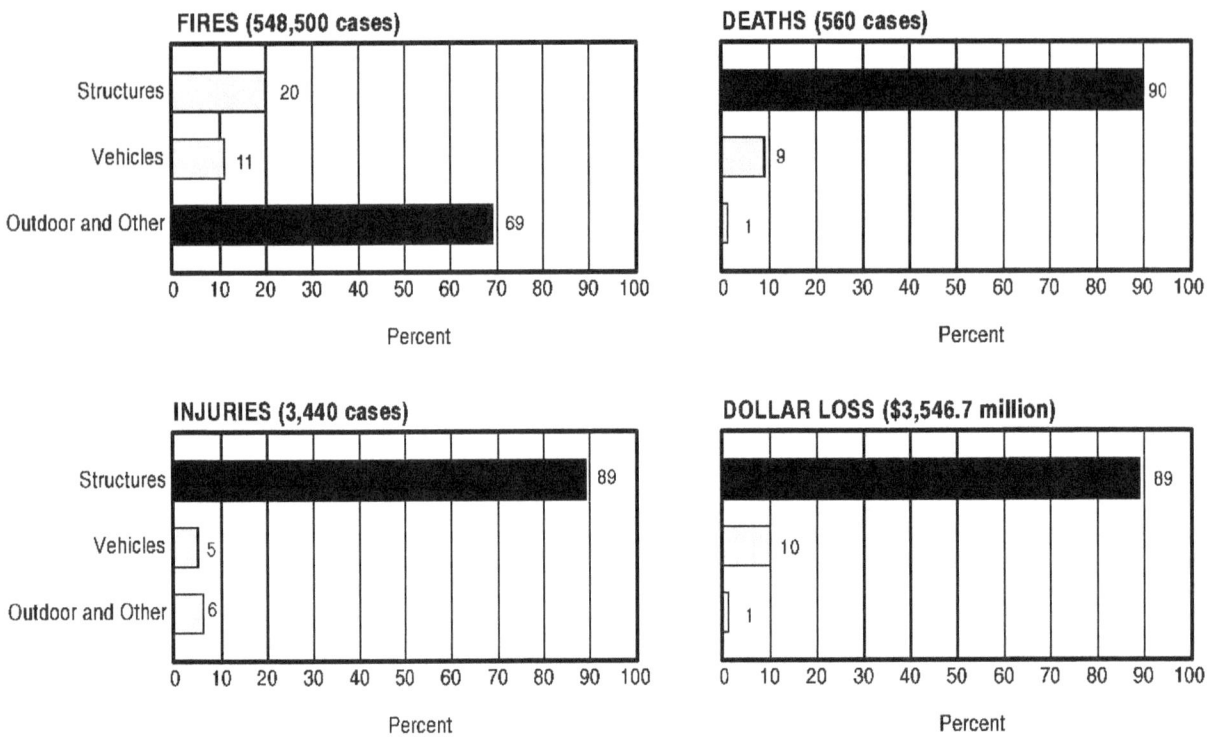

Sources: NFIRS and NFPA Annual Surveys

Figure 134. 1994 Arson Fires and Fire Losses by Major Occupancy Type

ARSON FIRES IN RESIDENTIAL STRUCTURES. The category "structures" includes both residential and non-residential properties. Analyzing these property types separately reveals that arson fires in residential structures account for higher losses than non-residential structures in terms of both fire deaths and fire injuries. Of all the people killed in arson structure fires in 1994, 96 percent were killed in residential structures according to NFIRS data. Similarly, of all injuries in structure fires caused by arson, 82 percent were injured in residential properties.

TRENDS IN ARSON FIRES. Figure 15, Chapter 2, shows that, overall, the number of fires occurring each year in the United States has fallen 19 percent over the past 10 years. Although the absolute number of arson fires occurring in structures has also fallen, the trend lines in Figure 136 show that arson in residential structures is decreasing slower than arson in non-residential structures (a 19 percent decrease versus a 34 percent decrease over 10 years). Consequently, the trend is for an increasing proportion of arson fires in any given year to be in residential structures. This is an important trend to watch given that higher losses, in terms of both lives and property, are associated with arson in residential structures than in non-residential or commercial structures.

While the number of residential deaths due to arson each year can vary substantially, Figure 137 reveals that over the past 10 years the overall trend has declined about 17 percent in the number of residential arson fire deaths. This is compared to a 32 percent decline in the overall number of residential fire deaths over the same period. Deaths in 1994 dropped to a 10-year low.

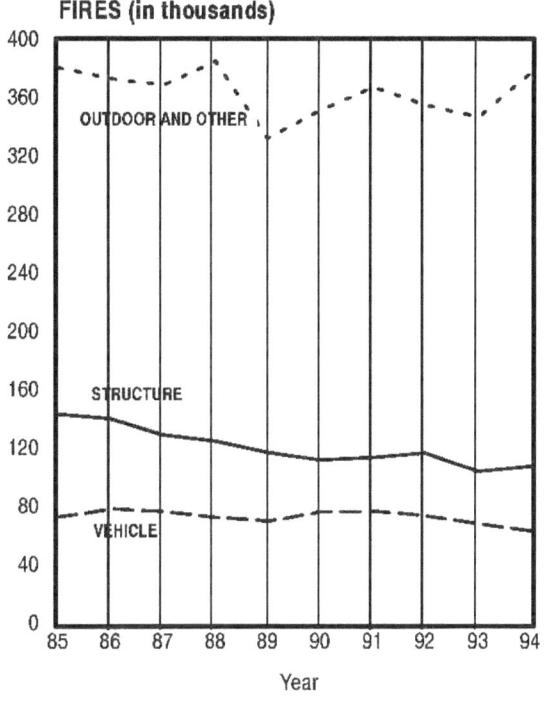

FIRES (in thousands)

	Structure	Vehicle	Outdoor and Other
1985	143.6	72.9	381.0
1986	140.5	78.6	373.5
1987	120.1	76.5	368.9
1988	125.1	72.5	385.8
1989	117.0	70.0	332.6
1990	111.9	76.2	351.4
1991	113.9	76.8	367.4
1992	116.6	73.5	355.1
1993	104.4	68.2	347.2
1994	107.8	63.1	377.6

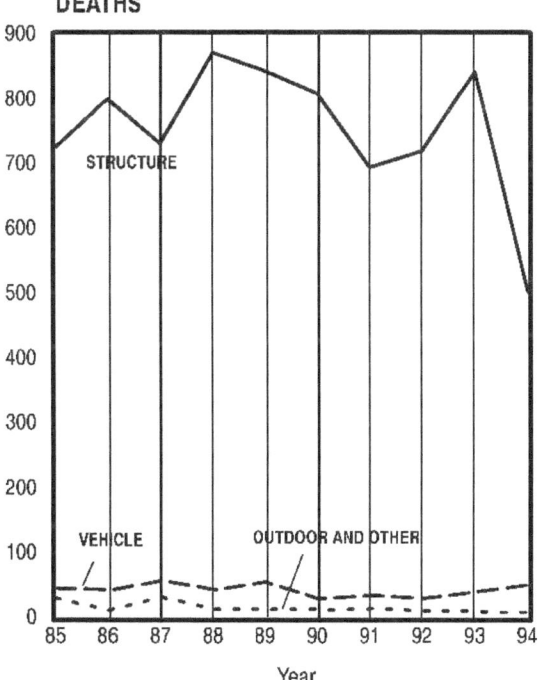

DEATHS

	Structure	Vehicle	Outdoor and Other
1985	724	45	30
1986	800	42	11
1987	731	57	31
1988	871	43	13
1989	843	55	13
1990	809	30	12
1991	694	34	14
1992	720	30	10
1993	841	40	8
1994	502	51	7

Sources: NFIRS and NFPA Annual Surveys

Figure 135. Trends in Arson Fires and Deaths by Major Occupancy Type

FIRES (in thousands)

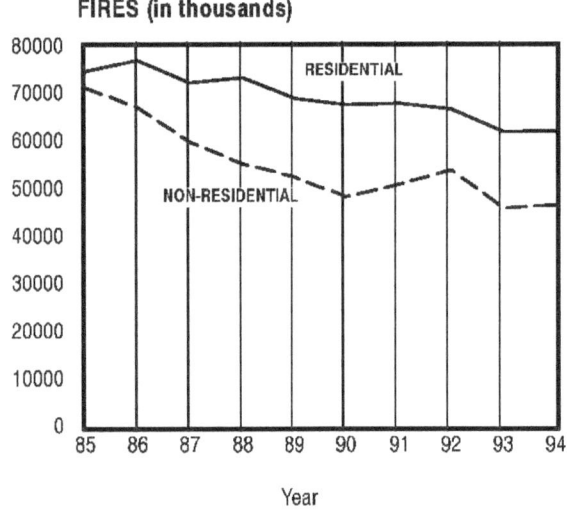

	Residential	Non-Residential
1985	74.6	71.1
1986	76.9	67.0
1987	72.2	60.0
1988	73.2	55.0
1989	68.8	52.4
1990	67.5	48.2
1991	67.7	50.9
1992	66.6	53.8
1993	61.9	45.9
1994	61.9	46.5

Residential 10-Year Trend = –34.0%
Non-Residential 10-Year Trend = –18.9

Sources: NFIRS and NFPA Annual Surveys

Figure 136. Trends in Non-Residential vs. Residential Structure Arson Fires

DEATHS

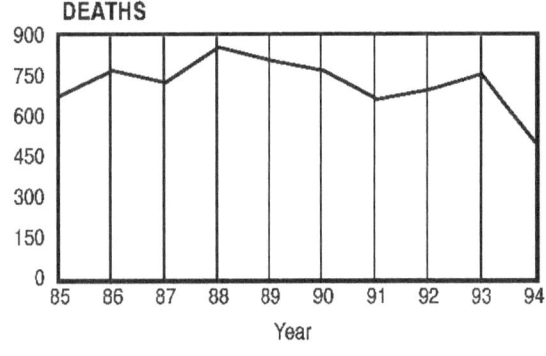

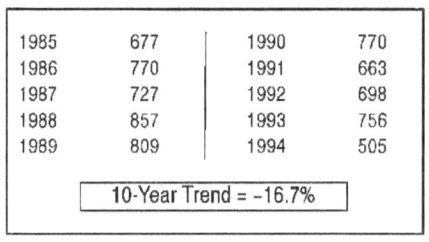

1985	677	1990	770
1986	770	1991	663
1987	727	1992	698
1988	857	1993	756
1989	809	1994	505

10-Year Trend = –16.7%

INJURIES

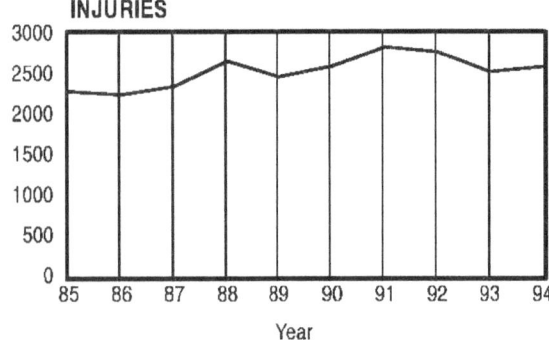

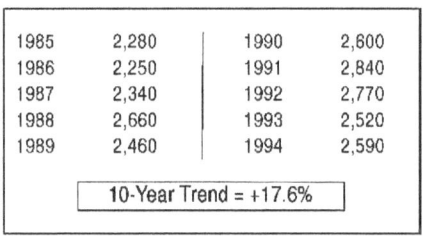

1985	2,280	1990	2,600
1986	2,250	1991	2,840
1987	2,340	1992	2,770
1988	2,660	1993	2,520
1989	2,460	1994	2,590

10-Year Trend = +17.6%

Sources: NFIRS and NFPA Annual Surveys

Figure 137. Trends in Residential Structure Arson Casualties

In contrast to the trend in arson fire deaths, the trend in residential arson fire injuries is upward—about 18 percent over the past 10 years. This is similar to the overall level of residential fire injuries, which have increased by 20 percent over the past 10 years.

METROPOLITAN VS. NON-METROPOLITAN RESIDENTIAL ARSON FIRES. While arson is among the leading causes of all residential fires, the data suggest that arson fires represent a higher proportion of all residential fires in metropolitan areas than in non-metropolitan areas. In 1994, arson fires comprised 14 percent of all residential fires with known causes and were the third leading cause of all residential fires. However, in metropolitan areas, arson fires accounted for 24 percent of residential fires and were the leading cause of residential fires. Cooking fires were second, accounting for 21 percent of metropolitan residential fires.[5]

This relationship between the character of place and the incidence of arson is borne out in other studies. A 1995 study by the National Fire Protection Association shows that for the period 1990–1994, the rate for incendiary and suspicious fires in cities of 250,000 or more was greater than twice the rate for communities of 5,000 to 10,000 or 10,000 to 25,000 (Figure 138). For rural areas with under 2,500 residents or with 2,500 to 5,000 residents, the rates were higher than for communities of 5,000 to 10,000, but lower than the rates for the largest places.

In the same study, NFPA reported that arson fires represented a higher proportion of all fires reported in large cities than in smaller cities for the period 1990–1994. The proportion of all fires attributable to suspicious and incendiary causes is more than two times higher in large cities than in smaller communities, as shown in Figure 139.

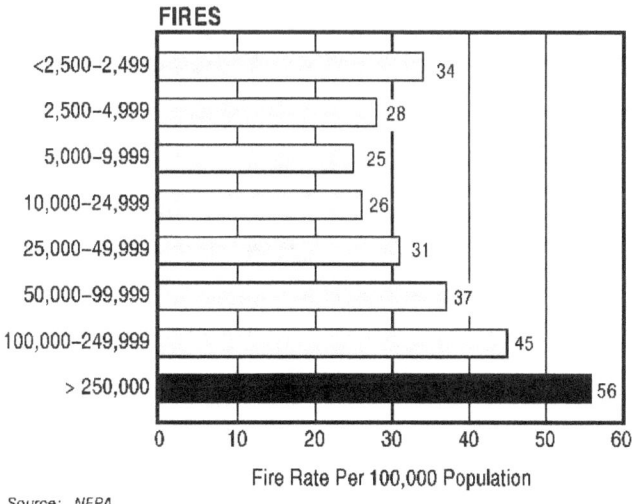

Figure 138. Severity of Arson Fires by Area Population Size (1990–94)

---

[5] United States Fire Administration. *Fire in the United States, 1983–1990*, Eighth Edition, p. 355. This information is intended to give readers a general idea of how metropolitan and national data compare, but it must be interpreted with care. Due to the way the data are collected and reported in NFIRS, some areas that are largely rural in character are included under metropolitan areas.

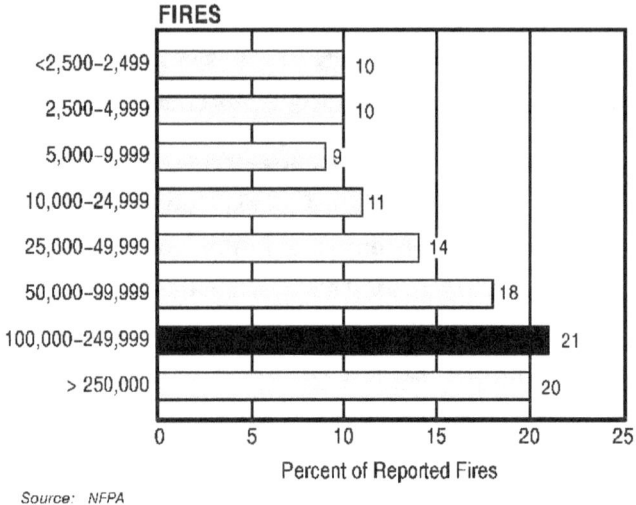

FIRES

Source: NFPA

Figure 139. Arson Fires by Area Population (1990–94)

In another study, NFPA analyzed the types of structures that experienced arson fires and the distribution of arson losses between 1990 and 1994. The results appear in Figure 140. Residential structures experience the highest number of arson fires and account for the highest proportion of total arson dollar losses of all structure types. Important to note, however, is that while stores and offices experience only 6 percent of all arson fires, they account for 20 percent of all dollar losses resulting from arson fires.

Characteristics of the Arson Problem

TIME OF DAY. Figure 141 shows the distribution of arson fires throughout the day for 1994. The two time periods with the highest proportion of arson fires are from 4 p.m. to midnight. This period accounts for almost half of all arson fires.

DAY OF THE WEEK. Figure 142 shows the distribution of arson fires throughout the week for 1994. Although arson fires are fairly well distributed throughout the week, there are proportionately more on Saturday and Sundays. Together, these weekend days account for almost one-third of all arson fires.

FIXED PROPERTY USE. Figure 143 shows the distribution of the types of properties that experienced arson fires in 1994. The six categories of fixed property that experienced the highest number of arson fires are listed. Together, these properties accounted for 80 percent of the fixed property use types that experienced arson fires.

Motives of the Firesetter

As with any crime, people set arson fires for varied and complex reasons. For the purposes of prosecuting criminals in other types of criminal cases, motive is often a secondary consideration and

190

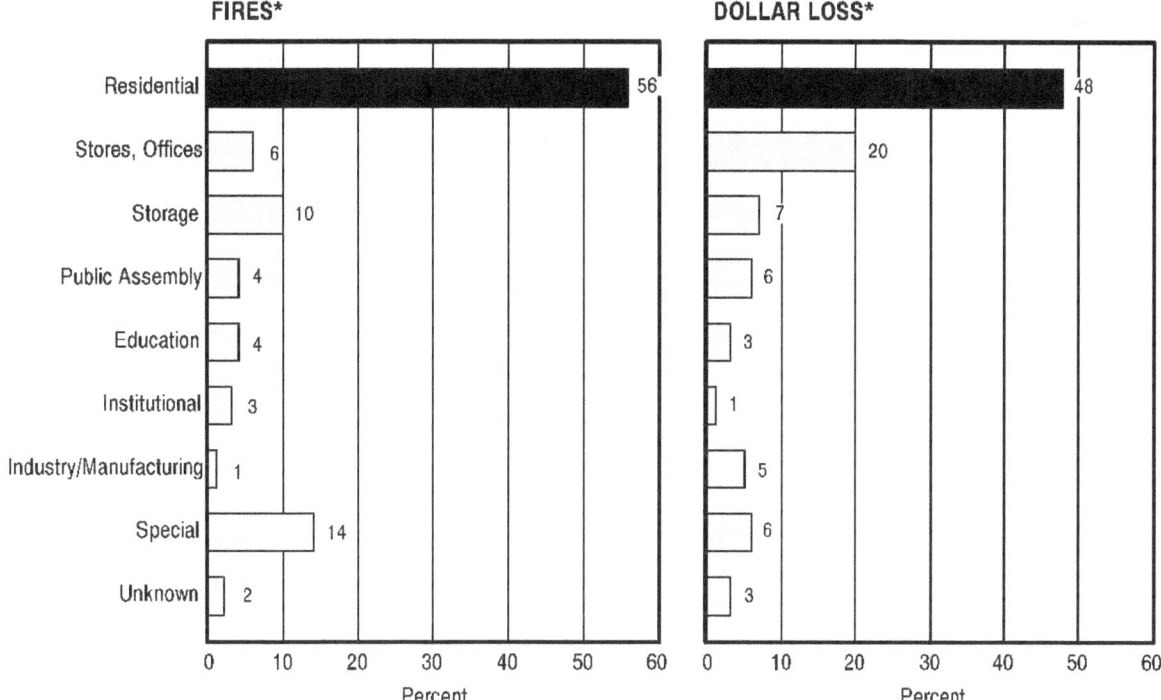

*Numbers of cases and actual dollars are not reported by NFPA.

Source: NFPA Annual Surveys

Figure 140. 1990–1994 Arson Fires and Dollar Loss by Property Type (1990–94)

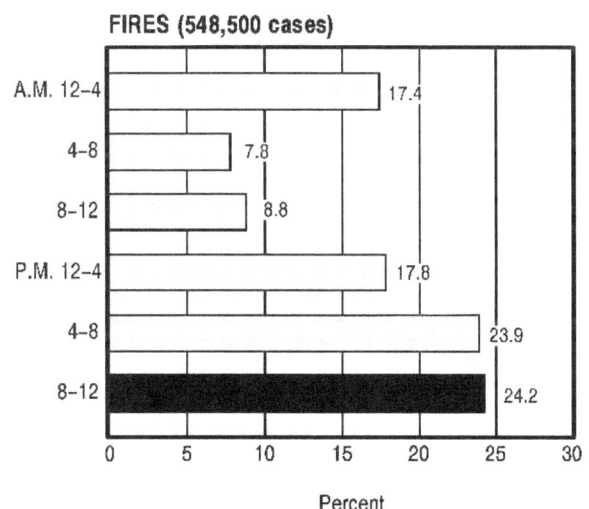

Source: NFIRS

Figure 141. Time of Day of 1994 Arson Fires

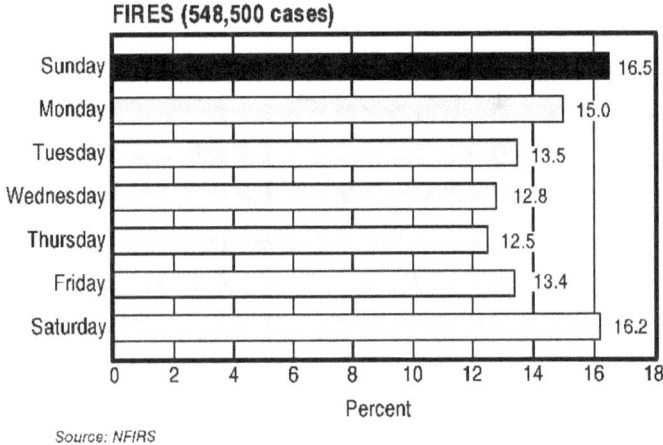

Figure 142. Day of Week of 1994 Arson Fires

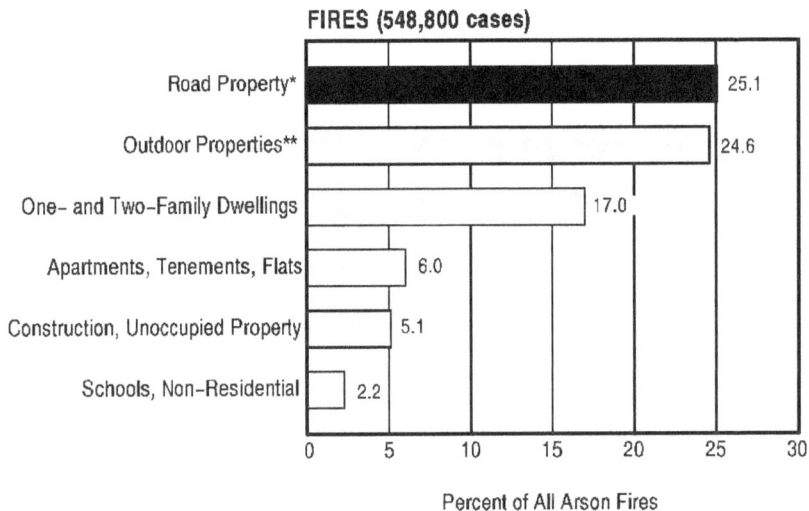

* Refers to vehicle fires.
** This is one of the NFIRS categories that refers to outdoor fires.

Source: NFIRS

Figure 143. Fixed Property Use of Locations Experiencing 1994 Arson Fires

is not necessarily crucial for conviction. But because arson is a clandestine crime where witnesses are rare and some or most of the direct evidence may burn in the fire, motive becomes a critical element in prosecuting firesetting cases. Pinpointing the motivation for setting a fire helps identify suspects and helps convince a jury of the accused's guilt. The most common motives behind firesetting are:

- Vandalism
- Spite and revenge, including angry spouses, ex-spouses, and boyfriends and girl-friends; disgruntled employees or former employees; and gang-related retaliation

- Intimidation

- Concealment of a crime

- Economic motives, including insurance fraud, debt removal, direct monetary gain, elimination of unwanted ownership, land assembly for development, and removal of business competition

- Emotional/psychological dysfunction, including juveniles in crisis; vanity ("hero" and attention), especially fires set by firefighters; and pyromania.

Of particular concern is the involvement of juveniles in intentionally set fires. The NFPA reports that in 1994, for the first time, juvenile firesetters accounted for a majority (55 percent) of those who were arrested on arson charges.[6] The growing involvement of youths in setting fires is cause for serious concern and deserves more attention from policymakers.

In a U.S. Fire Administration-sponsored management assistance project, investigators from nearly 60 state and local fire investigation units ranked spite and revenge as the most frequent motive behind incendiary fires. This is a common motive for both adult and adolescent firesetters. A fire set by someone bent on revenge is a premeditated act directed at an individual he or she knows and wants to hurt, such as a former spouse or business partner. Consequently, these tend to be the most dangerous fires in terms of casualties because volatile emotions motivate the crimes. Other spite and revenge arsons include gang-related fires that are set to exact revenge against rivals or their rivals' sympathizers.

Fires set for the sport of vandalizing property were ranked high as motives in the USFA study. Juveniles were responsible for the majority of these arson cases.

Some arsonists use fire to conceal companion crimes such as murder, embezzlement, or burglary. For example, fire death victims have been found with bullet wounds, which is why it is important to conduct autopsies on all fatalities found in fires. Medical examiners can determine whether fire or other causes resulted in the death.

There is some debate as to the relative frequency of fraud as a motive for arson. Although arson done for direct profit or to eliminate debt remains a problem, there are signs that this is less frequently behind firesetting today than it had been previously. One of the complications in assessing the frequency of arson cases motivated by fraud is that they tend to be more complex than other types of cases and require more investigation time.

Given the destructiveness of arson fires, a better understanding of their occurrence and nature is needed. Particularly important is more data on the incidence of arson fires and the motivations of arsonists in setting them. The National Fire Incident Reporting System only collects "cause" of fire data; no additional information is collected on incendiary fires. Unless all suspected cases of

---

[6] Hall, John R. Jr., *U.S. Arson Trends and Patterns—1994*. Quincy, MA: National Fire Protection Association, 1995, p. 3.

arson are pursued and classified as incendiary or suspicious via reporting systems such as NFIRS, our understanding of the arson picture in the United States will remain incomplete. One possibility is to develop a national protocol on collecting data on incendiary and suspicious fires. This would allow collection of data critical to understanding the nature of the arson problem, including types of firesetters, firesetter motives, and prior histories of firesetting among firesetters.

There are many challenges to collecting quality data, and primary among these are circumstances at the local level that may complicate data collection processes. Inadequate investigator training in some areas, particularly rural areas, means that many fires of an incendiary or suspicious nature are not identified and do not get fully investigated. Also, budgetary constraints or other demands on staff have reduced the level of fire and police resources available to investigate suspected arson cases. These circumstances exist even though many are aware that closer cooperation between fire and law enforcement officials is critical to stemming the tide of arson fires. Without thorough investigations and, as importantly, the willingness of local prosecutors to pursue cases against suspected arsonists, it is difficult to make headway in the effort to reduce the incidence of arson fires in the United States.

WILDLAND

Wildland fires have long been a concern of the rural residents of the United States, impacting the culture in many ways. Fire is a tool used by farmers to clear their fields after harvest and is an important ecological factor in wilderness areas. The general public's awareness of fire has grown over the last 25 years as the country's urban areas continue to expand into formerly rural and wilderness areas. The zone where communities and wildland fire characteristics overlap is now described as the wildland/urban interface. The last three decades have seen significant growth of the population living in the interface zone, and several catastrophic fires that have occurred there demonstrate the seriousness of this growing problem.

History

Catastrophic fires are not a new phenomenon in this country. In the fall of 1871, more than 800 people were killed and all but one structure destroyed in Peshtigo, Wisconsin, in what we would describe today as a wildland/urban interface fire. The early twentieth century saw a continued debate about the role of fire and the value of deliberate fires for clearing land and for lessening the chance of larger catastrophic fires by clearing fuel—a prescribed burn. The birth of the U.S. Forest Service in 1905 marks the beginning of a concerted and focused effort on the part of the government to extinguish wildfires. The commitment to fighting fires and the country's belief that they should be fought at nearly any cost was reinforced by the dramatic and historic fires of 1910 in which 5 million acres burned and 78 firefighters lost their lives. Internally, the Forest Service set itself the objective of having a fire under control by "10:00 a.m. the next morning."

The efforts to control fire were so effective over several decades that there has been a build up of fuels in many wildland areas that normally would have been consumed by periodic fire. The federal agencies recognized this growing ecological and fire hazard and modified their firefighting approach to "let burn" fires that were within a prescription for a given area. This translated into aggressive firefighting in areas abutting towns and communities and a less aggressive effort on fires that were burning in wilderness areas as long as the weather and other factors were within set limits, and there was a "natural" and desirable impact on the ecology. This was the case in Yellowstone in 1988 when fires that were originally "in prescription" went out of prescription and burned large areas of the park. The Yellowstone fires brought the issue of wildfire back into the realm of national debate, which continues today.

Many foresters advocate the use of "prescribed fire," that is, fire that is set during low fire danger periods, to remove fuels and thereby lessen the chance of large catastrophic fires. Using prescribed fire does entail the risk of it "escaping" and burning beyond the objectives set forth when it was lit. Also, smoke from a prescribed fire is the same as from a wildfire and is equally unpopular with residents of nearby communities. But despite the risks, prescribed fires are considered a tool for avoiding even larger risks from extremely large fires. The role of prescribed fire continues to be debated within the federal agencies that fight fire, at the state level, and by the public.

Interface Challenge

From 1985 to 1994, nearly 9,000 homes were destroyed by fire in interface areas.[7] The people who move into interface zones often do so because of the proximity to rural and wildland areas. They pride themselves on the integration of their homes into the surrounding vegetation. Unfortunately, the dictates of taste are often in contrast to the concept of "defensible space" and other steps homeowners can take to defend against interface fires.

The Wildland/Urban Interface Fire Protection Program has identified four common components of the interface fires it has examined. First, low relative humidity, high temperatures, and high winds often are in place before the fire starts. Second, human activity such as arson, debris burning, or downed electrical wires cause many interface fires. Third, many of the homes destroyed are constructed of combustible material or have especially vulnerable features such as wood shingle roofs. Fourth, a lot of combustible materials surround the home, such as woodpiles or fences.

As people relocate into the wildland/urban interface, they may bring with them a perception that, as is the case in urban areas, local fire departments are capable of handling virtually all of the fires that occur. The reality is that the interface presents a different set of risks than is usually faced by urban departments.

---

[7] *NFPA Journal*, March/April 1994, p. 84.

For example, in August of 1992 the Fountain Fire in Northern California burned 64,000 acres of forest land and destroyed 330 homes, 37 businesses, and more than 200 other structures.[8] Although overwhelmed, the system was not caught off guard, mutual aid agreements were in place, and all those called on responded quickly. But the fire overwhelmed the ability of the local fire jurisdiction. The suppression costs were over $20 million in tax dollars and losses were over $105 million, but no one died in the fire.

In 1991 the Oakland Hills fire overwhelmed the Oakland Fire Department. It destroyed 3,000 structures and killed 25 people in one of the most densely populated metropolitan areas in the United States. Fire officials in the Bay Area were well aware of the risk of such a disaster and had predicted that such a fire would eventually occur. No one predicted the over $1 billion worth of damage or how completely overwhelmed the fire suppression system would be. Residents abandoned their homes and fled by foot as firefighters and police officers retreated in the face of the oncoming walls of fire. For the first several hours, the fire was completely out of control and free to burn where it wished. Frantic callers who were able to reach 911 expecting to be told a fire engine was on its way were told instead to gather their family and evacuate the area.

It is estimated that $250–$300 million of the federal wildland fire suppression budget was spent in protecting the wildland/urban interface in 1994.[9] The role of federal government in wildland/urban fire protection is not completely defined or coordinated. Except for the National Park Service, the federal agencies involved in fire suppression do not have specific responsibilities to protect structures. And the Park Service structural responsibility is limited to only those structures located on NPS land. Although agency mission statements do not include responsibilities for structural protection, federal wildland resources are often expected to come to the aid of nearby communities during interface fires.

Mitigation

The lack of public enthusiasm for interface fire mitigation is a significant issue. Local communities are reluctant to enact zoning regulations and building codes that might decrease the risk of interface fires because of the associated increase in building cost and public opposition to further regulation. Some have suggested that property insurers should move to reward wildfire-awareness construction through reduced premiums. Insurance companies currently set premiums based on zone or neighborhood characteristics, not on individual homeowners' efforts to decrease the vulnerability of their own property. Some programs are underway, however, that might become models for interface mitigation. Summit County, Colorado, is incorporating hazard and risk assessment into zoning requirements. Other communities are working to educate property owners about the risk of interface fire and what they can do to reduce the risk. A state forester in Colorado observed that new

---

[8] *Fire Protection in the Rural American: A Challenge for the Future*, 1994, p. 16.
[9] "Interface," *Federal Wildland Fire Policy and Program Review Report*, 1995.

residents moving from wildland fire-prone areas of California are bringing their interface awareness with them.

Most of the structures destroyed in interface fires are destroyed in the first few hours of the fire when the fire suppression resources are overwhelmed. The public must recognize the potential of interface fire. Local governments need to educate their citizenry about wildfires and support efforts. Citizens must work with their local government to ensure that the risks are reduced before a fire starts. This is achieved by maintaining a defensible space around their property, using fire-resistant materials, and employing fire-wise landscaping.

Special Concerns for the Future

More people are moving into interface zones. And more forests have declining health and greater fuel buildup. The result is a growing number of potential wildland fires of high intensity that will threaten homes and communities. These fires are also a greater threat to firefighters, who not only have a more severe fire to contend with but will also probably have to fight the fire differently when it is in the vicinity of homes. Federal and state funding cutbacks have reduced the number of wildland firefighters, so more of the burden of initial attack on a wildland fire is shifting to rural fire departments. In many cases, these are small volunteer fire departments with inadequate wild-land fire training or equipment. In turn, this raises the risk to both the firefighters and the communities. Fire resources, public expectations, and prevention/mitigation efforts have to be better balanced to effectively address this growing fire problem.

TOTAL COST OF FIRE IN THE UNITED STATES

Over the last two decades, many attempts have been made to estimate the total annual cost of fire in the United States. The total annual cost is much greater than just the value of property destroyed by fire. The total cost includes the cost of fire services; the cost of fire protection built into buildings and equipment; the cost of fire insurance overhead; the many indirect costs, such as business interruptions, medical expenses, and temporary lodging; the value to society of the injuries and deaths caused by fire; the cost of government and private fire-related organizations; and the myriad of other related costs that add up to a very large economic impact. The same is true for any industrialized nation, but most nations have not estimated this total cost at all, and very few have done it in detail.

The total loss of fire is on much larger scope than people think, in the range of $60 to $120 billion a year, depending on how it is defined and the estimation methodology used. The first modern attempt to estimate the total cost of fire for the United States was undertaken by the U.S. Fire Administration as a project for a team of fire protection engineering students from Worcester Polytechnic Institute (WPI) circa 1980. This initial estimate was based on first-cut thinking about the problem and has been widely disseminated.

A more recent effort to estimate the total cost of fire was made in 1993 by William Meade, an economist for the National Institute of Standards and Technology.[10] It made initial estimates of some new cost areas that, though crude, have yet to be improved upon.

Dr. John Hall, the head of fire data analysis at NFPA (and a former USFA fire data analyst), has made a series of estimates of the total cost of fire built on the WPI and Meade estimates, with further development of the methodology. His most recent estimate of the total cost of fire was released in 1995.[11] The total cost of fire protection is much greater than most would suspect. In 1983 the total cost of fire was conservatively estimated by Hall to be on the order of $32.9 billion. By 1993, the estimate had grown to $45.5 billion, an increase of $12.6 billion over a 10-year period. Much of this increase can be attributed to inflation. Using a broader definition of costs—including an estimate of the value of volunteer firefighters labor—Meade estimated the total cost of fire to be between $92 to $139 billion (Table 17).

Table 17. Summary of Total Cost of Fire

Cost Component	Range of Cost Estimates ($ billions)	Most Likely Estimate ($ billions)
Category A: Losses		
Residential Property	4.0	4.0
Industrial Property	4.2	4.2
Other Property	0.7	0.7
Residential Interruption	0.6–1.0	0.8
Business Interruption	6.1–8.4	8.4
Product Liability	3.5	3.5
Category B: Insurance		
Product Liability	0.1	0.1
Net Fire Insurance	5.6	5.6
Category C: Fire Service		
Paid	9.6	9.6
Volunteer Conversion	16.2–36.8	30.0
Category D: Preventative		
Built into Structures	20.7	20.7
Built into Equipment	13.5–22.5	18.0
Standards Activity	0.1–0.6	0.2
Retardants/Testing	1.9–4.0	2.5
Fire Maintenance	4.3–16.6	6.5
Disaster Recovery	0.6	0.6
Total	91.7–138.9	115.4

Source: Meade, William, *A First Pass at Computing the Cost of Fire in a Modern Society*, The Herndon Group, March 1991; prepared for Center for Fire Research, National Institute of Standards and Technology, Gaithersburg, Maryland.

---

[10]Meade, William, *A First Pass at Computing the Cost of Fire in a Modern Society*, The Herndon Group, March 1991; prepared for Center for Fire Research, National Institute of Standards and Technology, Gaithersburg, Maryland.

[11]Hall, John, *The Total Cost of Fire in the United States Through 1993*, NFPA Report, October 1995.

Total Cost Components

When people talk about the cost of fire, the most common statistic quoted is *direct losses from fire*—what was burned up or damaged by fires—but this is only a small fraction of the total. Other categories of losses and costs must be taken into account to fully estimate the total cost of fire.

A second type of major cost category is the *cost of the fire service*. People often focus on the fires themselves and forget the cost to the public and government to maintain a "ready army" of firefighters, equipment, and stations. This includes the cost of local paid and volunteer departments, forest fire management, and even the portion of the municipal water supply attributed to the cost of fire protection. For example, some water mains are sized to meet firefighting needs and are larger than would be necessary for household water supplies.

Insurance overhead is the cost paid by the public for insurance less what is returned to the public in payments for insured losses (which are accounted for as part of direct losses).

There is also an *indirect loss from fire*, including business interruptions, costs of temporary lodging, tax losses, loss of market share, legal expenses, and many other categories. This is one of the most difficult categories to estimate with a high degree of accuracy.

Another category of costs is the cost of *fire protection built into buildings, equipment, infrastructure, and business operations*. Built-in fire protection in buildings is hard to quantify. Much of the built-in protection against fire also provides protection from other hazards (e.g., thick walls offer resistance to fires and provide structural integrity, electrical safety features reduce the hazard of electrical shock as well as fires).

The cost of fire protection built into equipment is even more difficult to estimate because there are so many more types of equipment than buildings. Equipment ranging from color TVs to portable space heaters to cigarette lighters have special features to prevent them from becoming the "equipment involved in ignition."

The cost of operations affected by fire considerations includes the training of employees in fire safety, cost of special transportation for flammables, the use of special containers for flammables, and work time lost evacuating buildings from false alarms.

Finally, there is the *cost of deaths and injuries* to society. Part of these costs are conceptually clear if difficult to estimate, such as the cost of medical treatment, funeral expenses, and time lost from work. Other, more conceptually difficult and, to some, distasteful costs to include are the value of a life and of pain and suffering. Estimating these aspects of losses is often done as part of cost effectiveness studies in other fields.

The most recent list of components of the total cost of fire is in Table 18. This was done for a study of the cost of fire in Canada.

Table 18. Elements of the Total Cost of Fire

Type of Cost (and Major Components)	Percentage of Total Cost
I. Direct Dollar Losses from Fires	15
• Fires reported to the states by the fire service or insurance companies • Fires that go unreported to the fire service and insurance companies • Wildland fires	
II. Cost of Fire Services	20
• Municipal career or part-paid fire departments – Personnel (including benefits/social costs) – Hardware: Fire apparatus, supplies, equipment, vehicles – Stations – Water system (fire-related cost) • Volunteer departments – Hourly or per-call wages – Equipment – Pensions – Attributed cost of replacing volunteers (option to include) • Industrial fire brigades • State and national fire forces (including state fire marshal offices) • Military firefighting forces • Management of forest fires	
III. Cost of Fire Protection in Structures (Buildings and Other Engineered Structures)	28
• Active fire protection systems – Detection, alarms, sprinklers, halon, and other suppression agents – Extinguishers, standpipe systems, smoke control systems • Passive fire protection (above structural needs) – Fire-related construction elements (e.g., fire-rated ceilings, floors, walls, doors, cladding, compartmentation, trenches for containing liquids in chemical plants) – Extra exiting – Extra spacing between structures (land value) • Fire protection in building systems – Fire-rated components and design of permanent electrical and mechanical systems • Infrastructure costs (e.g., wider roads and turnarounds for fire vehicle access)	
IV. Cost of Fire Protection in Equipment, Vehicles, Goods, and Industrial Operations *(Beyond what is needed to function or for shock protection, such as tipover switches in kerosene heaters, gas shutoff valves, protection around fuel tanks in cars)*	20
• Equipment and vehicles – Civilian – Military (ships, planes, etc.) • Industrial operations (e.g., fire safety training, fire drills; electrical, gas, and oil industry safety operations)	
V. Insurance Overhead and Profit	3
• Total cost of premiums less payouts	

Continued on next page

Table 18. Elements of the Total Cost of Fire (continued)

Type of Cost (and Major Components)	Percentage of Total Cost
VI. Indirect Losses From Fires	3

• Indirect losses to businesses, including:
  – Business interruption losses
  – Temporary displacement expenses
  – Long-term losses in market share
  – Secondary losses in dependent businesses
• Indirect losses to residences, including:
  – Temporary lodging, vehicles, and other living expenses
• Litigation expense
  – Legal costs (before and after a fire)
  – Settlements
• Tax losses
• Environmental impacts of fires and fire protection (e.g., halon impact on ozone layers; aquifer damage from runoff of contaminated water; air pollution)	

VII. Attributed Cost of Lives Lost and Injuries From Fires	10

• Civilian and firefighter fatalities (reported and unreported)
  – Medical costs, funeral costs, attributed value of life lost
• Civilian and firefighter injuries (reported and unreported)
  – Medical expenses, attributed cost of pain, suffering, and lost income	

VIII. Miscellaneous Costs	0.9
• Regulatory, research, and testing • National and state fire agencies and associations • Disaster recovery	
Total	100

Source: Schaenman, Philip, et al., *Total Cost of Fire in Canada,* The National Research Council of Canada Fire Research Laboratory, December 1994.

Conclusion

The total cost of fire is very large. Different sources estimate the cost at perhaps as much as 6 to 12 times higher than the direct losses alone. This puts fire protection up among the larger national problems in terms of its economic impact. The full magnitude of the problem is probably under-appreciated by the general public, media, and elected officials. The total cost is important to estimate when considering priorities across programs.

It also is important to consider each major cost element and tradeoffs among them when making fire protection policy. For example, the size of the fire service affects losses; the extent of built-in protection affects the cost of fire services and the losses; insurance overhead should be affected by the number and size of losses, etc. Analysis of changes in incremental costs of the major components of the total cost of fire should be given more consideration in setting priorities than it is usually is.

Appendix A

DIFFERENCES BETWEEN NFPA AND NFIRS ESTIMATES

The National Fire Incident Reporting System collects data from nearly 14,000 fire departments. The National Fire Protection Association annual survey of fire departments collects data from more than 3,000 fire departments. Neither is a perfect random sample; not all fire departments asked to participate do so. As one might expect, the distribution of fire departments is not the same in the two samples. And the NFPA survey collects tallied totals whereas NFIRS collects individual incident reports. Not surprisingly, therefore, there are differences between the NFPA annual survey results and the NFIRS results. In each of the 10 years examined (1985–94), the deaths reported to NFIRS represent a larger fraction of the NFPA national estimate of deaths than the NFIRS number of fires is of the NFPA estimate of fires. NFIRS injuries and dollar loss are even larger fractions of the NFPA totals than are deaths or fires (Figure A–1). Not only are the ratios different for fires, deaths, injuries, and dollar loss, but the ratio for fire deaths has sharply increased over time.

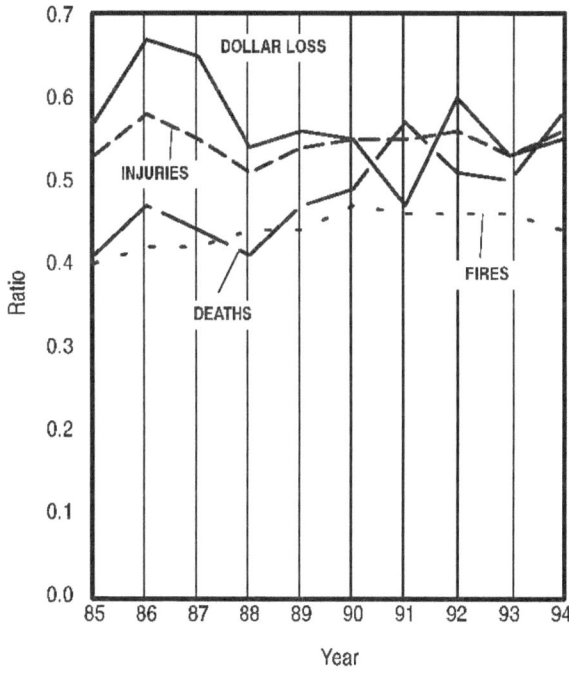

	Dollar Loss	Injuries	Deaths	Fires
1985	0.57	0.53	0.41	0.40
1986	0.67	0.58	0.47	0.42
1987	0.65	0.55	0.44	0.42
1988	0.54	0.51	0.41	0.44
1989	0.56	0.54	0.47	0.44
1990	0.55	0.55	0.49	0.47
1991	0.47	0.55	0.57	0.46
1992	0.60	0.56	0.51	0.46
1993	0.53	0.53	0.50	0.46
1994	0.55	0.56	0.58	0.44

Sources: NFPA Annual Surveys and NFIRS

Figure A–1. Ratio of Raw NFIRS Sample to NFPA National Estimates

Looking at the problem another way, Figure A–2 shows the number of deaths per fire, injuries per fire, and dollar loss per fire from NFIRS and NFPA from 1985 to 1994. Deaths per fire are quite similar for NFIRS and NFPA, with a maximum difference of 32 percent in 1994. Injuries and dollar loss per fire are consistently lower in the NFPA sample than in the NFIRS sample, but the difference has narrowed in the last few years.

The reasons for these differences are not known. One possibility is that some departments that report summary data to NFPA may undercount their casualties and losses when reporting on the NFPA survey forms. Another possibility is that there are data entry errors in NFIRS, with larger numbers of deaths, injuries, and dollar loss creeping into the database despite edit checks at state and federal levels. (It appears that at least some of the dollar loss difference is due to this.)

A third possibility for the differences is that fire departments might not report some minor fires to NFIRS that they include in their own totals that are reported to NFPA. We know that some departments do not fill out NFIRS forms for minor fires such as food on stove or chimney fires, but we are unsure whether these fires are or are not included in the department's report to NFPA nor the extent of the problem.

Resolving the differences between the two major sources of fire statistics in the United States is important to prevent confusion among the users of the data. As NFIRS gets more complete data on the population protected by its participating departments, the NFIRS estimates will be able to be made independent of other sources, which should improve consistency, too.

Figure A–3 represents the NFPA survey trends for non-residential property fires and dollar loss.

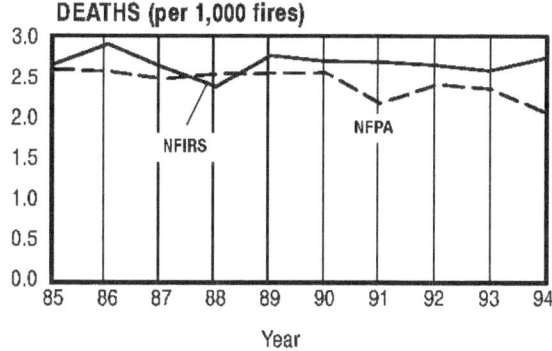

	NFIRS	NFPA
1985	2.67	2.61
1986	2.92	2.58
1987	2.64	2.49
1988	2.39	2.55
1989	2.78	2.56
1990	2.71	2.57
1991	2.70	2.19
1992	2.66	2.41
1993	2.59	2.37
1994	2.75	2.08
10-Year NFIRS Trend = –1.4%		
10-Year NFPA Trend = –15.8%		

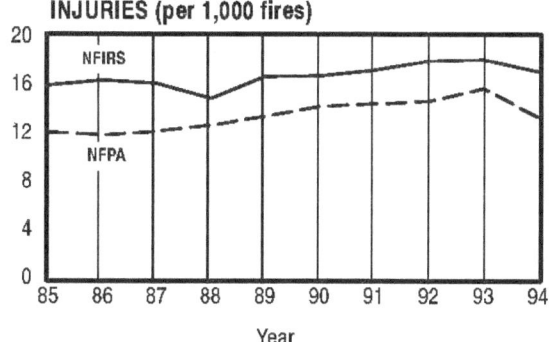

	NFIRS	NFPA
1985	15.87	11.99
1986	16.29	11.81
1987	16.02	12.11
1988	14.77	12.64
1989	16.62	13.36
1990	16.68	14.17
1991	17.12	14.39
1992	17.92	14.61
1993	18.01	15.61
1994	17.02	13.26
10-Year NFIRS Trend = +13.7%		
10-Year NFPA Trend = +26.1%		

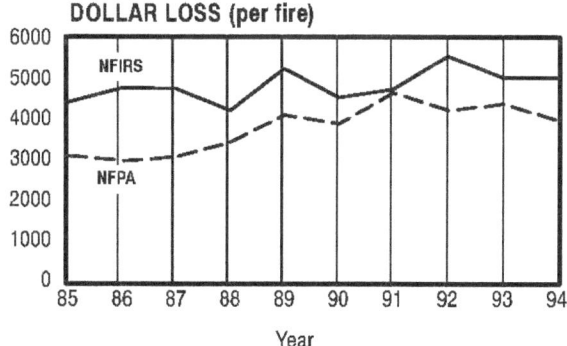

	NFIRS	NFPA
1985	$4,403	$3,089
1986	4,762	2,954
1987	4,744	3,073
1988	4,211	3,428
1989	5,247	4,092
1990	4,532	3,872
1991	4,734	4,637
1992	5,560	4,222
1993	5,035	4,377
1994	5,031	3,967
10-Year NFIRS Trend = +15.2%		
10-Year NFPA Trend = +48.6%		

Sources: NFPA Annual Surveys and NFIRS

Figure A–2. NFIRS vs. NFPA Survey: Severity of Losses

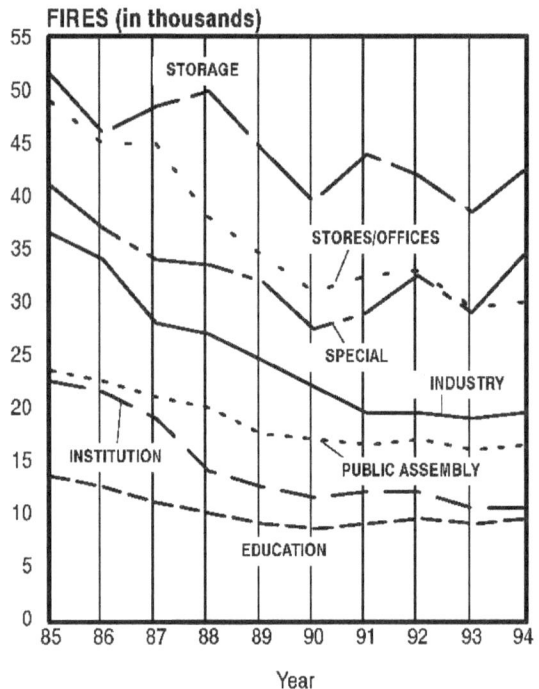

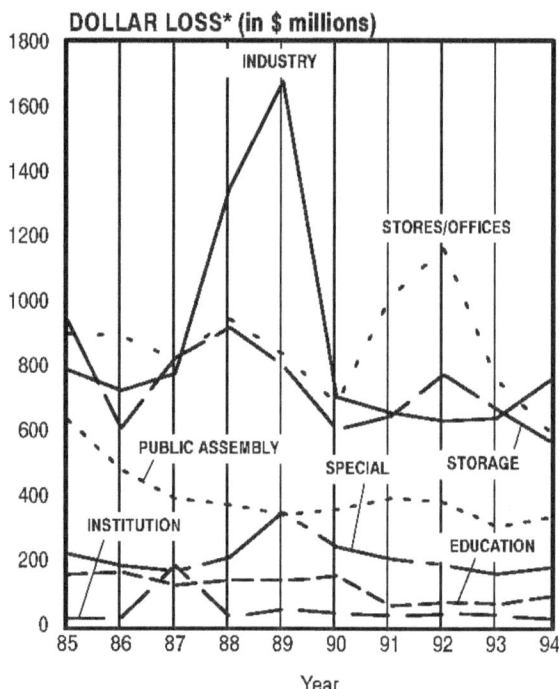

	Public Assembly	Education	Institution	Stores/Offices	Industry	Storage	Special
Fires							
1985	23,500	13,500	22,500	49,000	36,500	51,500	41,000
1986	22,500	12,500	21,500	45,000	34,000	46,000	37,000
1987	21,000	11,000	19,000	45,000	28,000	48,500	34,000
1988	20,000	10,000	14,000	38,000	27,000	50,000	33,500
1989	17,500	9,000	12,500	34,500	24,500	44,500	32,000
1990	17,000	8,500	11,500	31,000	22,000	39,500	27,500
1991	16,500	9,000	12,000	32,500	19,500	44,000	29,000
1992	17,000	9,500	12,000	33,000	19,500	42,000	32,500
1993	16,000	9,000	10,500	29,500	19,000	38,500	29,000
1994	16,500	9,500	10,500	30,000	19,500	42,500	34,500
Dollar Loss* ($ millions)							
1985	$636	$157	$ 23	$ 901	$ 788	$941	$222
1986	481	166	24	894	725	608	185
1987	391	125	184	824	778	826	168
1988	372	140	29	945	1,352	921	210
1989	342	139	51	840	1,680	804	350
1990	359	154	37	684	706	605	243
1991	394	63	30	1,013	658	646	207
1992	381	72	37	1,167	631	775	187
1993	306	66	31	760	639	668	160
1994	334	91	20	598	758	568	181

* Adjusted to 1994 dollars.

Source: NFPA Annual Surveys

Figure A–3. Trends in NFPA Non-Residential Structure Fires and Dollar Loss by Property Type

Appendix **B**

DATA TABLES

The six tables in this appendix provide data used in developing the 10-year trend charts in Chapters 3−5 where they were too crowded to present the actual numbers with the graphic.

Table B–1. Residential Fires and Fire Losses

Cause	1985	1986	1987*	1988	1989	1990	1991	1992	1993	1994
Fires										
Incendiary/Suspicious	74,586	76,928	72,202	73,236	68,847	67,467	67,738	66,589	61,888	61,948
Children Playing	29,100	28,615	27,890	27,880	25,952	23,207	24,102	25,463	25,215	25,872
Careless Smoking	46,536	44,780	41,291	40,439	35,533	31,707	30,559	28,814	27,931	26,470
Heating	183,309	149,729	133,623	125,945	116,939	88,339	89,163	87,728	88,695	77,366
Cooking	116,481	115,146	113,723	115,056	109,183	109,741	111,140	118,605	116,919	107,455
Electrical	51,900	49,227	47,222	49,162	44,197	43,662	45,091	43,084	45,089	44,545
Appliances	39,865	39,002	37,542	38,581	35,777	35,395	35,401	34,748	35,322	35,089
Open Flame	37,146	33,973	34,372	35,050	34,253	28,590	27,386	27,797	28,232	30,188
Other Heat	8,597	8,194	7,691	8,072	7,670	6,263	6,318	6,417	6,641	6,973
Other Equipment	6,013	6,554	6,170	6,401	5,547	5,584	5,885	6,564	6,020	7,009
Natural	9,687	10,985	10,569	9,895	10,413	8,820	9,985	8,951	10,164	10,182
Exposure	18,780	18,367	19,206	22,784	19,187	18,226	25,232	17,241	17,885	17,973
Total	622,000	581,500	551,500	552,500	513,500	467,000	478,000	472,000	470,000	451,000
Deaths										
Incendiary/Suspicious	677	770	727	857	809	770	663	698	756	505
Children Playing	409	397	519	526	474	353	450	398	423	426
Careless Smoking	1,567	1,356	1,368	1,500	1,103	1,062	823	994	913	817
Heating	1,050	648	691	783	636	635	542	516	570	503
Cooking	365	420	360	417	382	382	299	280	371	314
Electrical	436	614	472	382	538	382	329	354	353	409
Appliances	101	176	126	175	129	122	121	99	112	147
Open Flame	220	190	252	257	186	218	169	245	168	179
Other Heat	104	119	50	74	88	93	80	80	35	62
Other Equipment	22	37	54	47	20	42	36	29	70	73
Natural	19	7	14	12	37	10	18	13	14	15
Exposure	56	37	25	35	34	48	44	61	38	15
Total	5,025	4,770	4,660	5,065	4,435	4,115	3,575	3,765	3,825	3,465
Injuries										
Incendiary/Suspicious	2,280	2,119	2,342	2,655	2,460	2,595	2,836	2,772	2,521	2,586
Children Playing	2,144	2,246	2,501	2,567	2,513	2,391	2,786	3,019	2,997	2,781
Careless Smoking	3,445	3,153	3,334	3,681	3,095	3,000	2,793	2,767	2,907	2,430
Heating	2,876	2,330	2,489	3,079	2,627	2,211	2,287	2,278	2,595	2,005
Cooking	4,419	4,535	5,019	5,435	4,975	5,573	5,564	5,499	6,143	5.040
Electrical	1,323	1,260	1,490	1,671	1,444	1,477	1,836	1,730	1,854	1,545
Appliances	1,055	1,117	1,033	1,058	1,191	1,186	1,286	1,180	1,300	1,186
Open Flame	1,364	1,301	1,316	1,586	1,471	1,190	1,375	1,352	1,377	1,514
Other Heat	376	374	290	332	330	381	381	343	332	285
Other Equipment	230	227	307	183	265	245	304	277	245	358
Natural	167	163	169	178	193	189	212	153	154	147
Exposure	148	199	150	175	186	211	189	228	175	147
Total										
Dollar Loss (millions of dollars)[†]										
Incendiary/Suspicious	1,085.2	1,160.7	1,088.7	1,080.1	1,032.2	1,109.6	1,334.8	902.0	913.7	906.6
Children Playing	256.5	252.9	270.5	269.3	278.5	243.0	319.2	232.7	289.0	280.9
Careless Smoking	394.0	378.2	391.0	363.9	324.1	337.9	404.5	231.8	295.9	271.3
Heating	1,070.9	818.0	830.9	885.7	836.9	723.1	902.9	618.6	656.8	631.7
Cooking	424.4	453.9	508.4	508.1	481.8	480.2	548.8	393.1	503.8	531.8
Electrical	839.0	657.8	651.9	754.1	633.0	655.5	813.0	541.2	734.3	612.8
Appliances	295.8	366.7	284.4	296.6	282.0	282.4	351.5	224.0	288.2	263.5
Open Flame	281.5	265.7	299.7	289.4	297.6	338.2	368.7	216.8	299.5	283.8
Other Heat	83.9	86.6	55.3	85.5	121.1	65.3	66.6	53.4	59.9	61.9
Other Equipment	57.6	75.3	81.0	99.7	75.4	103.2	97.0	147.9	103.4	116.6
Natural	123.4	117.2	136.9	124.6	133.9	139.6	214.4	131.8	184.4	169.5
Exposure	285.7	175.3	227.0	279.1	281.9	344.6	619.8	405.2	638.0	186.7
Total	3,774.0	3,556.0	3,699.0	4,020.0	3,998.0	4,253.0	5,552.0	3,880.0	4,843.0	4,317.0

Note: These data support the Figure 36 chart. Columns may not add exactly to the totals due to rounding.
* Adjusted for $150 million questionable fire loss
[†] Adjusted to 1994 dollars

Sources: NFIRS and NFPA Annual Surveys

Table B-2. One- and Two-Family Dwelling Fires and Fire Casualties

Cause	1985	1986	1987	1988	1989	1990	1991	1992	1993	1994
Fires										
Incendiary/Suspicious	51,096	53,744	49,795	51,276	47,660	47,762	46,686	46,977	43,838	43,811
Children Playing	21,997	21,529	20,795	20,994	19,276	17,004	17,567	18,722	18,519	18,981
Careless Smoking	27,005	25,816	23,572	23,119	20,194	18,055	17,564	16,761	16,339	16,057
Heating	182,323	150,314	130,894	123,452	115,185	85,443	84,208	84,132	84,730	72,554
Cooking	77,359	78,125	73,941	74,652	69,445	70,160	69,447	73,887	72,053	66,283
Electrical	43,332	41,256	39,129	40,366	37,319	36,530	37,094	35,855	37,539	36,904
Appliances	32,276	32,667	30,496	30,930	28,929	28,592	28,216	27,746	28,284	27,824
Open Flame	30,496	27,683	27,938	28,597	28,315	23,257	21,834	22,562	23,264	24,332
Other Heat	6,615	6,725	6,208	6,664	6,251	5,070	5,102	5,122	5,419	5,484
Other Equipment	4,534	4,859	4,629	4,809	4,262	4,208	4,304	4,682	4,478	5,403
Natural	9,061	10,462	9,911	9,307	9,972	8,375	9,309	8,498	9,669	9,659
Exposure	15,358	14,821	15,694	18,335	15,693	14,543	21,670	13,055	13,868	13,708
Total	501,500	468,000	433,000	432,500	402,500	359,000	363,000	358,000	358,000	341,000
Deaths										
Incendiary/Suspicious	500	543	536	691	548	583	541	554	653	400
Children Playing	315	332	401	413	335	230	376	338	310	270
Careless Smoking	1,157	1,064	1,064	1,079	703	814	592	745	653	608
Heating	972	669	625	772	661	631	524	485	514	444
Cooking	279	349	270	264	314	314	205	248	269	236
Electrical	338	564	448	384	557	306	311	338	318	383
Appliances	81	156	93	168	126	119	96	86	65	120
Open Flame	216	147	215	216	151	206	133	212	147	174
Other Heat	72	114	38	67	67	87	65	61	24	48
Other Equipment	14	42	55	34	17	36	24	29	53	68
Natural	23	8	17	10	46	12	17	16	16	17
Exposure	54	17	17	29	21	32	21	49	12	17
Total	4,020	4,005	3,780	4,125	3,545	3,370	2,905	3,160	3,035	2,785
Injuries										
Incendiary/Suspicious	1,213	1,252	1,297	1,737	1,477	1,529	1,643	1,375	1,498	1,441
Children Playing	1,752	1,617	1,736	1,856	1,969	1,772	1,864	2,229	2,009	1,976
Careless Smoking	2,270	2,044	2,003	2,209	1,797	1,837	1,687	1,581	1,569	1,370
Heating	2,768	2,335	2,370	2,954	2,493	2,141	2,120	2,093	2,327	1,783
Cooking	3,271	3,573	3,788	4,065	3,535	3,928	3,819	3,693	4,083	3,340
Electrical	1,134	1,057	1,311	1,378	1,158	1,245	1,557	1,450	1,480	1,320
Appliances	930	995	812	902	945	1,012	950	960	1,055	874
Open Flame	1,084	1,021	1,076	1,307	1,039	941	1,085	1,057	1,055	1,146
Other Heat	286	263	250	289	275	275	286	287	235	231
Other Equipment	236	232	278	148	243	234	242	234	169	311
Natural	174	144	155	132	160	172	195	161	128	124
Exposure	133	116	123	148	133	166	153	156	92	83
Total	15,251	14,649	15,191	17,125	15,224	15,252	15,601	15,276	15,700	13,999

Note: These data support the Figure 49 chart. Columns may not add exactly to the totals due to rounding.

Sources: NFIRS and NFPA Annual Surveys

209

Table B-3. Apartment Fires and Fire Casualties

Cause	1985	1986	1987	1988	1989	1990	1991	1992	1993	1994
Fires										
Incendiary/Suspicious	18,047	17,202	17,775	17,764	16,494	15,913	16,943	16,014	14,571	14,693
Children Playing	6,512	6,364	6,569	6,568	6,066	5,786	6,134	6,263	6,230	6,400
Careless Smoking	14,705	13,934	13,787	13,565	11,345	10,526	10,475	9,306	9,228	8,573
Heating	7,367	6,274	6,695	6,244	5,847	5,647	5,999	5,647	5,940	5,362
Cooking	34,460	32,129	35,615	37,178	34,798	36,070	38,750	41,060	41,376	38,306
Electrical	5,691	5,106	5,637	6,023	4,980	5,250	5,678	5,406	5,775	5,700
Appliances	5,460	4,811	5,323	5,700	5,125	5,391	5,715	5,587	5,581	5,681
Open Flame	5,791	5,301	5,765	5,831	5,241	4,828	4,863	4,794	4,643	5,274
Other Heat	1,739	1,345	1,311	1,398	1,276	1,133	1,153	1,225	1,198	1,347
Other Equipment	1,189	1,320	1,223	1,347	1,024	1,066	1,288	1,539	1,211	1,321
Natural	731	785	783	684	694	646	775	631	719	641
Exposure	2,807	2,926	3,018	3,698	3,109	3,243	3,725	3,528	3,526	3,701
Total	104,500	97,500	103,500	106,000	96,000	95,500	101,500	101,000	100,000	97,000
Deaths										
Incendiary/Suspicious	161	148	157	173	233	175	128	110	122	133
Children Playing	80	66	143	118	95	88	58	80	119	130
Careless Smoking	346	244	270	321	300	217	193	187	235	193
Heating	62	28	37	30	31	41	31	30	29	40
Cooking	68	59	50	79	49	41	70	37	58	59
Electrical	83	40	30	30	33	41	19	24	42	40
Appliances	22	16	23	15	8	12	27	11	16	6
Open Flame	22	23	53	42	26	32	24	32	23	25
Other Heat	12	12	13	15	3	15	19	19	10	14
Other Equipment	6	2	3	0	3	0	10	4	6	0
Natural	0	0	0	0	0	0	0	0	0	0
Exposure	3	12	10	6	10	17	15	11	26	0
Total	865	650	790	830	790	680	595	545	685	640
Injuries										
Incendiary/Suspicious	679	674	795	771	835	791	960	1,080	892	952
Children Playing	426	466	704	668	573	628	835	823	941	704
Careless Smoking	887	847	1,026	1,111	1,011	929	916	1,017	1,128	872
Heating	248	186	248	262	289	212	242	255	266	223
Cooking	1,002	956	1,221	1,289	1,384	1,557	1,638	1,685	1,965	1,616
Electrical	170	207	189	262	242	226	302	267	357	277
Appliances	125	145	197	164	186	201	270	200	224	255
Open Flame	261	244	242	284	343	232	309	291	293	385
Other Heat	78	93	49	63	60	93	89	74	91	52
Other Equipment	16	17	45	17	45	29	61	51	58	56
Natural	10	28	23	31	35	27	20	10	20	22
Exposure	22	62	26	29	58	49	32	72	64	62
Total	3,924	3,925	4,765	4,951	5,061	4,974	5,674	5,825	6,299	5,476

Note: These data support the Figure 60 chart. Columns may not add exactly to the totals due to rounding.

Sources: NFIRS and NFPA Annual Surveys

Table B–4. Hotel/Motel Fires

Cause	1985	1986	1987	1988	1989	1990	1991	1992	1993
Incendiary/Suspicious	1,608	1,666	1,620	1,519	1,307	1,287	1,417	1,077	912
Children Playing	125	140	133	123	119	85	97	106	121
Careless Smoking	1,731	1,779	1,689	1,806	1,388	1,186	1,120	1,084	941
Heating	751	767	786	799	658	509	638	569	598
Cooking	1,071	1,113	1,221	1,159	900	891	1,028	986	879
Electrical	575	679	642	693	521	531	561	552	507
Appliances	754	802	894	914	769	749	818	780	786
Open Flame	448	518	440	466	402	319	361	319	307
Other Heat	99	123	119	85	119	72	79	55	64
Other Equipment	140	164	172	161	126	170	144	161	160
Natural	94	123	166	134	109	74	126	113	114
Exposure	104	126	116	142	84	125	110	199	110
Total	7,500	8,000	8,000	8,000	6,500	6,000	6,500	6,000	5,500

Note: These data support the Figure 68 chart. Columns may not add exactly to the totals due to rounding.

Sources: NFIRS and NFPA Annual Surveys

Table B–5. Non-Residential Fire and Fire Losses

Cause	1985	1986	1987	1988	1989	1990	1991	1992	1993	1994
Fires										
Incendiary/Suspicious		66,985	59,720	55,063	52,390	48,222	50,871	53,794	45,888	46,472
Children Playing		8,029	7,714	6,955	5,553	4,522	4,887	5,235	4,550	4,934
Careless Smoking		13,760	12,783	11,297	9,673	8,713	8,658	7,933	7,410	6,794
Heating		14,903	13,903	12,780	12,688	10,113	10,350	11,024	10.798	9,927
Cooking		14,853	14,677	13,304	12,285	11,927	12,277	13,239	12,617	11,959
Electrical		23,188	22,647	20,977	18,159	17,674	17,618	17,198	16,765	15,871
Appliances		11,890	11,455	10,498	9,724	9,265	9,344	9,613	9,442	8,849
Open Flame		23,312	22,991	22,288	19,1813	15,805	16,053	16,435	14,447	15,934
Other Heat		4,084	3,883	3,668	,235	2,908	3,017	2,956	2,769	2,999
Other Equipment		16,246	15,445	14,158	13,163	11,404	11,734	11,782	11,650	11,879
Natural		7,555	7,315	6,801	6,308	5,722	5,900	5,133	5,277	5,417
Exposure		13,695	13,967	14,711	12,140	10,725	11,791	11,158	9,888	10,463
Total	237,500	218,500	206,500	192,500	174,500	157,000	162,500	165,500	151,500	151,500
Deaths										
Incendiary/Suspicious	52	64	65	33	46	73	63	63	57	21
Children Playing	12	9	8	2	4	6	26	4	4	0
Careless Smoking	40	48	37	42	26	34	24	20	15	24
Heating	27	16	17	16	17	17	15	6	21	15
Cooking	0	7	8	12	7	17	4	4	6	10
Electrical	25	9	8	21	4	14	15	20	15	12
Appliances	2	11	0	2	9	17	4	6	6	0
Open Flame	22	14	37	28	24	28	17	26	9	21
Other Heat	25	14	3	9	9	8	4	9	3	5
Other Equipment	22	14	31	40	61	17	13	15	9	17
Natural	0	7	3	2	4	56	4	0	7	0
Exposure	12	2	3	7	9	0	0	4	3	0
Total	240	215	220	215	220	285	190	175	155	125
Injuries										
Incendiary/Suspicious	670	761	499	615	643	735	740	556	811	577
Children Playing	102	95	80	96	140	122	70	65	102	76
Careless Smoking	250	275	215	389	268	231	286	214	285	182
Heating	334	307	357	312	214	263	211	255	326	227
Cooking	330	284	264	349	254	282	324	264	413	298
Electrical	395	416	390	397	344	393	305	340	478	353
Appliances	246	208	142	251	223	198	160	166	196	284
Open Flame	454	559	485	483	404	414	435	290	357	366
Other Heat	114	84	72	96	69	71	81	74	78	60
Other Equipment	472	550	685	523	457	506	351	378	746	498
Natural	120	147	165	113	198	168	144	90	122	156
Exposure	39	40	21	50	62	42	19	32	37	22
Total	3,525	3,725	3,375	3,675	3,275	3,425	3,125	2,725	3,950	3,100
Dollar Loss (millions of dollars)*										
Incendiary/Suspicious	1,538.9	1,291.0	1,311.9	1,413.1	881.6	1,019.8	1,139.9	1,903.7	844.8	905.2
Children Playing	36.0	22.9	28.0	37.8	16.4	22.7	21.1	19.6	28.7	22.9
Careless Smoking	99.8	91.0	93.6	122.4	166.3	64.1	94.6	53.7	52.0	70.6
Heating	282.2	222.8	179.6	246.2	202.6	203.2	196.9	176.8	160.3	194.9
Cooking	101.8	96.3	84.1	101.1	95.8	87.5	111.3	90.2	70.6	96.5
Electrical	438.5	415.0	424.3	592.4	341.9	443.5	356.6	306.5	412.7	341.8
Appliances	111,6	95.5	113.9	149.2	87.1	94.9	110.3	81.2	88.2	76.1
Open Flame	254.0	222.6	310.0	329.2	243.3	200.9	185.1	151.2	264.1	217.9
Other Heat	57.9	52.9	60.9	63.1	33.7	67.5	60.2	41.6	44.0	43.3
Other Equipment	348.7	282.0	305.8	362.9	1,884.6	265.7	327.4	193.5	284.9	345.8
Natural	194.8	119.7	139.9	211.7	108.7	172.1	253.4	96.2	129.3	143.4
Exposure	203.8	172.6	244.7	259.6	145.0	147.6	155.3	136.1	249.0	104.7
Total	3,667.8	3,084.4	3,296.7	3,968.7	4,207.0	2,789.4	3,011.9	3,250.3	2,628.6	2,563.0

Note: These data support the Figure 78 chart. Columns may not add exactly to the totals due to rounding.
* Adjusted to 1994 dollars

Sources: NFIRS and NFPA Annual Surveys

Table B–6. Firefighter Injuries in Non-Residential Fires

Type	1985	1986	1987	1988	1989	1990	1991	1992	1993	1994
Percent of Injuries in Fires										
Public Assembly	4.8	4.9	6.0	4.5	5.0	5.4	5.3	4.6	4.0	3.7
Eating, Drinking	7.5	7.5	5.6	6.3	5.8	7.6	5.5	7.4	5.5	6.0
Education	3.9	3.3	2.3	2.5	3.3	3.3	2.4	3.6	2.2	2.6
Institutions	1.9	2.1	2.5	2.1	1.5	2.1	3.0	2.8	2.7	3.1
Stores, Offices	23.3	21.6	22.9	19.5	24.3	22.3	19.9	19.1	24.4	21.1
Basic Industry	3.8	3.0	1.0	2.4	4.1	3.0	2.8	2.8	2.2	3.5
Manufacturing	13.0	14.8	15.3	12.6	11.3	12.9	12.9	10.9	11.5	11.4
Storage	20.7	18.7	18.1	20.5	18.2	16.4	13.8	14.5	12.7	15.7
Vacant, Construction	11.4	14.5	14.5	15.5	15.7	15.3	17.9	19.3	14.6	15.4
Outside Structure, Unknown	4.8	4.3	6.4	7.4	5.6	6.8	7.8	9.4	12.8	9.7
Number of Injuries Per 1,000 Fires										
Public Assembly	59.4	54.8	66.5	51.8	56.8	60.4	58.1	49.5	39.5	36.2
Eating, Drinking	50.1	47.0	33.9	39.1	36.2	46.7	34.1	43.4	29.8	32.6
Education	44.7	34.2	24.8	26.0	33.8	34.3	22.5	38.8	16.6	19.5
Institutions	13.2	13.0	14.6	12.4	8.8	12.7	18.6	17.2	15.0	18.1
Stores, Offices	70.6	58.6	59.9	50.5	61.9	58.3	51.8	43.2	57.0	49.4
Basic Industry	65.8	52.7	16.4	36.9	60.7	50.0	42.6	43.3	34.0	48.3
Manufacturing	69.0	71.7	71.1	56.9	52.3	64.5	68.1	57.4	55.0	52.8
Storage	72.1	58.9	52.1	55.3	51.9	50.4	39.2	36.9	33.2	54.9
Vacant, Construction	63.9	67.0	63.2	62.7	58.8	61.9	70.5	67.8	52.8	49.6
Outside Structure, Unknown	29.2	17.2	31.0	33.9	25.0	30.8	28.7	29.3	37.5	29.3
Average	52.8	46.2	45.2	49.6	49.1	51.3	44.2	43.7	39.6	38.3

Note: These data support Figures 116 and 120 charts.

Sources: NFIRS and NFPA Annual Surveys

Appendix **C**
NFIRS REVISION PROJECT

The current National Fire Incident Reporting System employs techniques of data entry, validation, transmittal, and analysis that represented the state of the art at the time of its original design in 1975. NFIRS, however, has become outdated by the fast pace of computer technology advances. Methods available when NFIRS was introduced are very cumbersome by today's standards, often resulting in delays in obtaining information at the local, state, and federal levels and receiving less-than thoroughly validated data. Survey feedback from participating departments, states, and vendors has resulted in an extensive list of valuable suggestions to improve the system, many of which cannot easily be implemented by way of minor alterations to the current system because of the vintage of its architecture. The suggestions fall into five broad categories:

- *Clarify the collection requirements:* Rid the system of those current data elements and codes that are no longer relevant or that are so confusing or burdensome as to diminish the likelihood of complete and accurate data entry. Also, some of the required data inputs can be easily derived from other sources.

- *Simplify the forms:* Clarify and simplify the rules for completion of paper and automated forms.

- *Accommodate local information needs:* Federal, state, and local information needs are not coincident. As many as possible of these diverse requirements need to be accommodated in a single system if participation at all levels is to be achieved.

- *Expand the breadth of the system to all incidents:* Since the introduction of NFIRS, it has become increasingly important to document the full range of fire department activities (e.g., NFIRS 4.1 does not address EMS incidents). Consequently, a revised NFIRS should encompass the full range of departmental activities.

- *Collect data relevant to incident suppression/mitigation:* Since the current system was not designed by the fire service or those who used incident data, parts of the data that are currently collected in the system are not used. Other parts are imperfectly designed, leading to poor utility for analysis or prevention programs.

As a result of these suggestions, NFIRS is under development to respond to these changes in technology and needs. The new NFIRS is a substantial revision of NFIRS 4.1. The new system will:

- Use newer computer technology.

- Delete some currently used data elements.

- Revise some currently used data elements.

- Add new data elements.

- Expand the scope of the system from fires to all incidents.

The new system will be hardware independent and will have the option of being operated by states and municipalities via the Internet. It is anticipated that these improvements will enable even more participation in the data collection system. The expected increased participation will substantially affect the completeness of the national incident database, particularly in light of the growing proliferation of computers at the state and local departmental levels. By taking advantage of newer technology, USFA can also help speed the process of collection, which will result in a more timely analysis, thereby enhancing the utility of the information.

The system under development is to be used at three governmental levels—local, state, and federal—each with distinct needs.

Participating metro fire departments (as well as all fire departments) require on-line data entry, edit, and analysis of all the information required by USFA and participating states. These departments also require information that USFA and state fire marshals do not need, such as remarks and local resource use details. The identification of specific individuals and apparatus deployed to each incident is one such example.

Participating states require on-line entry, edit, and analysis of only a subset of the data elements collected by local fire departments in order to capture state-required information submitted on paper forms by nonautomated, local departments within their jurisdictions. In addition, states require a facility to accept, edit, and merge modem and disk uploads of incident reports collected by automated local fire departments within their jurisdiction.

At the federal level, USFA accepts data from states and selected metro fire departments (i.e., fire departments not located in participating states). Validation of data is necessary at the federal level. This validation includes the ability to edit and merge incidents with the national database.

Fire departments that are not metro fire departments are responsible for procuring their own software. Those without computers must make do with paper forms that are keypunched at the state level. To maintain a national standard of fire reporting, a tight software specification will be developed as part of this project for the many software vendors who cater to the fire reporting market as well as a method of testing, registering, and tracking which vendors are in compliance with the specification.

INDEX

G

FIGURE/TABLE INDEX

The following Deaths, Dollar Loss, Fires, *and* Injuries *entries refer specifically to charts and tables presented in this document.*

Dollar Loss

Fires

Injuries